Arithmetic Counts:

Why Americans Have Trouble with Math and How to Fix It

CITI OF BOOKS

Paul Shoecraft
Coauthor of The Mad Minute

Paul Shoecraft
pjs6425@aol.com

Ordering Information:

Quantity sales. Special discounts are available on quantity purchases by corporations, associations, and others. For details, contact the publisher at the address below.

CITIOFBOOKS, INC.
3736 Eubank NE Suite A1
Albuquerque, NM 87111-3579
www.citiofbooks.com
Hotline: 1 (877) 389-2759
Fax: 1 (505) 930-7244

Printed in the United States of America.

ISBN-13:   Softcover   979-8-89391-596-9
           eBook   979-8-89391-597-6

Library of Congress Control Number: 2025905560

Introduction

No problem is too big to run away from. — Charles M. Schulz, creator of Peanuts

Dear Reader,

If, like me, you grew up in America, *Arithmetic Counts!* is a story about how you and I were taught arithmetic. It is about our adding and subtracting in grades 1 and 2 and multiplying and dividing in grades 3, 4, and 5. In other words, *it is about how arithmetic is taught today*. Except for the New Math in the 1960s, arithmetic has been taught that way for more than 100 years.

If you learned how to add and subtract in grades 1 and 2 and learned how to multiply and divide in grades 3, 4, and 5, you did well. Me too. We deserve praise for paying attention and following instructions. We learned the arithmetic we were supposed to learn in those grades, *but we were robbed*, as were all of our classmates. *Everyone was robbed!* even our teachers. We all were held back in math in elementary school as if we were math dummies, and it shows in international testing.

Since 1995 and every four years thereafter, the International Association for the Evaluation of Educational Achievement has tested fourth and eighth graders in different countries in the most substantial international study of student achievement in math and science ever conducted. The results through 2019, published as TIMSS[1] results, are shown below. The rank of the last country for each year of testing equals the number of countries tested that year. Only eighth graders were tested in 1999. The TIMSS results for 2023 are addressed on page 328 in the appendix. According to an overview of the 2023 results for math, America's 4th- and 8th-graders scored lower, on average, in 2023 than in 2019 by 18 and 27 points, respectively.

TIMSS Results for the U.S and Top Ten Countries in Mathematics, 4th Grade, 1995-2019

1995		2003		2007		2011		2015		2019	
Singapore	625	Singapore	594	Hong Kong	607	Singapore	606	Singapore	618	Singapore	625
South Korea	611	Hong Kong	575	Singapore	599	South Korea	605	Hong Kong	615	Hong Kong	602
Japan	597	Japan	565	Taiwan	576	Hong Kong	602	South Korea	608	South Korea	600
Hong Kong	587	Taiwan	564	Japan	568	Taiwan	591	Taiwan	597	Chinese Taipei	599
Netherlands	577	Flanders	551	Kazakhstan	549	Japan	585	Japan	593	Japan	593
Czech Rep.	567	Netherlands	540	Russia	544	N. Ireland	562	N. Ireland	570	Russia	567
Austria	559	Latvia	536	England	541	Flanders	549	Russia	564	N. Ireland	566
Slovenia	552	Lithuania	534	Latvia	537	Finland	545	Norway	549	England	556
Ireland	550	Russia	532	Netherlands	535	England	542	Ireland	547	Ireland	548
Hungary	548	England	531	Lithuania	530	Russia	542	England	546	Latvia	546
* * *		* * *		11th USA	529	11th USA	541	* * *		* * *	
12th USA	545	12th USA	518	* * *		* * *		14th USA	539	15th USA	535
* * *		* * *						* * *		* * *	
26 Kuwait	400	25 Tunisia	339	36 Yemen	224	50 Yemen	248	47 Kuwait	353	64 Phillippines	297

[1] TIMSS is an acronym for Trends in International Mathematics and Science Study.

TIMSS Results for the U.S and Top Ten Countries in Science, 4th Grade, 1995-2019

1995		2003		2007		2011		2015		2019	
South Korea	597	Singapore	565	Singapore	587	South Korea	587	Singapore	590	Singapore	595
Japan	574	Taiwan	551	Taiwan	557	Singapore	583	South Korea	589	South Korea	588
USA	**565**	Japan	543	Hong Kong	554	Finland	570	Japan	569	Russia	567
Austria	565	Hong Kong	542	Japan	548	Japan	559	Russia	567	Japan	562
Australia	562	England	540	Russia	546	Russia	552	Hong Kong	557	Chinese Taipei	558
Netherlands	557	**USA**	**536**	Latvia	542	Taiwan	552	Taiwan	555	Finland	555
Czech Rep.	557	Latvia	532	England	542	**USA**	**544**	Finland	554	Latvia	542
England	551	Hungary	530	**USA**	**539**	Czech Rep.	536	Kazakhstan	550	**USA**	**539**
Canada	549	Russia	526	Hungary	536	Hong Kong	535	Poland	547	Norway	539
Singapore	547	Netherlands	525	Italy	535	Hungary	534	**USA**	**546**	Lithuania	538
* * *		* * *		* * *		* * *		* * *		* * *	
26 Kuwait	401	25 Morocco	304	36 Yemen	197	50 Yemen	209	47 Kuwait	337	64 Phillippines	249

TIMSS Results for the U.S and Top Ten Countries in Mathematics, 8th Grade, 1995-2019

1995		1999		2003		2007		2011		2015		2019	
Singapore	643	Singapore	604	Singapore	605	Taiwan	598	S. Korea	613	Singapore	621	Singapore	616
S. Korea	607	S. Korea	587	S. Korea	589	S. Korea	597	Singapore	611	S. Korea	606	C. Taipei	612
Japan	605	Taiwan	585	Hong Kong	586	Singapore	593	Taiwan	609	Taiwan	599	S. Korea	607
Hong Kong	588	Hong Kong	582	Taiwan	585	Hong Kong	572	Hong Kong	586	Hong Kong	594	Japan	594
Flanders	565	Japan	579	Japan	570	Japan	570	Japan	570	Japan	586	Hong Kong	578
Czech Rep.	564	Flanders	558	Flanders	537	Hungary	517	Russia	539	Russia	538	Russia	543
Slovakia	547	Netherlands	540	Netherlands	536	England	513	Israel	516	Kazakhstan	528	Ireland	524
Switzerland	545	Slovakia	534	Estonia	531	Russia	512	Finland	514	Canada	527	Lithuania	520
Netherlands	541	Hungary	532	Hungary	529	**USA**	**508**	**USA**	**509**	Ireland	523	Israel	519
Slovenia	541	Canada	531	Malaysia	508	Lithuania	506	England	507	**USA**	**518**	Australia	517
* * *		* * *		* * *								* * *	
28th USA	500	19th USA	502	15th USA	504	* * *		* * *		* * *		12th USA	515
* * *		* * *		* * *								* * *	
41 S. Africa	354	38 S. Africa	275	45 S. Africa	264	48 Qatar	307	42 Ghana	331	39 S. Arabia	368	46 Morocco	388

TIMSS Results for the U.S and Top Ten Countries in Science, 8th Grade, 1995-2019

1995		1999		2003		2007		2011		2015		2019	
Singapore	607	Taiwan	569	Singapore	578	Singapore	567	Singapore	590	Singapore	597	Singapore	608
Czech Rep.	574	Singapore	568	Taiwan	571	Taiwan	561	Taiwan	564	Japan	571	C. Taipei	574
Japan	571	Hungary	552	S. Korea	558	Japan	554	S. Korea	560	Taiwan	569	Japan	570
S. Korea	565	Japan	550	Hong Kong	556	S. Korea	553	Japan	558	S. Korea	556	S. Korea	561
Bulgaria	565	S. Korea	549	Estonia	552	England	542	Finland	552	Slovenia	551	Finland	543
Netherlands	560	Netherlands	545	Japan	552	Hungary	539	Slovenia	543	Hong Kong	546	Russia	543
Slovenia	560	Australia	540	Hungary	543	Czech Rep.	539	Russia	542	Russia	544	Lithuania	534
Australia	558	Czech Rep.	539	Netherlands	536	Slovenia	538	Hong Kong	535	England	537	Hungary	530
Hungary	554	England	538	**USA**	**527**	Hong Kong	530	England	533	Kazakhstan	533	Australia	528
England	552	Finland	535	Australia	527	Russia	530	**USA**	**525**	**USA**	**530**	Ireland	523
* * *		* * *				11th USA	520					11th USA	522
17th USA	534	18th USA	515	* * *		* * *		* * *		* * *		* * *	
* * *		* * *											
41 S. Africa	326	38 S. Africa	243	45 S. Africa	244	48 Ghana	303	42 Ghana	306	39 S. Africa	358	46 S. Africa	370

United States Rank in TIMSS Results for Math and Science, 4ᵗʰ Grade, 1995-2019

4ᵗʰ Grade	1995	1999	2003	2007	2011	2015	2019
Math	12th	Not tested	12th	11th	11th	14th	15th
Science	3rd	Not tested	6th	8th	7th	10th	8th

United States Rank in TIMSS Results for Math and Science, 8ᵗʰ Grade, 1995-2019

8ᵗʰ Grade	1995	1999	2003	2007	2011	2015	2019
Math	28th	19th	15th	9th	9th	10th	12th
Science	17th	18th	9th	11th	10th	10th	11th

I was on my 80ᵗʰ trip around the sun in 2020 when I began writing this book. In earlier trips, I became a mathematician, taught high school and college math, and became a teacher of teachers on how to teach math. Based on the comparatively weak performance of American fourth and eighth graders in math and science, I am alerting the nation about how its children are not on track to excel in the STEM disciplines: science, technology, engineering, and mathematics. Many finish high school so far behind their international peers in math and science that they never catch up.

What is America to do to ensure that its children can compete for STEM jobs in America, let alone worldwide? Nix foreign competition by restricting immigration? No, not unless we want to weaken America's economy by forcing businesses to "hire American," even though we know that our schools are not producing the talent they need. Besides, we cannot force businesses to do that. They will just leave the country if we try. A case in point was provided by Microsoft.

In 2007, Microsoft opened a new development center in Vancouver, Canada. They could have located it in the U.S., but they situated it in Canada because there they could *"recruit and retain highly skilled people affected by immigration issues in the US"* (workpermit.com, 2007). Data from an early study by the International Association for the Evaluation of Educational Achievement shows why such issues mattered to Microsoft:

> *Data from the Second International Mathematics Study (1982) show that the performance of the top 5 percent of U.S. students [in math] is matched by the top 50 percent of students in Japan. Our very best students — the top 1 percent — scored lowest of the top 1 percent in all participating countries.* — National Research Council, 1989

The path to expertise in the STEM disciplines begins with arithmetic, then algebra, geometry, calculus and beyond using the skills, concepts, thought processes, and *language* of arithmetic and algebra. Language? Yes, although a *written* one for business, government, science, and people like you and me to describe quantitative matters concisely and precisely.

Arithmetic counts! *It matters.* Bill Gates, former chairman of the board and chief software architect for Microsoft, made clear a belief central to Microsoft's choice for a new development center in Canada: *"I have never met the guy who doesn't know how to multiply who created software"* (Friedman, 2007). Borrowing from Mr. Gates' quote, *I have never met the scientist, technician, engineer, or mathematician who didn't know arithmetic.*

In the analogy about how an arrow must leave a bow on target to hit its mark, if the arrows are children and the mark is algebra, being on target means being competent in arithmetic. A corollary of that analogy is that an arrow that is off target by even a small amount may miss its mark "by a mile." Children who do poorly in arithmetic tend to do poorly in algebra:

> *Too many students in middle or high school algebra classes are woefully unprepared for learning even the basics of algebra. The types of errors these students make when attempting to solve algebraic equations reveal they do not have a firm understanding of many basic principles of arithmetic.* — National Mathematics Advisory Panel, 2008

Can we agree that taking six years — *all of kindergarten through grade 5* — to only teach *some* children how to just add, subtract, multiply and divide is too many years for too little math? Can we then agree on the need for a way to teach arithmetic in less than six years that works for *all* children so more math can be taught in elementary school? How else can America expect its children to catch up with their international peers in math and science?

> *The reconstruction and restructuring of the school mathematics curriculum is among the most important problems in education reform. Without mathematics reform, there can be no science reform — without science reform, no real education reform.* — Robert Nielsen, 1990

In reading this book, you will acquire a new perspective on arithmetic and how it can be meaningfully taught, joyfully learned, and a source of inner pride and self-esteem for *all* children. In particular, you will learn about the proven effectiveness of MOVE IT Math, a methodology I developed for a summer math camp[2] I ran 50 years ago when I was teaching math and math methods courses as an Assistant Professor in the math department at Arizona State University.

You will also learn why you probably never heard of MOVE IT Math and how, for half a century, America has been denied a methodology to counter the underperformance of its schools in math and science. However, for America's elementary schools to adopt MOVE IT Math's methodology, millions of Americans will have to agree to two changes in how you, they, and I were taught arithmetic. Both changes would be highly controversial, so I will hold off on listing them until I tell you about MOVE IT Math.

According to Albert Einstein, *"a problem cannot be solved with the same consciousness that created it."* Thus new thinking is required to improve how arithmetic is taught in America's schools. That new thinking is sound judgement in practical matters, namely, *common sense.*

Peace and tolerance,

Paul Shoecraft

Paul Shoecraft, PhD
Professor of Mathematics Education
Architect of MOVE IT Math™ aka Multimodality Math

P.S. The references for the citations in the introduction are listed with the references for Chapter 1.

---

[2] You may read about the summer math camp, called 'Rithmetic in Residence, at moveitmath.com. Just click on the train at the top of the home page.

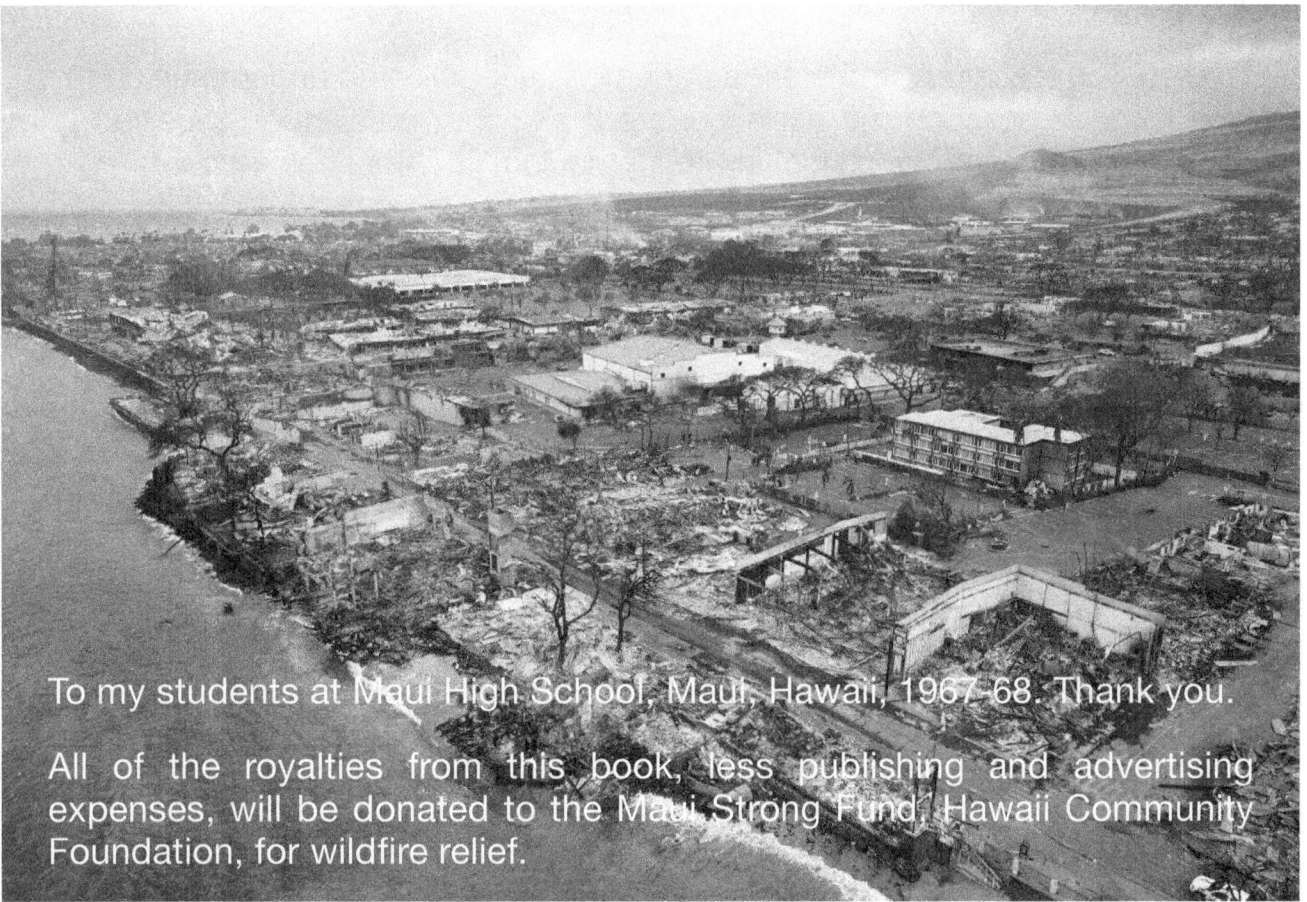

To my students at Maui High School, Maui, Hawaii, 1967-68. Thank you.

All of the royalties from this book, less publishing and advertising expenses, will be donated to the Maui Strong Fund, Hawaii Community Foundation, for wildfire relief.

An aerial image shows destroyed homes and buildings burned to the ground on Maui, Hawaii. *Los Angeles Times*, August 10, 2023

This book is testimony to the encouragement, support, and understanding my wife Linda heaped on me while I wrote it and the assistance I received to reform the elementary school math curriculum in American schools from elementary school teachers and administrators, university administrators, and especially Lynne Shoecraft, Ann Richards (1933-2006), governor of Texas (1991-1995), and Jim Ketelsen (1930-2017), former CEO of Tenneco (1978-1992), a Texas-based corporation. Others who helped are listed below. Their expertise, professionalism, and commitment to better educate America's children was extraordinary. I treasured working with them.

Rodana Aguirre, Judy Anderson, Mary Arnold, Carrie Baker, Peggy Bartlet, Donna Beach, Elizabeth Bell, Amy Boyd, Marilyn Bratcher, Hellene Brisco-Mieth, Keleigh Brooks-Muska, Pam Brown-Johnson, Gail Burlingame, Vanessa Burns, Jenny Button, Deanna Callahan, Kim Camarillo, Kim Carr, Linda Cartwright, Esther Castaneda, Terry Cavazos, Connie Cliffe, Elma Constancio, Sherise Davis, Kay Divan, Pam Dolozal, Susan Duke, Francis (Tina) Dusek, Glenn Goerke, Karen Earwood, Donna Edleman, Patricia Enlow, Sally Fabrygel, Linda Fielder, Carol Frederick, Diana Freudensprung, Shannon Gerik, Melodie Gohn, Kathy Gomez, Jeanne Greive, Carol Hall, Sammye Harper, Cheri Hart, Sharon Harton, Ann Hilborn, Patty Hirsch, David Holubec, Yvonne House, Deborah Hubbard, John Johnson, Lisa Johnson, Ruth Joslyn, Helen Jurries, Dot Kidd, Karen King, Dianne Kleinkauf, R. Kocurek, Denise Kofron, Edie Kost, Glennie Kraatz, Ann Kuester, Cheryl Laas, Toni Lahodny, Connie Lankford, Therese Leinger, Judy Leinius, Sharon Lempa, Mary Lester, Phyllis Malone, Susan Martin, Mike Maxwell, Pauline Maxwell, Joan McCadden, Paula McCauley, Betty McDaniels, Maria Mendoza, Evelyn Miller, Debbie Morrill, Brenda Motley, Lina Mutcheler, Robert and Geri Nielsen, Tammy Nobles, Kwame Opuni, Lisa Perkins, Patty Pittman, Kelly Pool, Betty Jean Ramsey, Shanedria Ridley, Toni Roeder, Suzanne Rogers, Tracy Rogers, Sandra Gayle Roome, Marilyn Routh, Mari Russ, Sue Schmaltz, Teresa Sellers, John Sharp, Anne Shaw, Deborah Sheback, Shari Shields, Donna Shimek, Mary Sklar, Jana Smith, Karen Soehnge, Charlene Stevens, Ann Stiles, Joyce Taylor, Sue Taylor, Mary Beth Thielen, Linda Thornton, Kathy Tipton, Yen Tran, Debbie Tumis, Pam Ulrich, Jovita Viallarreal, Pam Ward, Ellen Weeks, Eileen Weinstein, Judy Wenske, Lara Wenzel, Donna Wick, Rose Williams, Dell Wise, Amy Witten, Saralee Wittmer, Genevieve Wolter, Norman Woolsey, Dianna Ybarra, Wanda Zabransky.

Contents

Chapter 1: This Really Happened!

Move It Math Whizzes Cheer Governor Richards on City School Tour

Don Brown, *The Victoria Advocate*, Victoria, Texas, March 23, 1991

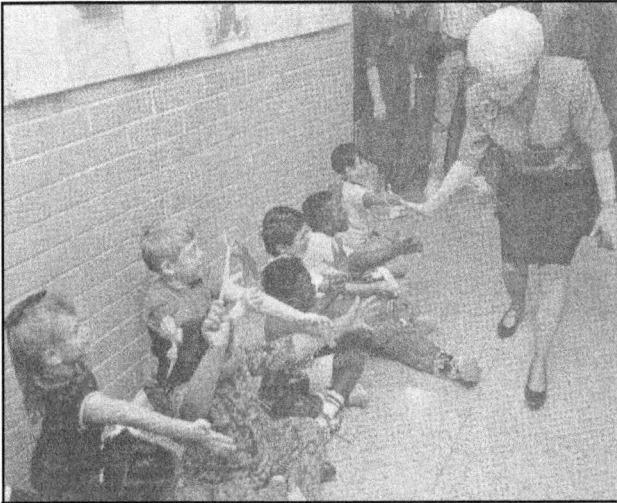

Governor Ann Richards has to bend over to give students a high five as she walks down the corridor at O'Connor Elementary Friday to see kindergarten and first-grade students demonstrate a new math program developed by UHV Professor Paul Shoecraft.

Governor Ann Richards found more than she bargained for while visiting classrooms at O'Connor Elementary School Friday.

Richards was in Victoria at the invitation of university of Houston-Victoria President Glenn Goerke to see firsthand the Multimodality Math/Move It Math program developed by Paul Shoecraft, a professor at UHV.

She admitted she knew very little about Shoecraft's concept before arriving in Victoria, but before leaving, she said she was "extremely impressed."

Richards received a briefing and watched a short video at the UHV Petroleum Training Institute, then went to the O'Connor campus to see first-graders breeze through "monster" addition, subtraction, multiplication and division problems.

MOVE IT, an acronym for math opportunities, valuable experiences, innovative teaching, focuses on the skills and thought processes children must acquire to succeed in mathematics. *"Move It Math is a quantum leap ahead of traditional math instruction concepts now taught in Texas classrooms,"* Richards said.

"I was impressed by the fact these children have not only learned a great deal about numbers, but also can work independently in the classroom," she said. *"I think it's remarkable. Obviously, these children are excelling far beyond the standardized requirements of Texas public schools."*

The governor is no stranger to the classroom. Prior to entering politics, Richards taught at the intermediate and high school levels. *"I have been in a lot of classrooms, but never in one where independent work was going on as successfully as what I have seen taking place here,"* she said.

She said she plans to urge newly appointed Education Commissioner Lionel "Skip" Meno and members of the State Board of Education to *"do as I did and come to Victoria to see a program that really works. Any exposure to this program would certainly convince people at the Texas Education Agency that it's worth emulating statewide. The question will be how much of it we can do and how fast we can put this program on track."*

When Richards arrived at the O'Connor campus, she was greeted by Principal Susana Mathis, Superintendent Bob Brezina and the school choir, directed by Lucille Araj. Students lined the sidewalk and hallways to get a glimpse of the governor.

Richards said she always asks children how they are doing in school and if they like math. *"Younger children usually say they like math, but the older they get, the more they say they're not good in math or don't understand it."* She said the education system *"has to assume blame when something happens to change a child's opinion of math from 'I can do it' to 'I can't do it.'"*

Richards said that while state leaders have been focusing a lot of attention on how Texas will finance public education, she is equally concerned about the "quality" of the product. *"I think the quality discussions about what children are actually learning and what we're turning out of the public school system will be the next big debate in Austin,"* she said.

Richards praised Shoecraft and the support given his math concept by the UHV faculty. *"The University of Houston–Victoria is doing a remarkable job, and their campus here is truly important in the higher education system of this state."*

She also commended teachers at schools like O'Connor *"who are willing to go out on a limb to allow their students to excel. A lot of times the big thing that holds back education is the fear of trying something new."*

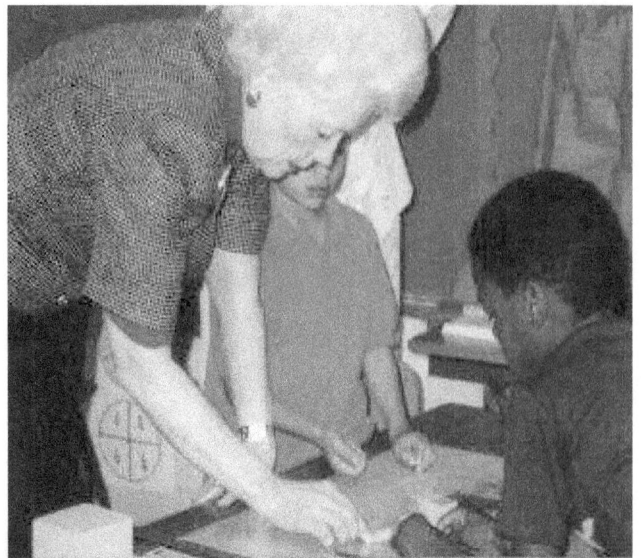

Port O'Connor Math Program at Cutting Edge of a Revolution in Education

George Macias, *Swisher Electric Edition*, Texas Co-op Power, March 1992

A small rural school on the Texas Gulf Coast is helping revolutionize the way educators think about mathematics and education in general.

The Port O'Connor School is a public school with classes from kindergarten through sixth grade. The school has a modest budget and limited resources, and students are typically from working class families, mostly the sons and daughters of shrimpers and farmers.

For two years in a row, third graders at Port O'Connor have scored 100 percentile on the state's standardized third grade math test. These young mathematicians not only add, subtract, multiply, and divide bigger numbers than much older students in other schools, they also enjoy the learning process. Astonishingly, they are grasping and excited about algebra concepts as early as kindergarten.

"Nothing sets these kids apart except for this new way of instruction. Any child from any background can profit from this way of being taught," says Mrs. Marilyn Bratcher, the principal of Port O'Connor School. She says the Port O'Connor experience is proving that what schools and teachers usually expect of students is far, far below what they can accomplish.

"Math books in the state of Texas do the same thing over and over. They're boring," says Mrs. Bratcher. *"We don't even use textbooks except as a resource because they're behind. In our country, intelligence is equated with math. Good math skills open doors for so many good occupations,"* she says.

The significance of the Port O'Connor "experiment" became clear to Principal Bratcher one day while sitting on the floor with kindergarten students in Judy Anderson's class. *"I was assessing Judy for the career ladder. When I realized what was happening, I was so excited I crawled up to her and said*

Second graders Tasha Jones and Ashley Hensley work math problems on the chalk board at the Port O'Connor School.

'Judy, you are teaching algebra to kindergarten kids — and it's only October!'"

Mrs. Anderson began rethinking her approach to teaching after attending a three-week class presented by Dr. Paul Shoecraft at the University of Houston – Victoria. The program was funded with federal assistance made possible by Eisenhower Math/Science grants and the Education for Economic Security Act.

Dr. Shoecraft promotes a revolutionary system of teaching math that is shaking the foundation of the educational hierarchy and drawing the interest of prestigious think tanks such as the National Science Foundation. His program, or parts of his program, have been tried or adopted by several school districts, including Calhoun, Comal, Hallettsville, Gonzales, Waelder, Victoria, Beeville, El Campo, Bay City, and Palacios ISD.

Dr. Shoecraft, who has been working on his educational revolution since the early 1970s, says several South Texas schools are part of this new cutting edge reform movement, but the Port O'Connor School was the first school to fully commit an entire school to this innovative teaching style. Port O'Connor's Mrs. Anderson was one of the first teachers to take elements of Dr. Shoecraft's math program and adapt them to a kindergarten classroom and make them work on a sustained basis.

Mrs. Anderson has also developed a program called "Math Buddies," which brings older children from higher grades together with younger kids. Principal Bratcher explains that Math Buddies serves several needs, such as reducing the amount of physical work for the teacher and reinforcing basic mathematical concepts for older students.

The math program at Port O'Connor includes many different modes of learning, using objects (called "manipulatives"), such as arithmetic blocks. The key to this learning process is that it attracts the student's attention by asking the child to do more than just listen to a teacher talk. Instead, the students are actively involved in the [learning] process.

Mrs. Anderson says the program grew out of total frustration: *"We had to do something to help these kids want to stay in school. So few graduated from high school. We had to make learning challenging and enjoyable."*

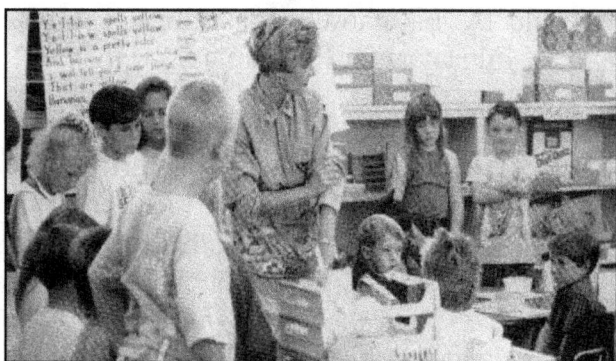

Mrs. Anderson leads a math discussion with children in her kindergarten class.

Her excitement for the program is apparent, both in the classroom and just talking about it. *"I like the fact that my students like math, but more important than that is that they love to study. They love to learn."*

In Texas, the general guidelines for kindergarten are that students should be able to compare things, to recognize numbers one through 20, and to be familiar with circles, squares, and triangles. [In contrast,] by the end of kindergarten, most of the Port O'Connor kids are familiar with addition and subtraction of whole numbers, fractions, coordinate graphing, positive and negative integers, solving for one unknown in an algebraic equation, area and perimeter of polygons, symmetry, and place values of numbers into the thousands.

Dr. Robert Nielsen, a [former math professor], member of the Mathematical Sciences Education Board in Washington, DC and an assistant to the president of the American Federation of Teachers, has visited the Port O'Connor School twice and is a strong advocate of both the Port O'Connor School and of Dr. Shoecraft's work.

Dr. Nielsen says, *"It's the direction that school mathematics ought to go everywhere. Port O'Connor has demonstrated that all kids can learn math. I hadn't realized how dramatically things can change until I visited Port O'Connor School. I'm convinced that this stuff is so radically different from the way you and I learned math that the only way to convince people is for them to see it happen. If you haven't seen it, you can't believe it."*

He says, *"There's something wrong when math is a subject that almost all kids learn to hate. The Port O'Connor School is dramatic proof that it doesn't have to be that way,"* and he encourages parents and teachers to ask questions about their own math programs in their local areas. He says, *"If you walk into the classroom and the kids are having fun, then the school has got the right stuff; if not, then it's just teaching math the old-fashioned way."*

Dr. William Kirby, an educational consultant who was the head of the Texas Education Agency for many years, says he thinks the program is *"super because it sets high expectations for kids, and it utilizes teachers with adequate training who believe in themselves."*

Dr. Kirby says students, through the use of concrete objects, are actually *"experiencing mathematical concepts first hand. They are actively engaged with their hands and their eyes, not just their ears. Too often kids sit passively, listening to a lecture."*

He says, *"If we're going to get where we need to be in this country, we need to change the model that's existed for the last 200 years of the teacher as merely a dispenser of facts."*

Dr. Kirby is also impressed with Port O'Connor's "Math Buddies" program because of the team effort it inspires: *"Sometimes older kids are helping younger kids, and sometimes the reverse is true."*

He says the Port O'Connor School has convinced him that teachers should be guides, mentors, and assistants to students. *"Teachers should be directors of learning, like symphony conductors. Not everyone in a symphony gets to sit in the first chair, but everybody participates and has an important role."*

For education reform to work, he says, it must come from local schools, unlike traditional "top down" reform. He says, *"Give people at local districts the freedom to be creative."*

Principal Bratcher is convinced that the math program at the Port O'Connor School is the tip of the iceberg, and teachers in all grades at the school are applying the successful methods used in Mrs. Anderson's class, such as crossing grade levels and using manipulatives, in teaching other subjects.

Parent volunteer Kim Beaty works with kindergarten student Tracy Romine using a math balance, which helps students understand mathematical concepts.

Mrs. Bratcher says, *"This has really changed the way we do things in our whole school, and because Port O'Connor has hardly any student turnover, the long-term success of the school will be easy to monitor."*

So far, with kindergarten students regularly testing out at second and third grade levels, and some as high as junior high levels, the success is fairly obvious.

The University of Houston – Victoria honored the young mathematicians at the Port O'Connor School in 1989 by offering UHV scholarships to all of the students on the condition they finish junior college. Dr. Shoecraft says the university wanted to acknowledge the important role these students played in making this type of innovative program work.

One Port O'Connor student, Devon Vasquez, wrote President Bush while attending second grade at the Port O'Connor School: *"Teachers from all over Texas have come to watch us do our work because they want their kids to do the fun things we do. We would be happy if you came to Port O'Connor, Texas, to our math class so we can show off our brains to you."*

Closing Remarks

By 2004, 135,000 students were enrolled in MOVE IT Math classes in 217 schools in seven states (Project GRAD, 2004). Nonetheless, you probably never heard of the program. How can that be considering its proven effectiveness and widespread use? The answer to that question may trouble you after you learn more about MOVE IT Math.

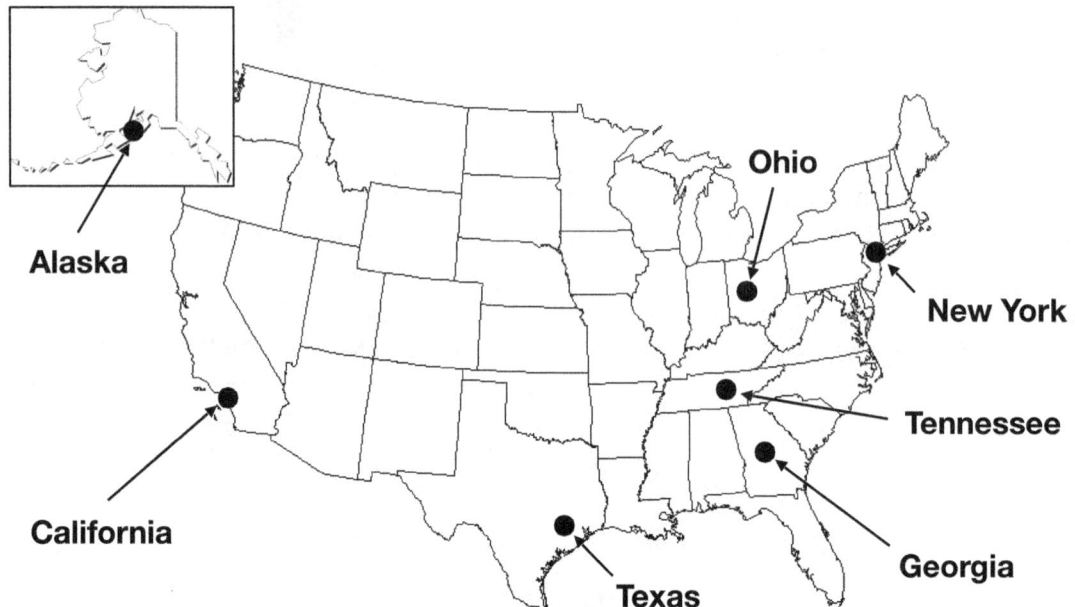

Alaska

Ohio

New York

Tennessee

California

Georgia

Texas

References (including those for the citations in the introduction)

Friedman, Thomas. *The World Is Flat: A Brief History of the Twenty-first Century*, Farrar, Straus, and Giroux, 2007.

"Move It Math Whizzes Cheer Governor Richards on City School Tour," Don Brown. *The Victoria Advocate*, Victoria, Texas, March 23, 1991.

National Mathematics Advisory Panel. *FINAL REPORT*, U.S. Department of Education, 2008.

National Research Council, *Everybody Counts: A Report to the Nation on the Future of Mathematics Education*. National Academy Press, Washington, D.C., 1989.

Nielsen, Robert. "Anyone Can Learn Math: New Programs Show How," *American Educator*, American Federation of Teachers (AFT), Spring 1990.

"Port O'Connor School Math Program at Cutting Edge of a Revolution in Education," George Macias. *Swisher Electric Edition*, Texas Co-op Power, March 1992.

Project GRAD, 2004: *Overview, Results, and Indicators.* A PowerPoint presentation from Project GRAD's website when active.

Schulz, Charles M. Creator of Peanuts. BrainyQuote.com.

TIMSS Results, nces.ed.gov/timss.

workpermit.com, 2007.

Chapter 2: America's Past and Present Arithmetic Curricula

The most durable way to improve schools is to improve curriculum and instruction rather than endlessly squabbling over how school systems should be organized, managed and controlled. — Diane Ravitch, 2010

A curriculum is like a table of contents. It is a list of topics for a subject that are to be taught in the order given. It may be just that or may include suggestions on how to teach the topics. In either case, as "the law of the land," so to speak, *it determines the instructional input for a subject*, which makes it the main variable affecting student output for the subject. Therefore, in keeping with the common-sense observation that it is insane to expect different outputs with the same input, to change student output, the curriculum must be changed.

Changing a curriculum is an immense endeavor. For a major subject like math, the curriculum for K-12 math has been changed nationally only three times in the last 140 years: once around 1880 as a result of the so-called progressive education movement (that was not progressive) that lasted for about 60 years (Wikipedia/Progressive Education), once in the 1960s with the New Math developed by university mathematicians, and once in 1989 with a call for new standards for K-12 math by the National Council of Teachers of Mathematics (NCTM).

Each of these nationwide changes in the K-12 math curriculum will be addressed in this chapter after an important observation about the nature of arithmetic, that portion of the math curriculum that is the centerpiece of this book.

The Nature of Arithmetic

Arithmetic encompasses three distinct but interrelated aspects of the subject that children must know in order to excel in it:

o **Basics:** The number facts (e.g., 5 + 8 = 13, 13 − 5 = 8, 8 x 6 = 48, 48 ÷ 8 = 6) and how to add, subtract, multiply, and divide whole numbers, fractions, and decimals with paper and pencil.

o **Concepts:** The big ideas that make arithmetic understandable, like knowing why the 9s in 25999.9 on the odometer of a car roll over and turn into zeros when the car is driven another tenth of a mile, that is, why 25999.9 + 0.1 = 26000.0.

o **Applications:** How to solve arithmetic word problems and the kinds of problems that occur in daily living that can be solved with arithmetic.

The reason for listing the three aspects of arithmetic is to provide a basis for comparing the presentation of arithmetic in the K-6 math curriculum before 1880 with the three curricula that followed it.

Before the progressive education movement, *all three aspects of arithmetic were integrated and taught concurrently,* as will be shown in Chapter 9 in examining two basic math books that were used before the movement. Since then, just one of the three aspects of arithmetic has been emphasized in each of the three curricula that followed with the assumption that the other two aspects would be learned as well — that focusing on just concepts, for example, would automatically teach basics and applications, a dubious assumption as noted by Hung-Hsi Wu, Professor Emeritus, University of California at Berkeley, as reported by Klein (2003):

The separation of conceptual understanding from basic skills in mathematics is misguided. It is not possible to teach conceptual understanding in mathematics without the supporting basic skills, and basic skills are weakened by a lack of understanding.

Professor Wu's assertion is confirmed by a common finding in educational research about the efficacy of most teaching methodologies: *no transfer*. Most students learn only what they are taught, *regardless of how it is taught*. Most do not learn a deeper understanding of what they are taught on their own. For instance, if the instructional input for arithmetic is focused primarily on basics, like adding and subtracting small numbers, most students will learn only that. Most will not make the connection between adding and subtracting small numbers and the concept of place value in base ten numeration or how to solve arithmetic word problems or problems in daily living that involve addition and subtraction.

America's Arithmetic Curricula Since 1880

The following are the mindsets that have governed the development of the three arithmetic curricula since the progressive education movement that began in 1880. As noted earlier, each mindset is based on just one of the three aspects of arithmetic: basics, concepts, or applications. Which mindset would you associate with utilitarians — those who shun the theoretical in favor of the practical? Which would you associate with math professors? With the NCTM?

o **Focus on Basics.** Concepts and applications will be discovered as natural consequences of learning how to compute with paper and pencil.

o **Focus on Concepts.** Basics and applications will be discovered as natural consequences of understanding place value in a base ten system of numeration and knowing that addition and multiplication satisfy the commutative, associative, and distributive properties.[3]

o **Focus on Applications.** Basics and concepts will be discovered as natural consequences of learning how to solve arithmetic word problems and their real-life equivalents with calculators.

The first mindset was harbored by utilitarians in the progressive education movement who saw the need for every citizen to know arithmetic whether they understood it or not. The second one was held by the math professors who developed the New Math, and the third one was adhered to by the NCTM and the educational establishment at large who provided guidelines for developing a curriculum that met their vision for K-12 math. An overview of the arithmetic curriculum that evolved from each of the three mindsets follows.

---

[3] The commutative, associative, and distributive properties for addition and multiplication are stated algebraically as $a + b = b + a$ and $a \times b = b \times a$, $a + (b + c) = (a + b) + c$ and $a \times (b \times c) = (a \times b) \times c$, and $a \times (b + c) = a \times b + a \times c$, respectively.

The Arithmetic Curriculum During the Progressive Education Era

Like most movements, the progressive education movement was a reaction to the status quo at the time. It was a rebellion against the elitist "Euro-American curricula of the 19th century, which was rooted in classical preparation for the university" (Wikipedia/Progressive Education).

In recognition of the small percentage of Americans that the pre-1880 curriculum served at the time, progressives opted for a utilitarian curriculum, which was anti-intellectual, not progressive, but served the needs of most people. For elementary school math, that meant focusing on computation — how to add, subtract, multiply, and divide whole numbers, fractions, and decimals.[4] Considering that calculators would not be available for another 100 years, focusing instructional input back then on just computation is understandable.

I was a student in this curriculum from 1946 when I began school in the first grade in Indianapolis, Indiana until 1958 when I graduated from high school in Globe, Arizona. I remember learning how to add and subtract in the first grade and nothing beyond that until the fourth grade in Arizona when I had to memorize the times table — the 100 multiplication facts from 0 x 0 = 0 to 9 x 9 = 81 — and learn how to multiply and divide. After that, I do not remember learning any math of consequence until I got to high school. Nonetheless, I never complained. I enjoyed the A⁺ I routinely got in math on my report card and the annual review of Roman numerals that enabled me to read the origin dates on public buildings. I was content to stagnate in math in grades 1-8, as were my classmates.

Then came algebra in the ninth grade! The first two weeks were among the most confusing I had ever experienced in school. *"What's this x, y, z stuff, and whadda ya mean by 5 – 8? Ya can't take 8 from 5."* In all the years I had spent in school up to then, I had never been exposed to literal (letter) numbers like x, y, z and had never seen a "backwards" subtraction problem like 5 – 8. All through elementary school, the slash for subtraction had meant "take away," then, just like that, it meant "to the left of zero on a *number* line" — a line I had never heard of before.

The motivation to change the math curriculum I experienced occurred on October 4, 1957 when the Soviet Union launched Sputnik 1 — the first human-made object to be placed in orbit around the earth. I was a senior in high school at the time and remember the spooky sight of the satellite that looked like a star racing across the night sky.

Fearing that the launch of Sputnick signaled that the United States was losing the Cold War with the Soviet Union, the U.S. Congress enacted and signed into law the National Defense Education Act (NDEA) on September 2, 1958. The act made available hundreds of millions of federal dollars in grants to colleges and universities to develop new curricula for K-12 math and science and provide scholarships for middle school and high school math and science teachers to advance their knowledge of the subjects they taught.

---

[4] According to Baker and others (2010), except for the New Math in the 1960s, an analysis of 141 textbooks for elementary school math published between 1900 and 2000 revealed that 85 percent of math instruction in K-6 was devoted to basic arithmetic.

I was one of many math teachers who benefited from the NDEA. I was dumbstruck when I learned that I could get *paid* to go to graduate school *tuition free*. Wow! Compare that to the obscene cost of higher education in the 21st century! So I quit my first teaching job as a math teacher at Arcadia High School in Scottsdale, Arizona and went back to school, acquiring a Master of Science degree in pure mathematics from Northern Arizona University in 1967 and a PhD in mathematics education from the University of Michigan in 1971. I owe my fulfilling career to the investment America made in me. I am forever grateful for that and offer this book as partial payback.

The New Math Arithmetic Curriculum

When World War II (WWII) ended in 1945, America's economy was booming. American manufacturing was at an all-time high, and 40 percent of America's workforce was unionized with good-paying jobs with cost-saving, stress-reducing benefits, like health insurance. For the first time ever in the nation's history, its middle class was becoming affluent. I remember my dad remarking back then about the increasing number of second cars parked in the driveways of the one-car garages that were common at the time.

Indulging in its newfound prosperity and role as a world leader right after the war, America languished as the rest of the world rebuilt. By the late 1950s, the world was catching up. Foreign brands, like SONY and Volkswagen, were becoming synonymous with quality, performance, and reliability. The competitive edge that American products had in the marketplace right after the war was tapering off.

During recruiting for WWII, it had become scandalous that most army recruits *"knew so little math that the army itself had to provide training in the arithmetic needed for basic bookkeeping and gunnery"* (Raimi, 2000). Nonetheless, America ignored the math illiteracy the war had revealed about its citizenry. All was well until the launch of Sputnik 1, which woke America to its inferior status in technology and forced it to address the underperformance of its schools in math and science.

A measure of America's world standing in math at the time was obtained by the newly formed International Association for the Evaluation of Educational Achievement. In 1964, said organization conducted the First International Mathematics Study. The subjects were eighth graders from 12 countries, one of which was the U.S. Imagine the shock when America learned that its eighth graders ranked *next to last* with a total math score that barely exceeded half that for eighth graders from the top ranked countries: Israel (#1), Japan (#2), and Belgium (#3).

No worries, though, or so America thought. With the passage of the NDEA in 1958, the nation was assured that its children would soon catch up with their international peers in math. With funding provided by the NDEA, math professors from notable universities were released from their usual duties to develop new and improved curricula for K-12 math. Surely, they would get it right, and they did, sort of. They got the math right, but their presentation of it, dubbed the New Math, was too much of a departure from the "old math" for it to be gracefully accepted by the general public.

The New Math was fueled by an academic think tank called the School Mathematics Study Group (SMSG) that was formed soon after the launch of Sputnik 1. The endeavor was directed by Edward G. Begle (1914-1978), a topologist and renowned Professor of

Mathematics at Yale University, and financed by the National Science Foundation. SMSG's mission was to reform the K-12 math curriculum, which it did, by creating curricula that made sense out of the math, at least to mathematicians.

Resistance to the New Math was immediate and justifiable. For arithmetic, all of the programs that emerged emphasized concepts at the expense of basics and applications and were known for their pedantry — their concern with detail of questionable importance, like differentiating between number and numeral and validating the obvious, like claiming that 3 + 4 = 4 + 3 and 5 x 6 = 6 x 5 because addition and multiplication are commutative. Duh. Formalizing self-evident truths about arithmetic in elementary school interferes with children's intuitive grasp of the subject.

> *Abstraction is not the first stage, but the last stage, in a mathematical development.* —
> Morris Kline, 1973

By the mid 1970s, the New Math had been rejected by the American public. Objections to the new terminology — like sets, union, and intersection — and computing in different bases were voiced around the dinner table and made the butt of jokes. Ever mindful of the need to sell books, the publishing industry responded with a new round of textbooks that were a hybrid of the "old math" with some of the terminology from the New Math, like "regroup" instead of "borrow" and "carry," which amounted to replacing two mathematically meaningless words with one mathematically misleading word.[5]

> *The result, after twelve years [of the New Math], was total failure. School mathematics was worse off in 1975 than it had been in 1955. The idiocies of the older curriculum had in most places been removed, but often to be replaced with new ones.*
> — Ralph Raimi, 1995

Number or numeral?

The word "number" refers to an amount, like the number of pigs in a pen. Counting and computing are done with numbers. The word "numeral" refers to the spoken or written expression of a number. So is the answer to an arithmetic problem a number or a numeral? It can be either. It is a number while associated with the computation that generated it and a numeral if read or written down separate from said computation. What about telephone numbers? Are they numerals? Yes, but mathematicians do not ask acquaintances for their telephone numerals.

In fairness to the New Math, it was sparsely implemented and played out primarily in the nation's consciousness, not its schools, so its potential to improve math education was never realized, which is unfortunate. In spite of how confusing it was to the general public, it was a coherent explanation of arithmetic, algebra, and beyond and a clear departure from the anti-intellectualism associated with the progressive education movement that preceded it. One of the New Math's contributions was the introduction of calculus in high school, and for good reason:

---

[5] Regroup means to reassemble, which means to put or gather together again, like a family getting together around a dinner table after being separated during the day. *That is not what occurs in computing*, nor on the odometer of a car when a string of 9s "rolls over." What occurs is *trading*. If 15 ones, for example, are thought of as $1 bills, regrouping them would be putting a rubber band around ten of them, which is not what occurs in computing. What happens is that ten of the $1 bills are *traded* for one $10 bill, which is worth the same as a wad of ten $1 bills, but is a different entity.

Calculus has invaded every field of scientific endeavor and plays invaluable roles in biology, physics, chemistry, economics, sociology, and engineering. Calculus can be used to help explain the structure of a rainbow, teach us how to make more money in the stock market, guide a spacecraft, make weather forecasts, predict population growth, design buildings, and analyze the spread of diseases. — Clifford Pickover, 2009

In rejecting the New Math, America overdid it. Most of the furor about the New Math had to do with its presentation of arithmetic in elementary school, but the New Math for high school was scrapped along with it, which amounted to the old metaphor of throwing out the baby with the bath water. Key elements of the New Math for the college preparatory math courses in high school should have been retained.

I taught the New Math right after I graduated from Northern Arizona University (NAU), but not without some "serious" decision-making on my part. Shortly before I graduated, the math department at NAU offered me a job, which surprised me, even though I had taught classes for the department while working on my master's degree in math and, according to my students, had done a good job. Plus, I had distinguished myself in my course work and had scored in the 90s on the Graduate Record Exam in math. What surprised me was that I already had a job after I graduated: a job teaching high school math again.

I could have backed out of the high school job and accepted the more prestigious job the math department at NAU had offered, which I considered because of my longstanding relationship with the Chair of the department, Dr. Harvey Butchart (1907-2002), a relationship I will address later in this chapter, but I decided to stick with the high school job. My decision had nothing to do with the jobs, per se. It had to do with the *location* of the jobs. The high school where I would be teaching was in Hawaii on the island of Maui. The lure of sun-drenched beaches year round verses another Flagstaff (elev. 7,000 feet!) winter of six months of snow and freezing temperatures was a temptation I could not resist.

To my surprise, the job in Hawaii was fraught with human drama. Upon showing up for the job, I learned that I had been assigned to teach sixth grade social studies. I am kidding, right? Nope. I was recruited to teach high school math, but the school district needed a sixth grade social studies teacher, and I had minored in the subject for my bachelor's degree with a math major, so I got the job, and since mainland USA was a long swim away, I figured I did not have much choice in the matter. Besides, *I was in Hawaii!* so I hunkered down to prepare to teach social studies.

As I was seated in my classroom flipping through the pages of the textbook for sixth grade social studies, I was interrupted by the principal of the elementary school to which I had been assigned. He asked me if I knew anything about the New Math, to which I responded with a full throated *"You bet!"* Imagine my elation when I learned that I was being transferred to Maui High School (now defunct) to teach the New Math to juniors and seniors because the teacher who usually taught them was on leave for the first semester. Notably, I was not told that the switch would be for just one semester.

I did not learn of the temporary status of my job at Maui High until late in the first semester. I was astonished when administration told me that come second semester, the teacher I had replaced would be returning and that I would be assigned to the district office to do make-work jobs. When

I informed my students of this, I was astonished again. They organized and told administration that they would go on strike if I could not stay on as their math teacher. They even wrote a protest letter to the governor of Hawaii that I learned about when I received a letter from the then governor of Hawaii, the Honorable George Ariyoshi:

> *Dear Mr. Shoecraft, I often receive letters from students praising their teachers, but never one for a math teacher. ... Your students told me that for the first time, the book went through them instead of them going through the book.*

Mindful of the situation, administration allowed me to finish the year with my students.

Granted, I had put my heart and mind into my teaching at Maui High, and my students were eager and able to learn, but I attribute much of my success with them to the textbook series the school had me use: the New Math for grades 9-12 developed by the University of Illinois Committee on School Mathematics (UICSM) under the direction of Max Beberman, a celebrated mathematics educator. Dr. Beberman urged that math should be taught as a *language* (Miller, 1990), albeit a written one. Thus a noteworthy goal of the Illinois program was to teach students how to *read* math.

For example, an expression like $5 - 8$ would be read as *"positive 5 minus positive 8"* ($^+5 - ^+8$) to acknowledge the signage of the numbers that is unwritten but that mathematicians unthinkingly read into the expression. (The reason the signage is unwritten is because mathematicians established the convention of just knowing it was there so they would not have to write so much.) Then, because the subtraction sign in algebra may be read as *"plus the opposite of,"* $^+5 - ^+8$ would be read more fully as *"positive 5 plus the opposite of positive 8,"* which would be restated as *"positive 5 plus negative 8"* and recorded as $^+5 + ^-8$. Therefore, $5 - 8 = ^+5 - ^+8 = ^+5 + ^-8 = ^-3$.

Note how reading the unwritten signage in $5 - 8$ clarifies that the plus and minus signs have dual meanings in algebra, which can be confusing. Depending on the context, the plus sign can mean "add" or "to the right of zero" on a number line, and the minus sign can mean "subtract" or "to the left of zero" on a number line. Therefore, to read $5 - 8$ with understanding, algebra students must imagine it as $^+5 - ^+8$ and interpret that as $^+5 + ^-8$.

To teach students to perceive $5 - 8$ as $^+5 - ^+8 = ^+5 + ^-8$, the UICSM textbook for algebra included the signage of the numbers — the superscripted plus and minus signs — in the problem sets. Maintaining the signage while working with the numbers meant extra writing for my students, but it made the magic show of algebra meaningful to them. None of them ever complained to me about the extra writing.

Lessons that ensure that students know what math teachers often assume they know, *like all numbers are signed numbers in algebra whether visibly signed or not*, can make a big difference in student performance. I learned that about teaching *before* I became a math teacher. I learned it when I took geometry in high school. I got a C in the course, which was a major departure from the As I had been getting in math up to then. So why the C?

I got off to a bad start in high school geometry because my first impression of it was all wrong. I thought it was just a bunch of facts about pictures of points, lines, and shapes that I could tell were "true" just from looking at them, else the pictures would be screwed up.

So when my geometry teacher would do something like draw two parallel lines on the chalkboard, cross them with a transversal, and tell the class that he was going to *prove* that the alternate interior angles thus formed were congruent (the same size and shape), I used to think, *"Why? I believe you."*

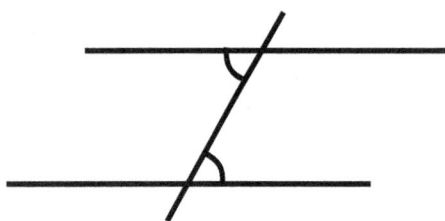

Alternate Interior Angles

My difficulty in high school geometry was in not realizing that in proving a theorem, there is no problem to solve, like in arithmetic and algebra. In geometry, the problem, so to speak, is to figure out which definitions and postulates go with a theorem in order to verify that your geometry teacher is not lying to you, but my geometry teacher never made that clear to me, so when he droned away at the chalkboard to prove a theorem that I thought was "obviously" true, I tuned him out. Instead of paying attention to what he was doing, I slumped in my seat and tried to be invisible so he would not call on me. I credit the C he gave me to my showing up for class.

Surprisingly, my fiasco with high school geometry may be common. Years later, in talking to numerous friends and acquaintances about their experiences with high school math, I learned that most of them had done well in algebra but not in geometry or vice versa and that hardly any of them, like me, had done well in *both*. In wondering about that, I think I know why.

Arithmetic and algebra are about solving problems, and that is what students do for ten years in K-9 before they take high school geometry. A potential side effect of that is to condition students to think that the sequence of events in math is problem first, answer second, which is *backwards* from what one does in high school geometry. There, if given a theorem, the task is to substantiate its veracity, but in my mind, the theorem was an answer when it was obviously true, so I did not know what to do with it because I had been conditioned to think that I was done when I knew the answer.

I suspect that those who struggled with geometry after doing well in arithmetic and algebra had been conditioned, as I had been, to think problem first, answer second, and were perplexed when given a theorem and told to prove that it was true when it was clear that such was the case just from looking at a pictorial representation of the theorem.

What I failed to grasp when I took high school geometry was the *purpose* of the course. I got hung up on all the definitions, postulates, theorems, and corollaries that were italicized, bold faced, and framed in rectangles in my textbook as if they were the Mona Lisa on display in the Louvre in Paris. I thought I was supposed to memorize all that stuff. Wrong! *I was supposed to use it as a context for practicing logical thinking* — reasoning from a set of givens to a conclusion that necessarily follows from the givens — which has been the reason for teaching geometry in high school since the 19th century (Committee of Ten on Secondary School Studies, 1892).

More than most other school subjects, mathematics offers special opportunities for children to learn the power of thought as distinct from the power of authority. — National Research Council, 1989

Geometry is pure thought. It conjures up a cathedral in the mind of necessary inferences. To move about in it, the theorems and corollaries are the archways, and the definitions and postulates are the keystones for the archways. Once I understood that about geometry, which occurred when I took a course in geometric constructions my junior year in college at Arizona State College (now Northern Arizona University), I excelled in the subject. I even received an award from the Chair of the math department, the aforementioned Dr. Harvey Butchart, for creating a novel way to construct a certain figure with just ruler and compass given a short list of assumptions about the figure.

Dr. Butchart was the teacher that we all hope for — a teacher who taught way more than what was on the final exam. I met him when I was registering for math classes my first week on campus as a freshman at ASC. When I signed up for set theory,[6] a course on the axiomatic foundation of mathematics that he would be teaching, he surprised me by telling me that I had scored the highest on the math exam that the incoming freshman class had taken.

After some small talk, Dr. Butchart surprised me again by inviting me to visit him at his home at "two to the tenth" on some street in Flagstaff, which probably had me thinking about how lucky he was to have such a "cool" street address. Two to the tenth (2^{10}) equals two times itself ten times, which can be figured out mentally by doubling two, doubling four, doubling eight, and so on seven more times: 2-4-8-16-32-64-128-256-512-<u>1024</u>. So Dr. Butchart had told me that his house number was 1024.

Although Dr. Butchart was much older than me from my perspective as an 18-year-old when I met him, we connected that day, and I became one of his most attentive students during my freshman and junior years at ASC and my year there as a graduate student after ASC had become a university. From a respectful distance, we became friends, and he mentored me. I never took him up on his offer to visit when I was an undergraduate, but I did after he retired and I had completed my doctoral work at the University of Michigan.

I was fortunate to have Dr. Butchart as a math teacher and mentor. He was a world-class mathematician and could have worked for Harvard, Yale, Princeton or any major university, like UCLA or the University of Michigan, but he chose Arizona State College, a nondescript school, for one reason and one reason only: ASC was in Flagstaff, which was just an hour and a half drive from the Grand Canyon.

Dr. Butchart taught math at ASC/NAU from 1945 until he retired in 1976. During that time, he became a leading authority on the Grand Canyon. According to Wikipedia, he kept a detailed log of more than a thousand excursions in the canyon in which he walked more than 12,000 miles (19,000 km), climbed 83 summits, and scaled the walls at 164 places, claiming 25 first ascents. In 2010, the U.S. Board of Names honored him posthumously by naming a 7,000-foot butte in the canyon "Butchart Butte." Now back to my wacky perception of high school geometry.

---

[6] The textbook for the course was *Naive Set Theory* by Paul Halmos. Ten years later, I concluded my pursuit of mathematics where it began. I took an advanced version of the *same* course with the *same* textbook at the University of Michigan except taught by Dr. Halmos himself.

The point of my story about my experience with high school geometry is that *nothing in math is obvious to a novice in the subject*, not even that 7 = 7, as you will learn in a later chapter. What was obvious to my high school geometry teacher about why I had to take geometry was not obvious to me. Had he told my class that the objective of the course was to initiate us to deductive thinking and then told us a Sherlock Holmes story to explain what he meant by that, I believe I would have been able to relate to his delight in the subject.

> *Mathematics, rightly viewed, possesses not only truth, but supreme beauty.* — Bertrand Russell, 1918

Later, when I became a high school math teacher and my karma required that I teach geometry, I used to stress to my geometry students that the purpose of the course was to teach them how to think logically — how to reach conclusions based on a set of givens or personal values without making stuff up or altering the givens or values in order to justify a particular outcome. When I gave them a test, I had them ask themselves if they valued honesty and pointed out that if they did, they should not cheat because that would be illogical. I remind myself of that every time I do my taxes.

Returning to my situation at Maui High, I could envision a showdown in the making. The high school had just one math teacher for juniors and seniors, so come next year, that teacher would be me or the teacher I had replaced for the year. If me, that meant humiliation and reassignment for the teacher I had replaced and notoriety for me with the teachers at Maui High for outing a popular colleague of theirs. If not me, that meant whatever administration decided to do with me.

My worries ended when I learned that an application I had submitted to the University of Michigan for a National Science Foundation scholarship had been approved. Since the scholarship paid for a year's tuition and fees and included a $5,000 stipend, I barely took a pay cut. So much for my adventure in Hawaii with the New Math. Now back to the general discussion about the New Math era.

By the late 1970s, enrollment in advanced math and science courses in American high schools had declined noticeably, and it was widely known that America's schools were not producing the talent in those subjects that American corporations needed in order to be competitive with their foreign counterparts, both locally and internationally. In response, the NCTM initiated a project called Priorities in School Mathematics (PRISM) to study the situation. Their findings were reported in 1980 in *An Agenda for Action: Recommendations for School Mathematics for the 1980s.* Initially, not much came of the report because it was soon overshadowed by another more compelling one.

In 1981, the U.S. Secretary of Education created the National Commission on Excellence in Education to respond to the widespread dissatisfaction with the performance of America's schools. Eighteen months later, the commission submitted its report entitled *A Nation at Risk: The Imperative for Educational Reform.* The report was brutal in its assessment of the nation's educational system:

> *Our nation is at risk. The educational foundations of our society are presently being eroded by a rising tide of mediocrity that threatens our very future as a Nation and a people. If an unfriendly foreign power had attempted to impose on America the mediocre educational performance that exists today, we might well have viewed it as an act of war.*

According to the report, *"between 1975 and 1980, remedial math courses in public 4-year colleges increased by 72 percent and amounted to one-fourth of all math courses taught in those institutions."* I could be wrong, but I imagine that or worse is still the case. Many of the math courses currently taught at community colleges are repeats of basic math and beginning algebra for zero credits.

When *A Nation at Risk* was released, I was an Associate Professor in the math department at Metropolitan State College (now a University) in Denver, Colorado. I learned of the report on my car radio while driving home from work. Although shocked, I was hopeful. I envisioned deliberations between mathematicians and K-12 math teachers that would culminate in sensible changes in the K-12 math curriculum. Instead, what ensued was political strife that came to be known, somewhat comically, perhaps, as the math wars of the 1990s:

> *The rhetoric surrounding [the curriculum for K-12 math] has been much more shrill, the policy differences more sharply drawn, the participants more diverse [than during the New Math era]. The so-called math wars of the 1960s were largely civil wars and were conducted primarily in journal articles and at professional meetings. Today's warfare [from 1980 on]ranges outside the profession and has a more strident tone; it is much less civil in both senses of the word.* — Jeremy Kilpatrick, 2001

As the math wars increased in stridency and fervor, the New Math faded in the minds of the American public. No more animated discussions about alienating jargon and the perceived pointlessness of computing in different bases. Gone also, though, was the goodnatured humor it had evoked: jokes about parents trying to help their children with their New Math homework, Peanuts cartoons like one with Charley Brown exclaiming *"How can you do 'New Math' problems with an 'Old Math' mind?"* and a song by Tom Lehrer about how to compute 342 − 173 in base ten and base eight, whose closing lines satirically summed up the public's attitude toward the New Math: *"It's so simple, so very simple, that only a child can do it!"*

The New Math era ended with a fizzle in 1977 with the termination of the School Math Study Group (SMSG). As reported by Rami (1995), its Director, Dr. Begle, had this to say at the conclusion of his directorship:

> *I see little hope for any further substantial improvements in mathematics education until we turn mathematics education into an experimental science, until we abandon our reliance on philosophical discussion based on dubious assumptions, and instead follow a carefully constructed pattern of observation and speculation, the pattern so successfully employed by the physical and natural sciences.*

As of the writing of this book, Dr. Begle's advice on how to improve the K-12 math curriculum has yet to be taken seriously.

The Current NCTM Standards/Common Core Arithmetic Curriculum

By 1980, the elementary school math curriculum, in particular, had become a hodgepodge — a confusing mixture of the New Math and "old math." Basic principles of arithmetic, like the associative, commutative, and distributive properties, had been retained, but topics that had caused mass confusion, like computing in different bases, had been deleted, even though necessary to fully understand base ten numeration. Said hodgepodge and the act-of-war report from *A Nation at Risk* prompted the National Council of Teachers of Math (NCTM) to expand on its release of *An Agenda for Action*. The result was announced in 1989: *Curriculum and Evaluation Standards for School Mathematics*, hereafter referred to as the NCTM Standards.

Contrary to the words "curriculum" and "standards" in the official title of the NCTM Standards, the document was neither a curriculum nor a set of standards against which student progress could be measured. Instead, it was a set of guidelines for developing a K-12 math curriculum based on the prejudices of the organization's leadership. For arithmetic, that meant focusing on applications (problem solving) — just one of the three aspects of the subject — and downplaying the other two: basics and concepts.

The NCTM Standards prescribed a radical departure from the two previous curricula for K-12 math. For elementary school, it championed the use of calculators as early as kindergarten and disparaged the actual teaching of arithmetic, claiming that children would discover how to add, subtract, multiply, and divide on their own just from solving wordy arithmetic problems with calculators — problems like 15 ducks and 28 ducks are how many ducks? instead of 15 + 28.

Children who solve arithmetic problems with calculators who have not been taught how to solve them with paper and pencil learn that calculators are "smart," which tends to shackle them to calculators for even the simplest of arithmetic problems. In contrast, children who have been taught how to solve arithmetic problems with paper and pencil learn that *they* are smart — a belief that is arguably *essential* for them to succeed in algebra and advanced mathematics.

What algebra teachers say: *"Students need to be better prepared in basic math skills and not be quite so calculator dependent."* — National Mathematics Advisory Panel, 2008.

For high school, the NCTM Standards downplayed offering calculus and discredited "learning math for math's sake" (learning math in order to learn more math), which denied the very nature of the subject, that it is like a ladder that is ascended rung by rung, where the rungs for arithmetic must be ascended in order to ascend those for algebra, in order to ascend those for calculus, in order to ascend those for the STEM disciplines:

> *Without adequate foundations in arithmetic skills and concepts from elementary school, entering middle school students will be unable to progress to algebra. Without strong foundations in algebraic skills and ideas, the doors to subsequent meaningful math courses will be closed.* — David Klein, 2003

Arithmetic counts! It matters in the pursuit of higher mathematics, and research about how learning occurs at the granular level of neurons[7] explains why.

Speaking for myself …

When I use a calculator, the only time I pay attention to what I am doing is when I input numbers and select the operation I want performed on them. After that, I press the equals key (=) to get the answer. Since all that usually takes just a few seconds, I usually do it twice to check the answer, but once it checks out, I just go with it. It never enters my mind to see if I could get the same answer with some paper-and-pencil procedure, but then I know how to compute with paper and pencil. What about children who do not know how to do that? Why should they even believe that the answers they get with calculators are correct if they cannot check a few of them by working them out mentally or with paper and pencil?

Neuroscience describes learning in terms of *neural circuits* — neurons that are interconnected to form networks that carry out specific functions when activated. It claims that learning occurs when the neurons for a network connect with other neurons to form a more complex network. Therefore,

---

[7] A neuron is a brain cell that transmits nerve impulses to other neurons.

since beginning algebra is generalized arithmetic — arithmetic with literal (letter) numbers, like x, y, z — it is reasonable to assume that the neuronal network for algebra must include the neuronal network for arithmetic. To assume otherwise, that students can learn algebra without first learning arithmetic is, to use an old metaphor, to assume that they can put the cart (algebra) before the horse (arithmetic) and that the horse will somehow catch up with the cart.

Backlash to the NCTM Standards was immediate, but there was no stopping its widespread acceptance, partly because of when it was released. For most of the 20th century, an eighth-grade level of literacy had been adequate for most jobs in the United States, but that changed in the 1970s when American markets were flooded with foreign products that had been made at a fraction of what it would have cost to make them in America. As a result, in the 1980s, *hundreds of thousands of low-skilled American jobs migrated to other countries*, and most of the workers who lost their jobs were unprepared to meet the educational requirements of the jobs that remained.

By the late 1980s, American business leaders were complaining about having to spend *billions* on new hires to remediate their deficiencies in the basics:

> *The American workforce is running out of qualified people. If current demographic and economic trends continue, American business will have to hire a million new workers a year who can't read, write, or count. Teaching them how and absorbing the lost productivity while they're learning will cost industry $25 billion a year for as long as it takes.* — David Kearns, Chairman and CEO, Xerox Corporation, 1987

The need to significantly upgrade the performance of American schools was indisputable. America's economy was at stake! State governors were the first to respond to the issue.

In 1986, the National Governors Association (NGA) devoted a gathering of all 50 state governors solely to education. The outcome was a commitment to a standards-based movement that would hold schools and teachers accountable for results. Three years later, President George H.W. Bush convened a summit with the NGA to establish national goals for education. Among them was that American students would rank at the very top in international testing in math and science by the year 2000.

By 1989, America was desperate for a way to improve student performance in math and science, and lo and behold, along came the NCTM Standards — a set of guidelines for creating a "new and improved" K-12 math curriculum that was based on *standard*s (acceptable levels of attainment), instead of mere goals, as in the curriculum it replaced. The main advocate for the NCTM Standards was the National Science Foundation (NSF), the same organization that had been the main advocate for the New Math, but this time their advocacy succeeded.

Throughout the 1990s, the NSF promoted what were called systemic initiative grants to state education agencies and strategically selected school districts to encourage states and certain school districts to align their math curricula with the curricular guidelines in the NCTM Standards. The NSF even supported the creation of commercial textbook series, like *Everyday Mathematics* by McGraw-Hill and *Math Trailblazers* by Kendall & Hunt, that were purposely aligned to the NCTM Standards.

By 2000, most states had accepted the NCTM Standards as gospel and had revised their standards for K-12 math to reflect the curriculum it outlined (Raimi, 2000). Although the NCTM claimed that the NCTM Standards was not a curriculum, within a decade of its release, it was well on its way to becoming one that was national in scope in spite of notable objections to it. One such objection was its offhanded treatment of fractions.

> *The most important foundational skill not presently developed [in the NCTM Standards] appears to be proficiency with fractions. The teaching of fractions must be acknowledged as critically important and improved before an increase in student achievement in algebra can be expected.* — National Mathematics Advisory Panel, 2008

The NCTM Standards essentially scrapped computation with fractions, probably because the NCTM was focused on solving arithmetic problems with calculators at the time, and fractions cannot be entered directly into most calculators. Instead, their decimal equivalents must be entered, and having to change fractions into decimals and write them down or remember them long enough to enter them into a calculator is a nuisance. (To experience the bother, solve $3/7 + 4/11$ on a calculator. The answer is 0.79, an approximation of $61/77$.[8]) Notwithstanding, demoting fractions to historical relics in the NCTM Standards was a crushing deterrent to mathematical growth (and cooking, as added by one of my editors, Ann Hilborn, a retired English teacher).

Math is a cumulative subject. It builds on itself, and one of its building blocks is fractions. In algebra, if $7x = 3$, then $x = 3/7$. In geometry, if figures are similar (have the same shape), like the two trapezoids below, their corresponding parts are in proportion ($a_1/a_2 = b_1/b_2 = c_1/c_2 = d_1/d_2$). In probability, the likelihood of getting a head when tossing a fair coin is $1/2$. In trigonometry, for a right triangle with sides a and b and hypotenuse h, the sine, cosine, and tangent of angle A are a/h, b/h, and a/b, respectively. In calculus, the derivative of $y = f(x)$ with respect to x is dy/dx. I could go on.

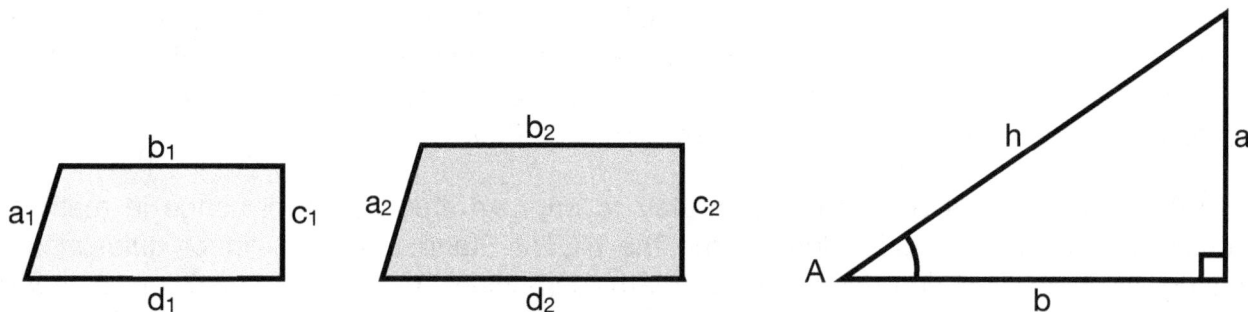

During the 1990s, controversy surrounding the NCTM Standards became so acrimonious that in 1998, the Secretary of the U.S. Department of Education (USDOE) called on the American Mathematical Society and the Mathematical Association of America to intervene. I am not aware of them doing that, but if they did, their effort was doomed to fail. Within a year of the Secretary's request of them, the USDOE escalated the dispute. It sent a letter to virtually every school district in the nation in which it recommended the use of textbooks from a list it had prepared of textbooks that were aligned with the NCTM Standards.

---

[8] $3/7 + 4/11 = 0.43 + 0.36 = 0.79$, an approximation of $3/7 + 4/11 = 33/77 + 28/77 = 61/77$.

Consternation with the list was immediate. A multitude of notable mathematicians sent a letter to the Secretary of the USDOE calling for the list to be withdrawn. According to Klein (2003), signatories to the letter included more than 200 mathematicians and *"seven Nobel laureates and winners of the Fields Medal, the highest international award in mathematics, as well as math department chairs of many of the top universities in the country."* Nonetheless, the list was not withdrawn.

In 2000, perhaps in response to the magnitude of the controversy surrounding the NCTM Standards, the NCTM released a revision of it entitled *Principles and Standards for School Mathematics* (PSSM). If the PSSM was meant to be a peace offering, it failed. It failed because it ignored many of the concerns that both parents and mathematicians had been expressing about the NCTM Standards ever since its takeover of the K-12 math curriculum in 1989. For instance, although the algorithms for the four basic operations in arithmetic (addition, subtraction, multiplication, and division) received additional emphasis in the PSSM, teaching them was still discouraged (Klein, 2003). Students were still expected to figure out how to compute on their own just from using calculators to solve arithmetic problems with words of things like ducks attached to the numbers.

Constructivism

The reason the NCTM did not budge on its insistence that students should discover how to compute was based on a flawed interpretation of a highly regarded learning theory called constructivism. Said theory asserts that learners should be actively engaged in the learning process, not passive recipients of information. It assumes that all learning stems from previous learning and that learning occurs *when and only when* learners <u>assimilate</u> new information with what they already know and construct new mental models with which to <u>accommodate</u> the new information. For example, in learning about dogs, assimilation is the process of taking in new information about dogs, and accommodation is reconciling the new information with one's prior knowledge of dogs.

Instead of having students memorize right answers or their teachers' explanations, constructivism declares that effective teaching is assisting students with the assimilation and accommodation of new information that they might understand it relative to what they already know. *So far, so good.*

> *Two of the key components which create the construction of an individual's new knowledge are accommodation and assimilation. Assimilating causes an individual to incorporate new experiences into the old experiences. This causes the individual to develop new outlooks, rethink what were once misunderstandings, and evaluate what is important, ultimately altering their perceptions. Accommodation, on the other hand, is reframing the world and new experiences into the mental capacity already present.* — Teach-nology.com, 2000

The problem with constructivism was how devotees of the NCTM Standards misinterpreted it. They believed that it implied that the *only* knowledge that is *truly* learned is that which the learner discovers, which was nonsense. Nonetheless, giving children calculators and expecting them to discover how to add, subtract, multiply, and divide on their own became the undisputed method for teaching arithmetic after the publication of the NCTM Standards.

Students need to construct their own understanding of each mathematical concept, so that the primary role of teaching is not to lecture, explain, or otherwise attempt to "transfer" mathematical knowledge, but to create situations for students that will foster their making the necessary mental constructions. — Math Forum, 2000

Some supporters of the NCTM Standards, such as Kamii and Dominick (1998), even claimed that it was *harmful* to teach arithmetic to children because teaching it would rob them of the opportunity to discover it, which was absurd. How can children be expected to discover how to compute in a numeration system with place value based on the exponential powers of some number, like 2 for binary arithmetic (e.g. $1011_2 = 1 \cdot 2^3 + 0 \cdot 2^2 + 1 \cdot 2^1 + 1 \cdot 2^0$) or ten for decimal arithmetic (e.g. $6053_{10} = 6 \cdot 10^3 + 0 \cdot 10^2 + 5 \cdot 10^1 + 3 \cdot 10^0$), just by fiddling with numbers on calculators? Neither the Egyptians who built the pyramids nor the Romans who constructed the aqueducts ever discovered place value numeration based on the powers of some number after thousands of years of "fiddling with numbers." What harms children is their knowing that they are not "getting it" when they see the raised hands of their classmates in response to a question that they cannot answer.

Insisting that arithmetic should be discovered, not taught, was misguided. Granted, discovering something about arithmetic all by oneself results in knowledge that by its very nature is meaningful; however, to insist that arithmetic *must* be discovered for it to be meaningful is going too far. For instance, the concept of percent may be taught meaningfully as a combination of two words, "per" and "cent," which together mean "per one hundred," like "per the number of *cents* in a dollar," because the word "cent" is derived from the Latin word "centum," meaning "hundred." That is why mathematicians settled on % to denote percent.

The percent sign is a scrambled 100 with the one between the two zeros like a fraction bar to suggest hundredths. Therefore, 25 percent, for example, written as 25%, equals 25/100, and if you just now learned how the word "percent" is linked to %, the percent sign, you did not discover that. You learned it by my *telling* you about a plausible connection between the two.

Bonafide discovery is usually a small step relative to a wealth of knowledge that has been acquired in the many ways that knowledge is acquired, as from reading books, listening to lectures, and watching documentaries on TV. When Isaac Newton created the calculus in 1665, he did not brag about how he had *discovered* it. Instead, he credited the "giants" whose "shoulders" he had stood on for his discovery. He thereby acknowledged how he had *learned* from the books and scholarly papers that his peers from times past had bequeathed to the scientific community. In doing so, he singled out an advantage that humans have over all other creatures — the ability to preserve information and transmit it from one generation to the next. Without that ability, humans would still be living in caves.

An unprepared mind discovers nothing, except maybe a $5 bill on the sidewalk that fell out of somebody's purse or wallet. Teaching, including chalk-and-talk, is a proven method for preparing minds, and the NCTM Standards claiming otherwise for arithmetic was malpractice. *It ensured that many children would not learn arithmetic*, at least not in school — not until they were grown and, if they had children, had to help them with their math homework.

Getting back to the NCTM Standards and its sequel, the PSSM, by 2005, the belief that arithmetic had to be discovered, not taught, finally reached a breaking point with the many who still believed that it should be taught. The NCTM had to either quit bashing the teaching of arithmetic or provide evidence that children were, in fact, discovering how to compute on their own just from using calculators to solve arithmetic problems embellished with words.

In 2006, the NCTM quit the bashing. It released *Curriculum Focal Points* (CFP), which listed the most critical topics for math in K-8, one of which was basic arithmetic *and the need to teach it*. The CFP was the NCTM's admission that it should not have squelched the teaching of arithmetic in the NCTM Standards and the PSSM. Never mind that it took them 17 years from 1989 to 2006 to admit their blatant disregard for computation, resulting in countless Americans who were supposed to discover how to compute when they were in elementary school who now have difficulty making change without a calculator.

Common Core State Standards for K-12 Math

By 2009, most states had adopted the NCTM Standards as modified by the PSSM and tweaked by the CFP as the basis for determining a set of benchmarks for measuring student progress in K-12 math. Unsurprisingly, the benchmarks differed from state to state, which caused problems for students transferring from one state to another who were out of sync in math with their new classmates. In 2009, the National Governors Association[9] (NGA) took the lead in addressing the problem and collaborated with the Council of Chief State School Officers[10] (CCSSO) to unify the benchmarks. The outcome was a 93-page booklet entitled *Common Core State Standards for Mathematics* (CCSSM) that became available in 2010. A copy of that portion of the CCSSM having to do with arithmetic in K-3 may be viewed in the appendix beginning on page 264.

Unlike the NCTM Standards and the PSSM, which explained how to *develop* a K-12 math curriculum that would reflect the NCTM's vision for such, *the CCSSM was a ready-made math curriculum that mirrored it.* Oddly, the NGA and CCSSO claimed otherwise. Nonetheless, *the CCSSM was a curriculum.*

> *How is the CCSSM different from the NCTM standards? While the documents share a vision, they do serve quite different purposes. The NCTM Standards put forth a broad vision for school mathematics, including many examples and advice for implementation, while the CCSSM provides a detailed set of grade-by-grade standards that can be immediately adopted as a state curriculum document.* — Association of Mathematics Teacher Educators, 2011

The creation of the CCSSM is the most significant event *ever* to impact the K-12 math curriculum in the United States! As pointed out in this chapter, the curriculum determines teacher input, which determines student output, so whoever controls the curriculum controls

---

[9] The National Governors Association (NGA) is a nonpartisan organization that is made up of the governors of all 50 states, three territories, and two commonwealths.

[10] The Council of Chief State School Officers (CCSSO) is a nationwide, nonprofit organization composed of the public officials who head the departments of elementary and secondary education in the states, five U.S. extra-state jurisdictions, the District of Columbia, and the Department of Defense Education Activity.

student output. Surprise! The NGA and CCSSO control the curriculum because they created the CCSSM and thereby own the copyright to it. *Thus the CCSSM is a nationwide takeover of the K-12 math curriculum by state governors*, albeit with good intentions.

As shown on the next two pages of covers of textbooks, workbooks, and testing materials that were in use in 2020, the CCSSM determines the instructional input for math in *every* textbook-based math program in *every* public, private, and charter school in America. As of 2020, 41 states — all but Alaska, Florida, Indiana, Oklahoma, Minnesota, Nebraska, South Carolina, Texas, and Virginia — have adopted the CCSSM (corestandards.org/math). In actuality, though, all 50 states have adopted it because they all draw from the same pool of textbooks, workbooks, and practice tests for K-12 math.

Common Core State Standards TEXTBOOKS for K-12 Math in 2020

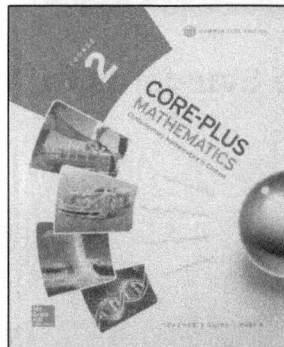

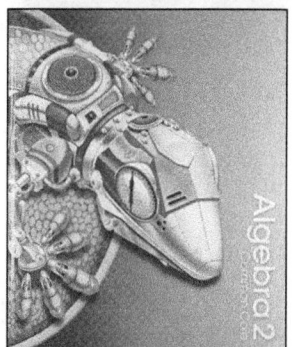

Common Core State Standards WORKBOOKS for K-8 Math in 2020

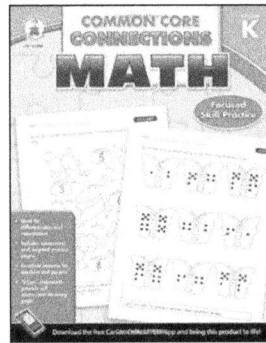

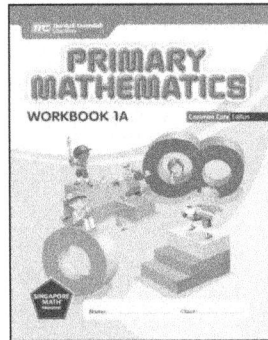

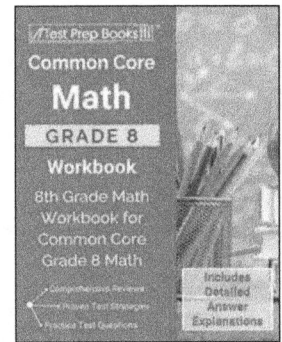

Common Core State Standards PRACTICE TESTS for K-5 Math in 2020

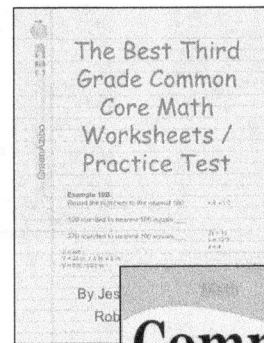

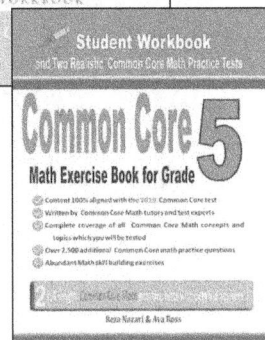

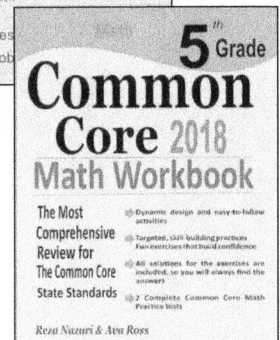

So there you have it. The CCSSM is America's national math curriculum for K-12 for the 21st century. Considering that it is a product of the NCTM Standards that have determined America's K-12 math curriculum since 1989, the TIMSS scores in the introduction clearly show that it is naive to believe that the CCSSM will someday result in American students performing as well or better than their foreign counterparts in math and science.

In 2019, nine years after the publication of the CCSSM in 2010 and 30 years after the publication of its progenitor, the NCTM Standards in 1989, the TIMSS scores in 2019 were just as dismal as those for 1995, and I predict that those for 2023 will be equally dismal when they become available in December, 2024. If I am right, but America's educational establishment just spins the results to make themselves look good and doubles down in defense of the CCSSM, what do you think the TIMSS scores for American students will be for 2027? 2031? ... 2099?

Closing Remarks

According to virtually all studies of the matter, textbooks have become the de facto curriculum of the public schools. — Harriet Tyson-Bernstein, 1988

The point of this book is not to have America disregard calculators and computers and return to paper-and-pencil arithmetic. Indeed, *trillions* of arithmetic computations are performed daily throughout the world with calculators and computers to solve a wide variety of problems in virtually every field of human endeavor. The point is that the root cause of the poor performance of American students in math and science in international testing, as shown in the introduction, is the curriculum and that the current one — the CCSSM based on the NCTM Standards — is a boondoggle, at least for arithmetic:

The textbooks today contain even less mathematics than they did in 1955. The books for grades 1-8 come packaged for teachers with mountainous "Teachers Guides," in which the mathematics is swamped into insignificance. — Ralph Raimi, 1995

According to Raimi (2000), the NCTM Standards was not the product of thoughtful, research-based deliberations to equip every U.S. citizen with the math used in daily living while simultaneously preparing him or her to advance in math and science in high school if they wished to do so. Instead, it was the creation of a "progressive" faction of the NCTM leadership in concert with like-minded "professional educational bureaucrats" from colleges of education, state education agencies, and the U.S. Department of Education who disallowed dissent, loudly and belligerently promoted so-called problem solving with calculators, and sneered at teaching computation. I can verify that as recounted later.

Criticism of the CCSSM for K-6 is not with the entirety of it. Much of it is great! Unlike the bland curriculum I experienced as a child in the 1940s and '50s or the stuffy New Math of the '60s, the CCSSM for elementary school math dictates a mathematical smorgasbord with side dishes of geometry, measurement, graphing, and more, but if elementary school teachers are directed to teach just tidbits of said subjects and mini-arithmetic as prescribed in the CCSSM (as will be shown in the next chapter), then children will only learn tidbits of said subjects and *piddly arithmetic*, which means that they will be unprepared to succeed in algebra, which they must if they are to succeed in calculus and the STEM disciplines.

References

An Agenda for Action: Recommendations for School Mathematics for the 1980s, National Council of Teachers of Mathematics, 1980.

Association of Mathematics Teacher Educators (AMTE). "How Is the CCSSM Different from the NCTM Standards?" *Frequently Asked Questions about the Common Core State Standards for Mathematics*, c. 2011.

Baker, David and Hilary Knipe, John Collins, Juan Leon, Eric Cummings, Clancy Blair, and David Gamson. "One Hundred Years of Elementary School Mathematics in the United States: A Content Analysis and Cognitive Assessment of Textbooks From 1900 to 2000," *Journal for Research in Mathematics Education*, 2010.

CCSSM. See *Common Core State Standards for Mathematics.*

Committee of Ten on Secondary School Studies (**1892**). Wikipedia.org.

Common Core State Standards for Mathematics (CCSSM), National Governors Association in collaboration with the Council of Chief State School Officers, 2010.

Curriculum and Evaluation Standards for School Mathematics, National Council of Teachers of Mathematics, 1989.

Curriculum Focal Points, National Council of Teachers of Mathematics, 2006.

International Association for the Evaluation of Educational Achievement, First International Mathematics Study. Appendix to School Organization and Student Achievement, University of Gothenburg (Sweden), Department of Education and Special Education, 1964.

Kamii, Constance and Ann Dominick. "The Harmful Effects of Algorithms in Grades 1-4," *The Teaching and Learning of Algorithms in School Mathematics*, National Council of Teachers of Mathematics, 1998 Yearbook.

Kearns, David T. *Learning to Be Literate in America*, Foreword, 1987.

Kilpatrick, Jeremy. *Journal for Research in Mathematics Education*, Volume 32: 421-27, 2001.

Klein, David. "A Brief History of American K-12 Mathematics Education in the 20th Century," Copyright by Information Age Publishing, 2003. Final version published in *Mathematical Cognition*.

Kline, Morris. *Why Johnny Can't Add: The Failure of the New Math*. New York, N.Y.: St. Martin's Press, 1973.

Math Forum. "Constructivism in Mathematics Education," Proceedings of the Ninth International Congress on Mathematical Education, Gravemeijer, Koeno et al. 2000.

Miller, Jeffrey W. *Whatever Happened to the New Math?* American Heritage, 1990.

National Commission on Excellence in Education. *A Nation at Risk: The Imperative for Educational Reform*, U.S. Department of Education, 1983.

National Mathematics Advisory Panel FINAL REPORT, U.S. Department of Education, 2008.

National Research Council, *Everybody Counts: A Report to the Nation on the Future of Mathematics Education*. National Academy Press, Washington, D.C.,1989.

NCTM Standards. See *Curriculum and Evaluation Standards for School Mathematics.*

Pickover, Clifford A. *The Math Book: From Pythagoras to the 57th Dimension, 250 Milestones in the History of Mathematics*, Sterling Publishing Co., 2009.

Principles and Standards for School Mathematics (**PSSM**), National Council of Teachers of Mathematics, 2000.

Raimi, Ralph. "What's at Stake in the K-12 Standards Wars: A Primer for Educational Policy Makers." Edited by Sandra Stotsky. *Judging Standards for K-12 Mathematics*, 2000.

Raimi, Ralph. *Whatever Happened to the New Math?* 1995. Slightly amended in 2005.

Ravitch, Diane. *The Death and Life of the Great American School System, How Testing and Choice Are Undermining Education*, Basic Books, 2010.

Russell, Bertrand. *Mysticism and Logic*, 1918.

Teach-nology.com/currenttrends/constructivism/piaget, Piaget's Theory of Constructivism.

Tyson-Bernstein, Harriet. *A Conspiracy of Good Intentions ... America's Textbook Fiasco,* Council for Basic Education, Washington, D.C., 1988.

Wikipedia.org/Butchart, Harvey.

Wikipedia.org/Progressive Education.

Wu, Hung-Hsi. "Basic Skills Versus Conceptual Understanding: A Bogus Dichotomy in Mathematics Education," *American Educator*, American Federation of Teachers, 1999.

Chapter 3: Why So Many Children Struggle with Arithmetic

We have inherited a mathematics curriculum conforming to the past, blind to the future, and bound by a tradition of minimum expectations. — National Research Council, 1989

This chapter examines the arithmetic curriculum for K-3 as prescribed in the 2010 release of *Common Core State Standards for Mathematics* (CCSSM) by the National Governors Association and the Council of Chief State School Officers. Said curriculum for arithmetic in those grades consists of 79 standards, from counting to 100 in kindergarten to relating area to the operations of addition and multiplication in grade 3. All 79 of them are listed in the appendix beginning on page 264: 17 for kindergarten, 18 for grade 1, 20 for grade 2, and 24 for grade 3. Except for highlighting, italics, and the alphanumeric code by which they are referenced, they are listed *exactly* as they appear in the CCSSM.

The Common Core standards for K-3 arithmetic are of two types: one about content, the other about performance:

Content standards are general statements about what students should know or be able to do as a result of instruction. They describe desired outcomes that should follow from instruction designed to achieve them.

Performance standards are measurable goals that must be met to supposedly ensure success with the subject. Of the 79 Common Core standards for K-3 arithmetic, 73 are content standards (noted by a circle in the list in the appendix), and six are performance standards (noted by a square in the same list), apportioned as follows: one for kindergarten, one for the first grade, two for the second grade, and two for the third grade.

To examine the Common Core standards for K-3 arithmetic, it is enough to examine the six performance standards for those grades that are listed below as well as in the appendix, because meeting them implies having met the content standards that are foundational to the performance standards. For instance, in order to meet the performance standard for addition of whole numbers, the content standards for base ten numeration must be met.

Common Core Performance Standards for K-3 Arithmetic

Kindergarten

■ K.OA.5

Fluently add and subtract within 5.

Grade 1

■ 1.OA.6

Add and subtract within 20, demonstrating **fluency** for addition and subtraction within 10. Use strategies such as counting on; making ten (e.g., $8 + 6 = 8 + 2 + 4 = 10 + 4 = 14$); decomposing a number leading to a ten (e.g., $13 - 4 = 13 - 3 - 1 = 10 - 1 = 9$); using the relationship between addition and subtraction (e.g., knowing that $8 + 4 = 12$, one knows $12 - 8 = 4$); and creating equivalent but easier or known sums (e.g., adding $6 + 7$ by creating the known equivalent $6 + 6 + 1 = 12 + 1 = 13$).

Grade 2

- **2.OA.2**

    Fluently add and subtract within 20 using mental strategies. (See performance standard 1.OA.6 for a list of mental strategies.) By [the] end of Grade 2, <u>know from memory all sums of two one-digit numbers</u>.

- **2.NBT.5**

    Fluently add and subtract within 100 using strategies based on place value, properties of operations, and/or the relationship between addition and subtraction.[11]

Grade 3

- **3.OA.7**

    Fluently multiply and divide within 100 using strategies such as the relationship between multiplication and division (e.g., 8 × 5 = 40 implies 40 ÷ 5 = 8, 40 ÷ 8 = 5, and vice versa) or properties of operations. By the end of Grade 3, <u>know from memory all products of two one-digit numbers</u>.

- **3.NBT.2**

    Fluently add and subtract within 1000 using strategies and algorithms based on place value, properties of operations, and/or the relationship between addition and subtraction.

Common Core Fluency Requirement for Arithmetic

To interpret the performance standards for arithmetic in K-3, my first task was to clarify the meaning of the word "fluency" as used or alluded to in the six such standards for arithmetic in those grades. According to the CCSSM, fluency means *"skill in carrying out procedures <u>flexibly</u>, <u>accurately</u>, <u>efficiently</u> and <u>appropriately</u>."* That is the only definition of fluency in the CCSSM, and the words "flexibly," "efficiently," and "appropriately" are not defined in the document in spite of their openness to interpretation. However, Gojak (2012), a former President of the National Council of Teachers of Mathematics, notes that fluency means, at the very least, *"fast and accurate."* She adds that it means more than that, as given below, but she, too, fails to explain what is meant by the words "flexibly" and "efficiently."

> *Students exhibit <u>computational fluency</u> when they demonstrate <u>flexibility</u> in the computational methods they choose/understand and can explain these methods, and produce <u>accurate</u> answers <u>efficiently</u>.* — Linda M. Gojak, 2012

Nonetheless, what fluency means in the CCSSM can be understood in how the word is used in the performance standards for arithmetic in K-3 and, as shown later, in how advocates for the CCSSM use the word in referring to computation.

---

[11] **Strategies based on place value** refer to the algorithms for addition and subtraction: the column-by-column procedures that are used to solve addition and subtraction problems. Those based on "properties of operations" and "relationships between addition and subtraction" refer to the mental strategies for figuring out the number facts that are listed in performance standard **1.OA.6**.

For the number facts, being fluent means being able to instantly recall the addition and multiplication facts from memory and being able to quickly figure out all of the number facts mentally with thinking strategies like the ones listed for 1.OA.6, the performance standard for the first grade.

For computation, it means being able to add, subtract, multiply, and divide whole numbers by rote (mechanically, without thinking) with the standard algorithms for such.

With that understanding of the word "fluency" and the assumption that the word "within" means "less than or equal to," the performance standards for arithmetic in K-3 can be restated as follows.

Common Core Fluency Standards for K-3 Arithmetic Restated

Kindergarten

■ **K.OA.5**

Kindergarteners must memorize the addition and subtraction facts with sums and differences less than or equal to **five** (e.g., 2 + 3 = 5 and 5 − 2 = 3). They must rehearse them until they can recall them automatically without hesitation, the same as they can recall their own names.

Grade 1

■ **1.OA.6**

First graders must memorize the addition and subtraction facts with sums and differences less than or equal to **ten** (e.g., 4 + 6 = 10 and 10 − 4 = 6). They must rehearse them until they can recall them automatically without hesitation, the same as they can recall their own names. Additionally, they must know how to count on[12] or use mental strategies to figure out all 100 addition facts from 0 + 0 = 0 to 9 + 9 = 18 and all 100 subtraction facts from 0 − 0 = 0 to 18 − 9 = 9.

Grade 2

■ **2.OA.2**

Second graders must know how to mentally figure out all 200 addition or subtraction facts. Additionally, they must memorize all 100 addition facts. They must rehearse them until they can recall them automatically without hesitation, the same as they can recall their own names.

---

[12] Counting on means to count forward a certain number of counts after some number. Its main use is to count out an addition or subtraction fact. For example, 5 + 8 can be counted out by counting on five from eight (9-10-11-12-13) or eight from five (6-7-8-9-10-11-12-13) to determine that 5 + 8 = 13, the last number in either count, and 14 − 6, for example, can be counted out by counting on from six to 14 (7-8-9-10-11-12-13-14) and keeping track of the counts on one's fingers or with tally marks (||||| |||) to determine that 14 − 6 = 8, the number of counts.

Note, however, that the leadership of the National Council of Teachers of Mathematics abhors counting out a number fact. They liken it to counting on one's fingers, which everyone is supposed to know is *bad*. That is probably why counting on is the only strategy in standard 1.OA.6 without an example. Were there one, it would show that counting out an addition or subtraction fact is meaningful and mathematically sound.

- **2.NBT.5**

  Second graders must know how to <u>add and subtract two-digit numbers</u> with sums and differences less than or equal to 100 (e.g., 36 + 64 = 100 and 72 − 18 = 54).

Grade 3

- **3.OA.7**

  Third graders must know how to use mental strategies to figure out all 100 multiplication facts from 0 x 0 = 0 to 9 x 9 = 81 and all 90 division facts from 0 ÷ 1 = 0 to 81 ÷ 9 = 9. Additionally, <u>they must memorize all 100 multiplication facts</u>. They must rehearse them until they can recall them automatically without hesitation, the same as they can recall their own names.

- **3.NBT.2**

  Third graders must know how to <u>add and subtract three-digit numbers</u> with sums and differences less than or equal to 1000 (e.g., 437 + 518 = 955 and 620 − 294 = 326).

Déjà Vu!

Except for having to memorize the number facts for multiplication in the third grade instead of the fourth grade, the Common Core arithmetic curriculum for K-3 is the same arithmetic curriculum that my classmates and I coasted through when we were in those grades more than 75 years ago.

As said in Chapter 2, I never complained about the redundancy in arithmetic when I was in elementary school in the 1940s. I was too naive then to do that. Now, however, I realize that my classmates and I were held back in arithmetic in K-3 and that America's children are still being held back in arithmetic in K-3. The Common Core arithmetic curriculum for those four years is just a perpetuation of how arithmetic has been taught in America since the progressive education movement that began in 1880. Said curriculum is not "spiraled," as publishers claim. It does not coil upwards from year to year, revisiting topics at more advanced levels.

> *In practice, although not in law, we have a national curriculum in math education. It is an "underachieving" curriculum that follows a spiral of almost constant radius, reviewing each year so much of the past that little new learning takes place.* — National Research Council, 1989

By the mid 1980s, the redundancy in the K-8 math curriculum was widely acknowledged. To assess the redundancy, Flanders (1987) examined the following textbook series for grades 1-8 and their ninth-grade algebra texts:

- *Addison-Wesley Mathematics* (1987) and *Algebra* (1986), Addison-Wesley Publishing Company

- *Invitation to Mathematics* (1985) and *Scott, Foresman Algebra: First Course* (1984), Scott, Foresman and Company

- *Mathematics Today* (1985) and *HBJ Algebra I* (1983), Harcourt Brace Jovanovich

To measure the redundancy, Flanders conducted a page-by-page review of the textbooks to count the number of pages from one grade to the next that contained content that had already been covered. *If a page contained even a fraction of new content*, the page was not included in the count. Thus by comparing the number of pages with no new content to the total number of pages in a textbook, he was able to measure the percent of redundancy from one grade to the next.

From kindergarten to grade 1, Flanders found that the redundancy in the textbooks in the three series averaged 26 percent. For grades 2-5, he found that it averaged 50 percent from grade to grade, and for grades 6-8, he found that it averaged 65 percent from grade to grade. In sharp contrast, he found that it averaged only 12 percent from grade 8 to the ninth grade algebra texts.

Flanders reported his findings in 1987 in the *Arithmetic Teacher,* an NCTM publication for elementary school teachers. The title of the report was "How Much of the Content in Mathematics Textbooks Is New?" The following are Flanders' words in describing his findings:[13]

> *Students can expect to see the majority of the mathematics they encounter in a given year again and again in the years to come.*

> *There should be little wonder why good students get bored: they do the same thing year after year.*

> *Even a cursory examination of tables of content shows that early chapters redo old arithmetic.*

> *On average, the first half of a grade 1-8 book has 35 percent new content, whereas the second half of the book has 60 percent new content.*

> *Early in the year, when students are likely to be more eager to study, they repeat what they've seen before. Later on, when they are sufficiently bored, they see new material — if they get to the end of the book.*

> *We say we want students to be active and creative problem solvers, yet we set up an environment that seems designed to discourage them from thinking about new ideas — in short, an environment designed to put them to sleep.*

> *The primary finding is that a relatively steady decrease occurs in the amount of new content over the years up through eighth grade, where less than one-third of the material is new to students. This decrease is followed by an astounding rise in the amount of new content in the texts of the most common ninth grade course, algebra.*

The NCTM's response to Flanders' disclosure of the redundancy in the textbook series he examined was not what I expected. The NCTM was obviously well aware of his findings from having published his article in the *Arithmetic Teacher*. Nonetheless, it ignored the redundancy he revealed in the textbook series he examined. Instead, *it embraced it in the NCTM Standards that it released in 1989,* two years after it published Flanders' article. It enshrined the redundancy with educational jargon (BS) about how repeating old ground is bearing down on the basics and codified it in supporting the Common Core State Standards math curriculum for K-6:

---

[13] If you think the current textbook series for math are not as redundant as Flanders reported more than 30 years ago, consider making your own assessment of their redundancy. Go to any elementary school that is using a textbook series for math and examine it for redundancy. You can probably estimate it just from reading the table of contents in the textbooks from one year to the next.

The Common Core calls for greater focus in mathematics. Rather than racing to cover many topics in a mile-wide, inch-deep curriculum, the standards ask math teachers to <u>significantly narrow and deepen</u> the way time and energy are spent in the classroom. This means <u>focusing deeply</u> on the major work of each grade. — Common Core website @ corestandards.org/Other Resources/Key Shifts in Mathematics, 2020

To "significantly narrow and deepen" the curriculum in order to "focus deeply" on what is left of it are false claims that dragging out the teaching of arithmetic by repeating a few topics year after year until memorized is the best way to teach arithmetic. Teaching arithmetic by dragging it out is like trying to teach a centipede how to walk a leg at a time. Assuming the centipede has the usual 15 pairs of legs for 30 legs all total, if taught to walk a leg at a time, the centipede would have to "assimilate and accommodate" (from the section on constructivism in Chapter 2) each of the 30 lessons with the ones already taught until all 30 were taught. Thus no surprise if the centipede never learns how to walk if taught to do so a leg at a time.

You CAN Tell a Textbook by Its Cover

In commenting on the redundancy Flanders noted in the textbook series he examined, he remarked that he could hardly tell the textbooks apart from one grade to the next without looking at their covers. I can attest to that.

In 1989 or thereabouts, I acquired a federal grant to work with a small elementary school near Victoria, Texas where I worked for a branch campus of the University of Houston. The grant provided for my salary to teach a course at the school on how to teach arithmetic effectively to all children. Additionally, it covered the cost of tuition for teachers to enroll in the course for three hours of graduate credit and paid to equip their classrooms with the instructional materials they would learn how to use in the course. In all, the grant was for about $30,000 in training and instructional materials for the school.

Prior to the first day of class, I had a preliminary meeting with the teachers. My goal for the meeting was to emphasize the need to enrich and accelerate their K-6 math curriculum. Anticipating an *"if it ain't broke, don't fix it"* attitude among the teachers, I planned to show them that their math curriculum was, in fact, "broken" and needed "fixing." To that end, I had asked the teachers to bring their textbooks for math to the meeting.

To begin the meeting, I asked a first grade teacher to read the table of contents in her textbook. (All of the grade school teachers were women.) I then asked the same of a second grade teacher, and so on. I soon noted that the teachers were becoming agitated. They were surprised to learn that so much of the math that they had been teaching was just a repeat of the math from the previous year. The discussion that followed was lively and encouraging. I left believing that I had achieved my objective.

The very next day, I received a call from my department Chair and was told that the school whose teachers I had met with the day before was opting out of the grant. *I got the message.* I had embarrassed the school's administration in the cloddish way I had revealed the redundancy in the school's textbook series for math.

I accept that I had been too brash in how I went about having the teachers in the above scenario realize that their math curriculum was "broken," but it was not my having them read the table of contents in their math texts that upset administration. It was the truth it revealed — *a truth too staggering for them to accept or for America to disregard.* Just by doing their jobs and teaching what they were told to teach from their math texts, their teachers were holding their students back in arithmetic, the same then as now if the school is still using a textbook series for K-6 or K-8 math.

Few facts stand undisputed in educational research, but the dependence of teachers on textbooks and of students on tests is as firm a finding as exists in this amorphous discipline; especially in mathematics, teachers teach only what is in the textbook and students learn only what will be on the test. — National Research Council, 1989

Teaching More than What Is in the Textbooks

Granted, elementary school teachers can offset the redundancy in their textbooks with enrichment material, and many do, but they are severely limited in how much they can veer from their textbooks. Just having to teach from a textbook makes a teacher a member of a grade K-6 or K-8 instructional team that is wedded to the rest of the textbooks in the series, and the team's playbook is a master list that tells each teacher on the team how much of their textbook to cover by a certain date. Thus if a teacher fails to teach what they are supposed to teach from their textbook, they will throw the team off track. *Worse, though, is if they teach <u>more</u> than what they are supposed to teach from their textbook.*

As pointed out in the Flanders' study, the redundancy in textbooks in a K-6 or K-8 series from one grade to the next is so pronounced that a teacher who teaches more math than they are expected to teach is liable to encroach on what the teacher next in line for their students is supposed to teach, which can lead to conflict with that teacher. A case in point is provided by Anne Shaw, an elementary school teacher I met about 30 years ago.

When I met Anne, she was teaching first grade at a private school in Victoria, Texas, and I was the Director of the Mathematics Education Initiative (MEI) at the University of Houston in Victoria (UHV). As the Director of the MEI, my self-imposed mission was to disseminate MOVE IT Math, the elementary school math program that I had developed, throughout Texas. To that end, I regularly taught courses in MOVE IT Math to practicing elementary school teachers.

I met Anne when she contacted me at UHV to learn about MOVE IT Math after the mother of one of her first graders recommended that she do so. To fulfill her request, I had her accompany me to a school in Port O'Connor, Texas that was using the program. We spent just the one day together, but I remembered her when she contacted me a few years ago through my website @ moveitmath.com.

Since spending the aforementioned day with me in Port O'Connor, Anne had obtained a master's degree in education and had made a business out of conducting workshops for teachers on how to implement project-based learning in their classrooms. After our reunion, she posted the following testimonial about MOVE IT Math on her website @ 21stcenturyschools.com:

I was teaching first grade in a private school when the mother of one of my students [told] me about a course she was taking at the University of Houston in Victoria, Texas taught by Paul Shoecraft. She urged me to contact him to find out about his math program.

I called Dr. Shoecraft to learn about his program. He stated that he was going to visit a pilot project taking place in Port O'Connor, Texas in a few days and invited me to accompany him.

We arrived at the school in Port O'Connor and entered a very large classroom/lab. There were no student desks, just math centers set up around the perimeter and down the center of the room. A fifth grade student was stationed at each center to act as facilitator, and small groups of KINDERGARTEN children were rotating through the centers.

A sort of clothesline was strung around the perimeter of the room above the math centers. Hanging from this line were many samples of the kindergartner's Monster Math[14] problems. The Monster Math problem you see here is actually much simpler and smaller than those I observed that day! These children were successfully — and excitedly — doing addition, correctly, with answers in the millions!

I continued to be astounded as I watched these five-year-olds go through the various math centers adding, subtracting, multiplying, dividing, doing Geometry (determining the area and perimeter of irregular polygons), Algebra (yes, solving for x), Graphing Coordinates, working with Fractions (converting fractions to decimals and vice versa; adding and subtracting mixed numbers; and more). They were developing their number concepts by working not only in Base Ten, but adding and subtracting in Bases 2, 3 and 4! And there was more!

I decided to immediately take this to my first graders. I bought [the hands-on materials and the first volume of Math Games and Activities to make the games that were at the instructional centers] and took this math to my classroom. Very soon my first graders were doing all this math, and they loved it! They were so very proud of themselves, and they walked around with their heads held high feeling like they were Einsteins!

There was a problem, though. It seems that the second grade teacher at this school was afraid that there would be nothing to teach the children when they got to her classroom the next year. She complained about this to the headmaster [who] called me into his office and told me about [her complaint]. Then he said, "You are teaching them too much. You are to stop it right now. I do not want to see your students using manipulatives or moving about the classroom. You are to put them in their seats, give them a timed math test every day and post the results in the hall."

I refused to do either of these things. I continued using MOVE IT Math and I did NOT put the children in their seats, give them timed tests and then post the results in the hall. That is against everything I believe in and know about teaching as well as child development.

At the end of the school year, the headmaster called me into his office again and said, "We do not want you to come back to this school next year." So I found a job teaching at a nearby school that embraced and supported my teaching style and MOVE IT Math!

---

[14] Pamela Dolezal, a first grade teacher in Hallettsville, Texas, created the "Monster Math" theme based on my use of the word "monstrous" in describing the size of the arithmetic problems that children could solve with low-stress algorithms if they understood base ten numeration. Thank you, Pam.

Tina Dusek, another First Grade Teacher Who Taught *"too much"* Math

I met Tina when she enrolled in a math course I taught at UHV for pre-service elementary school teachers. My impression of her at the time was that of a mature, confident woman who had raised her family and was now focused on "becoming," and for Tina, that meant becoming an elementary school teacher, which she voiced with conviction. Of all my students at the university, I believed she, at least, would override the dumbed down standard arithmetic curriculum and teach MOVE IT Math in its place.

After Tina graduated, I encountered her again when she enrolled in a graduate course I was teaching off campus. By then, she was a first grade teacher for Lamar Consolidated Independent School District in Rosenberg, Texas. In visiting with her after class, I asked if she were teaching MOVE IT Math. Imagine my shock when she said *"No."*

My surprise must have showed, because Tina teared up. I do not remember what we talked about after that, but her not teaching MOVE IT Math upset me. If not even Tina was able to stand up to an arithmetic curriculum that she knew was holding her students back, who then? If nobody, what was the point in my even teaching MOVE IT Math to prospective elementary school teachers?

Tina's response to our conversation was different than mine. She resolved to teach MOVE IT Math, which I now realize was just asking for trouble with administration, albeit "good trouble." If you saw the film Stand and Deliver, Tina got in trouble like Jaime Escalante, a high school math teacher in California who drew unwanted attention when his minority students scored *"too high"* in calculus on a state achievement test. As a result, he was suspected of cheating in administering the test and had to administer it again, except with "watchers" to make sure he did not cheat. Imagine how he must have felt when his students exonerated him when they again scored *"too high"* in calculus.

Within a year or so of my talk with Tina, I learned of the consequences of her resolve to teach MOVE IT Math to her first graders. What came of it led to my staying in touch with her ever since. I will leave it to Tina to tell you what came of it in a letter I asked her to write for this book.

Dear Paul,

I well remember the instance you're talking about. It was in Wharton, Texas. You were teaching a class for post graduate credit. I decided to take it. After the first class is when we had that discussion. Here's how I remember it.

I was looking forward to learning even more from you. In that first class you talked about how we must change the way we teach math to young children and that they were capable of so much more than we were allowing them to do. Listening to you, I vowed to do better when I got back to my first grade classroom. Then, after class, we talked and you asked me what I was doing in my class and I had to confess that I was doing the textbook. I was so ashamed! Yes, I got tears in my eyes and again vowed to do better.

The next day, I took all the textbooks from my students and put them in the closet and told the kids we were going to learn math in a new fun way and they were to let me know how they liked it. I started out teaching them the painless practice games, counting backward in as

53

many ways as I could think of, and counting on using the touchpoints. When I felt the kids were ready for a unit test from our Math textbook, I gave it to them and recorded it in their permanent record forms as I was supposed to. As the kids became proficient in the different areas covered by the textbook, I gave them the appropriate test for that unit. By April we had finished the textbook's tests and no one had scored less than a B on any Unit test.

I also instituted Fraction Friday. In first grade all they are supposed to do is recognize and color in various fractions. We used your Fraction Cakes and the kids were making all sorts of equivalents! I was amazed! So the next Fraction Friday I showed them how to add like fractions. The BIG rule was that to do so, the fractions had to have the same denominator or be the same color if using fraction cakes. So I would write a simple fraction addition problem on the board and using their fraction cakes they would solve it.

We did this for a few weeks and then I showed them how to write it for themselves. My first graders were adding fractions! They used manipulatives as long as they needed them, but most of them put them away in a few weeks, but they were always available if they were needed. Then we moved on to subtraction and it was a no brainer. The fun thing was at the end of each lesson I would write a problem on the board that had DIFFERENT denominators. They would tell me I couldn't do that because they weren't the same color. I would laugh and say they caught me and that I was proud of them that I hadn't been able to trick them.

This continued until April. After one such lesson I wrote $\frac{1}{2} + \frac{1}{4} = ?$ on the board, fully expecting them to "catch" me again. But a child in the back of class raised his hand and asked the wondrous question, "Why can't we make a trade? Can't we trade the $\frac{1}{2}$ for two $\frac{1}{4}$s? That's a fair trade isn't it? Then all the fractions would be the same color." I asked the class what they thought. They thought it was a fair trade, but we got out the Fraction Cakes to check it out, and they all agreed it was a fair trade.

My first graders were now adding and subtracting fractions with different denominators! The really interesting thing was that two years later when these kids were in 4th grade their teacher called me and asked if I had taught certain kids and she read out a list of students. I had taught every one of them. I was scared. What was wrong? She laughed and said she just knew it. She had given her class a pretest for fractions and those kids had tested out of that unit! They remembered how to add and subtract fractions despite not having any review whatsoever since first grade. They had truly internalized the concepts! The grade 2 and 3 teachers did not teach Move It Math in any form.

Later in April. My students were given the ITBS (Iowa Test of Basic Skills). 100% of my students mastered addition, subtraction, equality, and fractions. They did extremely well on everything else too. Evidently, a little too well. It caught the attention of our Deputy Superintendent and the District Math Coordinator. They had a meeting with my Principal thinking that as a first year teacher I had not given the test properly. Fortunately, my principal had a monitor in both 1st grade classrooms to help out. They talked to her and she said I followed the directions precisely and gave the test correctly.

Their next stop was me. I had to fess up. I showed them the textbooks still in the closet. They stared at me! "Well, what did you teach them?" they asked. I showed them what I had done. The students came in about that time and I asked them if they wanted to show my bosses how to add and subtract fractions. They answered with a resounding YES! So, we showed them how my first graders could add and subtract fractions. My bosses were agape! They went around to different students asking them to explain what they were doing and the kiddos did an excellent job of explaining fractions to them.

At the end of the year, we all sat down and discussed how they had learned math this year. They all agreed they liked "Dr. Shoecraft's" math the best!

"Now for the rest of the story," as said by Paul Harvey, a popular radio announcer a long time ago. Unlike Anne Shaw, Tina was not fired for teaching her first graders *"too much"* math. Instead, she was allowed to continue teaching MOVE IT Math, and she got the same outstanding results with her first graders her second year with Lamar Consolidated as she had her first year there.

In response to the success Tina was having with MOVE IT Math, Dr. Paul Slocumb, the Assistant Superintendent for Lamar Consolidated ISD at the time, asked her to apply for the job of Gifted and Talented (GT) Program Facilitator, even though applicants were to have five years of teaching experience, and Tina had only taught for two years. That may be why she told him that she was not yet ready to leave the classroom. However, when he asked her to apply for the job after her third year of teaching, she applied for it and got the job. For the next 21 years until she retired, she helped GT teachers by modeling how to elevate lessons to a gifted level, writing GT performance based units, and teaching K-5 GT pullout classes.

Math Ladder

To grasp the significance of restricting the arithmetic curriculum in K-3 to just addition and subtraction with small numbers through most of grade 3 and the memorization of the number facts for addition, subtraction, multiplication, and division, imagine a ladder whose rungs represent the major topics in math in kindergarten through four years of high school math. Beginning with arithmetic and assuming one rung per topic for the ladder, the first rung would be for counting, the second one for number/numeral.

To get to the first rung, kindergarteners must learn how to count to 100. To get to the second rung, they must know how to read and write the numerals from 0 to 100 and know that the numerals refer to amounts. For example, if asked how many eggs in a full carton of eggs, they must be able to count all of the eggs without counting any of them twice and express the number of eggs in the carton with the numeral 12.

After the first two rungs on the Math Ladder, there would be seven rungs for whole number arithmetic. The third rung would be for understanding base ten numeration. The fourth rung would be for the addition facts and related subtraction facts. The fifth rung would be for the addition algorithm. The sixth rung would be for the subtraction algorithm. The seventh rung would be for the multiplication facts (times table) and related division facts. The eighth rung would be for the multiplication algorithm, and the ninth rung would be for the division algorithm.

After the rungs for whole number arithmetic, there would be four rungs for adding, subtracting, multiplying, and dividing fractions intermixed with four rungs for the same with decimals. Finally, intermixed with the preceding 17 (2 + 7 + 4 + 4) rungs would be the 18th rung for percent, the 19th rung for rounding off, and the 20th rung for estimation. Thus arithmetic would take up the first 20 rungs of the Math Ladder and a few more, but 20 rungs is enough to appreciate the climb children must make to learn arithmetic.

After arithmetic, the Math Ladder would have at least 100 more rungs for the major topics in four years of college preparatory math courses in high school:

○ Algebra I in year 1

○ Geometry in year 2

○ Algebra II/trigonometry in year 3

○ Probability/statistics, pre-calculus/analysis, or calculus in year 4

Thus the Math Ladder would consist of a minimum of 120 rungs: 20 for arithmetic and 100 for four years of college preparatory math in high school. Yet according to the Common Core performance standards for K-3, America's children are only expected to reach the seventh rung of the Math Ladder by the end of grade 3, the rung designated by performance standard 3.OA.7 that mandates the memorization of the multiplication facts in that grade.

Now reflect on what it means for America's children to only reach the seventh rung of the Math Ladder by the end of the third grade. *It means that they still have at least 113 rungs to climb and have used up nearly a third (all of K-3!) of the 13 years from K through 12 to climb them.* That averages out to about 13 major topics per grade for grades 4 through 12 compared to just one or two major topics per grade for kindergarten through grade 3.

Small wonder that many fourth graders struggle with the change in pace in math and decide that they did not get the "math gene." By the time they get to high school, I imagine that as many as half of them are resigned to taking only enough math to graduate. They do not realize how quitting on math in high school may restrict their futures.

> *Over three quarters of the degree programs at most universities require courses in calculus, discrete mathematics, statistics, or other comparable mathematics. College students need this level of mathematical literacy in order to understand with precision the mathematical ideas that form the foundations for science, business, and engineering courses.* — National Research Council, 1989

The Most Important Consideration in Developing a Curriculum

Cognitive development — the blossoming of intelligence, conscious thought, and the ability to solve problems — begins in infancy and continues throughout adult life. Infants are born with a genetic potential, and their brains are primed to learn, but to reach their potential requires a wealth of experiences. Cognitive growth is maximized in an environment that is intellectually stimulating and flatlined in one that is intellectually stagnant. Thus the most important consideration in developing a curriculum is to ensure intellectual stimulation.

The current baby-talk arithmetic curriculum for K-3 is a mind-deadening experience for children. It is stifling and unimaginably boring. It drags out addition and subtraction with small numbers for four years! In doing so, it totally ignores the intellectual capability of young children as measured by their capacity to learn a *language* by the time they go to school.

With mainly just encouragement from family members, *all* children learn a language by the time they are in kindergarten, *which is an astounding intellectual achievement!* They learn how to use verbs to reflect the past, present, and future and how to arrange words and phrases to create sentences they never heard before — *all without formal instruction*. If raised in homes where two languages are spoken, they learn both of them without even trying and can speak them without an accent. *Never again will they learn so much with so little conscious effort*, so why the failure in the K-3 arithmetic curriculum to take advantage of such precociousness? No amount of inquiry reveals a satisfactory answer to that question:

> *Claims based on theories that children of particular ages cannot learn certain content because they are too young, not in the appropriate stage, or not ready have consistently been shown to be wrong. Nor are claims justified that children cannot learn particular ideas because their brains are insufficiently developed,[15] even if they possess the prerequisite knowledge for learning the ideas.* — National Mathematics Advisory Panel, 2008

Being deprived intellectually while growing up has *irrevocable* consequences as evidenced by how difficult it is for most adults to learn another language. Most Americans speak just English and at best a smattering of another language. That is because most were raised in families that spoke only English and thus were *deprived* of the opportunity to learn another language when they were children and would have learned it effortlessly. The older they get, the more they will struggle to learn another language, like me in trying to learn French and Spanish in college. Why should it be any different for children in learning arithmetic?

Arithmetic Is a Language

Arithmetic is a language, <u>*a written one for describing numeric events concisely and precisely*</u> — events like paying bills, preparing a budget, and buying a car.

o The numerals 0, 1, 2, 3, 4, 5, 6, 7, 8, 9 are the "alphabet" for this language and are used to "spell" numbers, including fractions and decimals with additional notation, and the numbers themselves are "nouns" because they are the names of things, namely, *amounts*.

o The operations of addition (+), subtraction (–), multiplication (x), and division (÷) are "verbs" because they convey actions to be performed with numbers.

o Equals (=), less than (<), and greater than (>) are also verbs because of how they are used to construct "sentences," namely, number sentences, like 9 + 3 = 12, 9 + 2 < 12, and 9 + 4 > 12.

To learn the language of arithmetic, children must be *immersed* in it, the same as for any language. In school, *they must be a member of a family of teachers and classmates who read and write arithmetic as the most informative and efficient way to record numeric events*, real or imagined. They must add, subtract, multiply, and divide numbers big and small, including fractions, *beginning in kindergarten*. The result is children who learn to "talk" (read and write) arithmetic *without an accent*.

---

[15] By the time children begin kindergarten, 95 percent of their brain development has already occurred (Roaten & Roaten, c. 2011).

Algebra Is a Language

Understanding arithmetic as a written language is the basis for understanding algebra as an extension of that language. Literal (letter) numbers, like x, y, z, are *pronouns*, like "she," "they," and "it," because they are the names of numbers that are not specified, and coefficients, like 5 in 5x = 30 or pi (π) in C = πd, are *adjectives*, because they modify the pronouns x and d. Thus to "talk" (read and write) algebra without an accent, *which children must if they are to succeed in the STEM disciplines* (science, technology, engineering, and mathematics), they must grow up in a school family of teachers and classmates who are conversant with algebra, which is made possible with concrete referents for the symbolics, as shown below with Algebra Tiles.

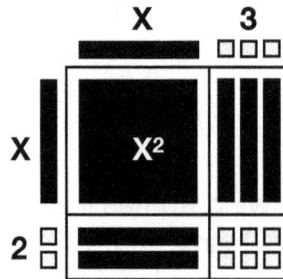

Area = Length x Width = (X + 3)(X + 2) = X² + 2X + 3X + 6 = X² + 5X + 6

Arithmetic without Pronouns Is Baby Talk

Imagine interacting at work, visiting with friends, or being around family for even five minutes if you were not allowed to use a single pronoun — a word referring to someone or something mentioned previously in a conversation. Below are some of the words you could not say:

I, Me, My, Mine, You, Your, Yours, He, Him, His, She, Her, Hers, It, Its, We, Us, They, Them, Our, Ours, Their, Theirs

Imagine how stilted you would sound in telling the story below about Al and Ben if you could not use any pronouns:

○ **Without pronouns:** Al and Al's brother Ben were playing outside. Al and Ben got so dirty that Al and Ben's mom told Al and Ben that Al and Ben could not come inside until Al and Ben hosed Al and Ben's selves off.

○ **With pronouns:** Al and his brother Ben were playing outside. They got so dirty that their mom told them that they could not come inside until they hosed themselves off.

Likewise, the written language of arithmetic is stilted without pronouns (literal numbers). For instance, consider X + 1 = 10. Equations like this show up in the first grade in the Common Core arithmetic curriculum as unknown-addend problems[16] to show that addition and subtraction are related. (See content standards 1.OA.4 and 1.OA.8 on pages 268-269 in the

---

[16] An unknown-addend problem is an algebraic equation of the form X + a = b where a and b are given and X is the unknown addend. For example, for X + 1 = 10, a = 1, b = 10, and X is the addend that must be determined.

appendix.) However, a literal number like X is not used to signify the unknown addend. Instead, a question mark, underlining, or a square is used. Thus instead of solving X + 1 = 10, children guess the answers to number sentences like ? + 1 = 10, ___ + 1 = 10, or □ + 1 = 10, all of which indicate that some number plus 1 equals 10, the same as X + 1 = 10. So why the avoidance of literal numbers and algebraic equations in elementary school math?

The motivation to shelter children from algebra is the long held but unsubstantiated belief that literal numbers are too abstract (scary!) for children, whereas a question mark, underlining, or a square are not because they just ask for an answer. That is ridiculous. Language is full of abstractions. Consider the word "love." How is what that word stands for less abstract than a letter like X standing for a number?

Elementary school mathematics leads up to algebra and, as such, it should prepare students for algebra. Using abstract variables whenever possible is great preparation. — W. Stephen Wilson, 2009

Hands-on Algebra Beginning in Kindergarten

Algebra beginning in kindergarten? *Absolutely!* Children must become familiar with algebraic equations early on because algebra cannot be avoided in arithmetic, as noted by the occurrence of unknown addend problems like X + 1 = 10 in the first grade.

To solve an equation like X + 1 = 10, young children use a math balance. As shown below, they load the beam with a weight on the one peg on one side and a weight on the ten peg on the other side. Then, with a single weight, as shown, they experiment with where to hang it until the beam is horizontal, that is, *balanced*. The answer is the number above the peg where they hang the weight to make the beam balance. Thus for X + 1 = 10, X = 9.

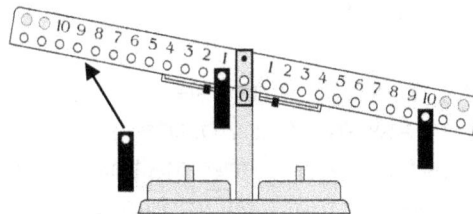

$$X + 1 = 10$$
$$X = \underline{}$$

In solving algebraic equations on a math balance, children learn that the equal sign (=) means "balanced" or "is the same as," not "show me the answer," as on a calculator. To be mathematically precise, they learn that "equals" is an *equivalence relation*, not an operation. It is not a command to do something, like add or subtract. It means that the two sides of an equation represent the same amount even though they may look different, as with equivalent fractions, like 2/3 = 6/9, and in algebraic equations, like X + 1 = 10.

Kindergartners do not balk at literal numbers like x, y, z in equations like I did in ninth grade algebra when I encountered them for the first time. They have not been in school long enough to be conditioned to just working with numbers, so they are not intimidated by the letters when they are told that they are numbers whose number names they have yet to know. Instead, they revel in discovering their number names with the algebra toys they get to play with that are listed in the appendix beginning on page 279. *Algebra counts, too!*

> *Algebra is a demonstrable gateway to later achievement. Students need it for any form of higher mathematics later in high school; moreover, research shows that completion of Algebra II correlates significantly with success in college and earnings from employment. In fact, <u>students who complete Algebra II are more than twice as likely to graduate from college compared to students with less mathematical preparation</u>.* — National Mathematics Advisory Panel, 2008

The path to success in the STEM disciplines and a multitude of careers, including ones in business, health care, and government, begins with *arithmetic*, followed by algebra as generalized arithmetic, followed by calculus using the language of algebra, followed by the math requirements for whatever career for which calculus is a prerequisite. The path is not sluggishly plodding through whole number arithmetic and zero algebra in K-6 except to claim that exposing children to *any* pattern, even one like red, blue, green; red, blue, green; and so on, is preparation for algebra. Publishers might as well claim that it is preparation for calculus.

The Automaticity Standard

> *Regardless of our age, we all share a common rite of passage in early education — the mastery of math facts.* — Rebecca Sholes, 2018

Children struggle with arithmetic because of the automaticity standard. To meet the standard, they must "automate" (memorize) arithmetic. They must be able to recall from memory all 390 number facts without hesitation and know how to add, subtract, multiply, and divide whole numbers, fractions, and decimals by rote.

The automaticity standard is time honored. It has set the benchmark for knowing arithmetic for as long as anyone can remember. Ask most any grade 3-6 elementary school teacher or junior high, high school, or college math teacher if the multiplication facts, in particular, should be memorized, and you will probably be told that they should be.

Automaticity has to do with how tasks are performed on a scale from <u>consciously</u> competent to <u>unconsciously</u> competent. If you play a musical instrument, you are somewhere on that scale for that instrument. To achieve automaticity with it if you have not done so, you would have to practice with it until you could play it without having to think about how to play it, which is what is meant by being "unconsciously competent."

Everybody attains automaticity for tasks that they perform regularly, like tying the laces on their shoes, walking, or driving a car. Very few, however, attain it with a musical instrument as played by a professional musician. Nonetheless, the automaticity standard requires every child to attain automaticity with arithmetic the same as Beethoven attained it for playing a piano. Adherence to the automaticity standard in the Common Core arithmetic curriculum is the reason for all the worksheets and speed drills on the number facts in K-3.

*Computational proficiency with whole number operations is dependent on sufficient and appropriate practice to develop **automatic recall** of addition and related subtraction facts, and of multiplication and related division facts. It also requires **fluency** with the standard algorithms for addition, subtraction, multiplication, and division.* — National Mathematics Advisory Panel, 2008

Oddly, the word "automaticity" does not appear in the CCSSM. Nonetheless, the automaticity standard is fully embedded in the document. As shown below, said standard is responsible for all six of the Common Core performance standards for arithmetic in K-3:

o **Kindergarteners** must achieve automaticity with the addition and subtraction facts with sums and differences less than or equal to **five** (K.OA.5).

o **First graders** must achieve automaticity with the addition and subtraction facts with sums and differences less than or equal to **ten** (1.OA.6).

o **Second graders** must achieve automaticity <u>with all 100 addition facts</u> by the end of the school year (2.OA.2). Additionally, they must be able to quickly and accurately add and subtract **two-digit** numbers (2.NBT.5).

o **Third graders** must achieve automaticity <u>with all 100 *multiplication* facts</u> (the times table) by the end of the school year (3.OA.7). Additionally, they must be able to quickly and accurately add and subtract **three-digit** numbers (3.NBT.2).

Thus the automaticity standard determines the Common Core arithmetic curriculum for K-3. It specifies 1) what will be taught and 2) how it will be taught. What will be taught are the number facts for all four operations and how to add and subtract small numbers. How they will be taught is with <u>repetition</u>. Granted, many teachers supplement their instruction on how to add and subtract with hands-on materials, as with arithmetic blocks and colored counters, but the main objective of the Common Core arithmetic curriculum for K-3 is to force the memorization of the number facts.

By <u>fact fluency</u>, we mean knowing a math fact with <u>automaticity</u>, much like students know their sight words without thinking about them. That is how we want students to know their math facts. — XtraMath, 2020

The goal in adding and subtracting just two-digit numbers in grade 2 and three-digit numbers in grade 3 is *not* to practice addition and subtraction. The goal is to avoid serious computation with paper and pencil until all 390 number facts have been memorized. Spending grade 2 on adding and subtracting two-digit numbers and grade 3 on the same except with three-digit numbers is just running in place after grade 2. Adding and subtracting three-digit numbers is no different than adding and subtracting two-digit numbers except for doing the same thing one more time. Thus it is fair to say that the only arithmetic children are expected to learn in the first four years of elementary school is the number facts.

Importance of the Number Facts in Arithmetic

The origin and staying power of the automaticity standard is due to the importance of the number facts. Unless students know them, they cannot compute because the number facts come up in each step of the step-by-step procedures that are used in the standard algorithms for addition, subtraction, multiplication, and division. To realize this, please verify in your mind the answers to the following four problems.

$$
\begin{array}{cccc}
& & \mathbf{2} & \\
73 & 96 & 29 & \\
+\ 25 & -\ 41 & \times\ 3 & 96 \div 4 = 24 \\
\hline
98 & 55 & 87 &
\end{array}
$$

To get the answers shown, you had to know from memory (or work out) the following number facts: $3 + 5 = 8$, $7 + 2 = 9$, $6 - 1 = 5$, $9 - 4 = 5$, $3 \times 9 = 27$, $3 \times 2 = 6$, $6 + \mathbf{2} = 8$, $9 \div 4 = 2$ R1, and $16 \div 4 = 4$. Thus the number facts and computation are inseparable. Students *must* know the addition facts to add, the subtraction facts to subtract, the multiplication facts to multiply, and the division facts to divide, _which is why many students falsely believe that they lack the ability to learn math_. They keep forgetting some of the 390 number facts.

Granted, memorizing the number facts has been made more meaningful in some textbook series, as with fact families in *Everyday Math* that show that addition and subtraction are the inverse (opposite) of one another, as are multiplication and division. For example, the fact family for an addition fact like $5 + 8$ is $5 + 8 = 13$, $8 + 5 = 13$, $13 - 5 = 8$ and $13 - 8 = 5$, and for a multiplication fact like 5×8, it is $5 \times 8 = 40$, $8 \times 5 = 40$, $40 \div 5 = 8$, and $40 \div 8 = 5$.

Fact families should definitely be taught, but to show how the operations are related, not as a novel way to promote memorizing the number facts with fact family worksheets. Nonetheless, making fact family worksheets for elementary school teachers to buy and duplicate, like the ones below by Edumonitor and Math-Aids, has become a cottage industry. To verify that, google "fact family worksheets." I did and got 27,500,000 hits!

Although arithmetic is regarded as baby math compared to the lofty college preparatory math courses in high school, the automaticity standard makes getting past the rungs on the Math Ladder for the number facts the toughest part of the climb to the top because the rungs are "slippery." Many children keep losing their footing on them because they keep forgetting some of the number facts after a break from practicing them, like after summer vacation, a school holiday, or even a weekend. That is why the arithmetic curriculum in elementary school is devoted to practicing the number facts. *The only remedy for forgetting is more practice.*

Importance of the Number Facts in Algebra

The number facts are too basic to success in arithmetic to entrust to memory without a backup if/when memory fails. In fact, they are too basic to success in *all* of math — *and therefore* science, technology, and engineering — to entrust to memory alone. For instance, note how the number facts come up in algebra.

o **Add** $(5X^2 - 8X + 82) + (7X^2 + 37X - 165)$. **Solution**: Work $5 + 7$, $37 - 8$, and $82 - 165$ to obtain $12X^2 + 29X - 83$.

o **Multiply** $(8X + 9)(7X - 6)$. **Solution**: using FOIL (an acronym for the "first, outer, inner, and last" coefficients), work $8 \cdot 7$, $8 \cdot 6$, $9 \cdot 7$ and $9 \cdot 6$ to obtain $56X^2 - 48X + 63X - 54$. Then work $63 - 48$ to obtain $56X^2 + 15X - 54$.

o **Solve for Y** given $12Y = 80$. **Solution**: Divide both sides of the equation by 12 to obtain $Y = 80/12$. Note that the notation $80/12$ in lieu of $80 \div 12$ requires an understanding of fractions. Work $80 \div 12$ to obtain $Y = 6\text{-}8/12 = 6\text{-}2/3$. Note again the need to understand fractions, in particular, *equivalent* fractions, improper fractions, and mixed numbers.

o **Solve for Z** given $3Z/4 = 5/6$. **Solution**: Using that the product of the "means" [4 and 5] equals the product of the "extremes" [3 and 6], work $4 \cdot 5$ and $3 \cdot 6$ to obtain $18Z = 20$. Divide both sides of $18Z = 20$ by 18 to obtain $Z = 20/18$. Reduce $20/18$ to obtain $Z = 10/9$. Convert $10/9$ to a mixed number to obtain $Z = 1\frac{1}{9}$.

Guessing in math is risky. It often results in wrong answers, and students who get a lot of wrong answers are likely to question their mental capacity, *because doing poorly in math tends to be internalized as being "dumb."* I imagine that guessing wrong for as few as ten number facts in algebra can result in enough wrong answers overtime to cause some students to drop out of math and even out of school.

> *Mathematics is the worst curricular villain in driving students to failure in school. When mathematics acts as a filter, it not only filters students out of careers, but frequently out of school itself.* — National Research Council, 1989

An alternative to guessing when a student forgets a number fact is to work it out, as promoted in 1.OA.6, the first grade performance standard for arithmetic; however, doing so would lack "fluency." So what should a student do if they forget a number fact while computing with paper and pencil?

Guess.

Skip it and go on to the next problem.

Work it out.

Suppose a student has to make one of those choices once or twice a week all through grade school. Which choice over time would best prepare the student for adulthood? Which choice or choices would you associate with dropping out of school before finishing high school?

Arithmetic Should Be Taught with Understanding

Imagine being taught how to read the same as you were taught arithmetic. In kindergarten, you would have been made to memorize the alphabet but not the sounds the letters make or when next to other letters. In grades 1-3, you would have been made to memorize 390 "basic" words. If you did as instructed, you would have dutifully memorized them without thinking about the sounds made by the letters in the words. If, however, you had figured that out and your teacher heard you sound one out, you would have been told not to do that because sounding out a basic word lacks "fluency." It wastes time, hinders comprehension, and is a "crutch" — a way to avoid the monotony of memorizing all 390 basic words.

You would not have been taught grammar, sentence structure, or any of the basic rules of language. Your teachers would have spent all of K-3 trying to get you to memorize the 390 basic words. Some students would have been able to do that, but many would have been told at some point that *"language is not your strong suit."* Does that sound familiar? Did anyone ever say that to you or to a friend of yours about math not being your or their strong suit, especially if you or your friend were girls? If so, did you or your friend decide that math made no sense and give up on it?

Fortunately, reading is not taught as just outlined. In K-3, children learn the alphabet, the sounds the letters make, and how to read by sounding out the letters in words, even words they have not seen before. By the end of grade 3, most children have learned how to read age-appropriate books, so why during the same time period have they not learned arithmetic? The answer to that question is the automaticity standard and how it holds children back in arithmetic, the same as it held me and my classmates back in the 1940s.

Understanding enables thinking, thus math teachers agree that subjects like algebra, geometry, and calculus must be understood because they are too vast in the topics they cover to be memorized, but acknowledging that does not imply that were they not so vast, they should be memorized. So why is arithmetic the only subject on the math ladder that is supposed to be memorized? <u>Requiring that denies children the intellectual independence and emotional certainty that comes from knowing they know how to figure out a number fact if forgotten and know how to think their way through a tedious calculation</u>.

Received Wisdom

As noted earlier, the automaticity standard is widely supported. Nonetheless, support for it should not be mistaken for validation. A bad idea can be passed on for generations as "received wisdom" without questioning its truthfulness, as illustrated in the following made-up story:[17]

---

[17] Ann Richards, while governor of Texas (1991-1995), told this story at a conference for educators about the need for change.

A young woman was preparing to cook a roast. Before putting it in the pan, she hacked two inches off the end of it. She did that because her mom had always hacked two inches off the end of a roast before putting it in a pan.

Wondering why one day, she asked her mom. Her mom said she did it because her mom had always hacked two inches off the end of a roast before putting it in a pan.

Now, both the daughter and mom were wondering why, so they asked the grandmother, who replied, the pan was too short.

Likewise, support for the automaticity standard without questioning its validity has led to foolishness in how it constricts teaching arithmetic in K-3, as will be addressed in the next chapter.

I realize that my railing against the automaticity standard is no match for the *millions* of authoritative voices that support it. Thus I do not expect you, the reader, to agree straightaway with my takedown of the standard. Regardless, I am compelled to speak out against it because I believe that the reasons for rejecting it that will be presented in this book cannot be denied. As an octogenarian in 2020 when I began writing this book, why else would I spend the next four years of my life writing this book except to get what is rattling around in my head on paper so I can quit thinking about it.

A lot of support for the automaticity standard just means that a lot of people are going to take convincing to believe that it is a bad idea and should be dismissed. My best effort at such convincing is forthcoming. My hope is that what became obvious to me during my career about the negative impact of the automaticity standard on the arithmetic curriculum will become just as obvious to *all* Americans that they might speak with one voice to demand that the standard be discarded. Until that occurs, how arithmetic is taught in American schools will continue to be a roadblock to success in the STEM disciplines — science, technology, engineering, and mathematics — instead of the super highway it can be if taught sensibly.

Closing Remarks

The NCTM should not have coined the word "fluently" in the NCTM Standards and supported its use in the Common Core performance standards for the K-3 arithmetic curriculum, as in K.OA.5: *Fluently* add and subtract within 5. The word "fluently" in these documents only applies to arithmetic and how it should be known. It means it should be memorized. Thus it creates a false equivalency between being fluent in one's native language and knowing arithmetic.

Arithmetic is a written language, not a spoken one. Tasked with memorizing it as dictated by the automaticity standard, children must rehearse each puzzle-piece of it in their minds until all of the pieces "stick." In school, they may recite the number facts as a group activity or explain to their class how they performed a computation, but except for events like that, they do not vocalize arithmetic. They do not indulge in small talk with classmates or friends and family about the number facts or the algorithms for adding, subtracting, multiplying, and dividing. Thus it is unrealistic to expect the number facts and how to compute with paper and pencil to flow from their minds as easily and unconsciously as words that they voice every day, seven days a week, in and out of school.

As for the redundancy in the textbook series for grade 1-8 math that Flanders revealed, the NCTM should have rebuffed it in the NCTM Standards. Instead, *it lauded it!* It cheered the redundancy as concentrating on the basics as opposed to covering multiple topics in a curriculum that it disdained as being *"a mile wide and an inch deep."* In so doing, it asserted that the mathematical needs of most of America's children were best served with calculators and a smattering of this-and-that math in grades 1-8 and pablum math (no calculus) in high school, which assumed that most of them were incapable of pursuing the STEM disciplines in high school. The malpractice in championing a curriculum based on low expectations is that student performance tends to match them.

The power of expectations to affect student performance was made clear to me in a conversation I had long ago with Mr. X, a man whose name I have forgotten. When I met him, he was the principal of an all-black high school in Dallas and had been its principal since the civil rights movement in the 1960s. In asking him what it had been like to be the principal of an all-black high school back then, he told me the following story.

During his first year as principal, he noted that the math department did not offer calculus. In asking why of the Chair of the department, he was told that the students were not capable of learning it. In response, he said, in essence,

There is <u>dignity</u> in failing calculus. There is only humiliation in failing basic math again.

He then added that he expected the math department to begin offering calculus and further added that he would fire the Chair if he did not make that happen. Within a few years of that conversation, the department had to offer *multiple* classes of calculus to account for student interest in the course. *Where there's a way, there's a will.*

Getting back to the arithmetic curriculum for K-3, it is not possible to accurately identify which children in those grades will choose to pursue the STEM disciplines in high school. Therefore, the only way to ensure that those who wish to pursue them are prepared to do so is to prepare *all* of them to do so, *which is doable*, as explained later, and better for all of them than having some of them think they are math dummies because they keep forgetting some of the number facts no matter how hard they try to remember them.

Any attempt to separate elementary school children into two groups, one group that will never have the option of becoming an engineer and another group that will be given that option, would seem grossly unfair. All elementary school children should have the option of choosing to try to be an engineer, so all children must be given the necessary mathematics in elementary school. — Stephen Wilson, 2009

The harder a standard is to meet, the greater the number of rejects, as zero-tolerance policies promptly reveal. Products that do not meet the standard can be discarded or sold as "seconds," but if children fail to meet a standard, they cannot be discarded, but if they cannot meet the automaticity standard that all children are expected to meet, they are likely to believe that they lack the innate ability to learn math. That being the case, they are likely to make decisions that negatively impact their futures, like the elementary school teacher who told me the following after completing the first course in MOVE IT Math: All Children Can learn Arithmetic.

When I was in elementary school, I dreamed of being an architect when I grew up, but I gave up on that dream because I was terrible in math. If I had learned arithmetic the way I just learned how to teach it, I would not have had to give up that dream.

Arithmetic counts! It mattered to that teacher.

References

CCSSM. See *Common Core State Standards for Mathematics.*

Common Core State Standards for Mathematics (CCSSM), National Governors Association in collaboration with the Council of Chief State School Officers, 2010.

Curriculum and Evaluation Standards for School Mathematics, National Council of Teachers of Mathematics, 1989.

Dusek, Tina. First grade teacher and GT [gifted/talented] Program Facilitator, Lamar Consolidated Independent School District, Rosenberg, Texas. Retired.

Edumonitor, Fact Families, 4th Grade Math Worksheet, theeducationmonitor.com.

Everyday Math, University of Chicago School Math Project, © 2008.

Flanders, James R. "How Much of the Content in Mathematics Textbooks Is New?" *Arithmetic Teacher*, 35(1), 18–23, 1987.

Gojak, Linda M, NCTM President (2012-14). Summing Up, November 1, 2012.

Math-Aids, Complete Each Family of Facts, math-aids.com.

Math Games & Activities, see Shoecraft, Paul.

MOVE IT Math, moveitmath.com. MOVE IT is an acronym for Math Opportunities, Valuable Experiences, Innovative Teaching.

National Mathematics Advisory Panel, U.S. Department of Education, 2008.

National Research Council, *Everybody Counts: A Report to the Nation on the Future of Mathematics Education*. National Academy Press, Washington, D.C.,1989.

NCTM Standards. See *Curriculum and Evaluation Standards for School Mathematics.*

Roaten, G.K. and D.J. Roaten. "Adolescent Brain Development: Current Research and the Impact on Secondary School Counseling Programs," Texas State University, c. 2011. Available online.

Shaw, Anne. 21st Century Schools at 21stcenturyschools.com.

Shoecraft, Paul. *Math Games & Activities*, Volumes 1 & 2 (more than 600 pages of backline masters for making hundreds of math games out of cardstock), Dale Seymour Publications, 1984.

Sholes, Rebecca. "Is Automaticity a 21st-century Math Skill?" *Victory*, victoryprd.com. 2018.

Wilson, W. Stephen. "Elementary School Mathematics Priorities," *AASA Journal of Scholarship & Practice*, 2009.

XtraMath. "Fact Fluency, Integrating Best & Next Practices into Common Core Math," XtraMath Website, 2020.

Chapter 4: The Automaticity Standard Is Pseudo-Intellectual Flimflam

A single fact can overturn an entire system. — Frederik Van Eeden

As noted in Chapter 3, the six performance standards for the Common Core arithmetic curriculum for K-3 hold children accountable to the automaticity standard:

o **Kindergarteners** <u>must memorize the addition and subtraction facts with sums and differences less than or equal to **five**</u> (K.OA.5).

o **First graders** <u>must memorize the addition and subtraction facts with sums and differences less than or equal to **ten**</u> (1.OA.6). Additionally, they must be able to use the counting and thinking strategies that are listed in the standard to work out all 100 addition facts and all 100 subtraction facts.

o **Second graders** <u>must memorize all 100 addition facts</u>. Additionally, they must be able to figure out all 100 addition facts and all 100 subtraction facts <u>with some mental strategy that does not involve counting</u> (2.OA.2). Moreover, they must know how to add and subtract two-digit numbers with sums and differences less than or equal to 100 (2.NBT.5).

o **Third graders** <u>must memorize all 100 multiplication facts</u>. Additionally, they must be able to figure out all 100 multiplication facts and all 90 division facts <u>with some mental strategy that does not involve counting</u> (3.OA.7). Moreover, they must know how to add and subtract three-digit numbers with sums and differences less than or equal to 1000 (3.NBT.2).

Justification for the Automaticity Standard

For the automaticity standard to have determined the arithmetic curriculum in American schools for as long as anyone can remember, there must be a good reason for that. Accordingly, this chapter examines the rationale for holding children accountable to the standard. In particular, it scrutinizes the justification for requiring children to memorize the number facts.

The gist of the automaticity standard is that arithmetic must be <u>automated</u>. It must be performed mechanically without conscious thought. Thus it dictates that the number facts and the algorithms for adding, subtracting, multiplying, and dividing whole numbers must be memorized.

The justification for imposing the automaticity standard on arithmetic is the assumption that having to work out a number fact or otherwise think while computing is a <u>cognitive distraction</u>,[18] a distraction that will cause children to 1) lose their place in the computational process and make mistakes and 2) hinder their understanding of the math at hand as they advance in the subject.

> **Make mistakes:** *Doing arithmetic, whether mental or written, is a complex cognitive task. If the [number] facts are not available at the level of automatic recall, this means that the student has to shift cognitive attention from the arithmetic task to obtain an answer for such facts questions*

---

[18] The term "cognitive distraction" is used to differentiate between a distraction that involves cognitive activity, like working out a number fact, as opposed to a distraction like a loud noise.

as 3 + 4 by counting or using some other slow procedure. Because the short-term memory of early years students is quite limited, this disruption of focus can negatively affect performance and learning. In other words, the student can too easily lose the train of thought concerning an arithmetic task [and make mistakes] because too much cognitive attention is being paid to a minor component of that task. — Automaticity of the Basic Facts of Arithmetic, 2020

Hinder understanding: *Students must learn counting and acquire instant recall of the single digit number facts for addition and multiplication (and the related facts for subtraction and division). ... In later courses, the student who has to quickly do the single digit computations, even if in their head, rather than just recall the answers, will find they are unable to focus completely on learning and understanding the new mathematics in their course.* — W. Stephen Wilson, c. 2006

Basing the requirement for young children to memorize the number facts on the assumption that their short-term memories are "quite limited" is indefensible. In reviewing more than 100 studies on the short-term memory limitations of children, Chi (1976) was unable to find conclusive evidence that *"the capacity or the rate of information loss from STM [short-term memory] varies with age."* Instead, she found that what appeared to be STM limitations in young children were caused by insufficient knowledge and a lack of strategies for processing information.

For older children, whose short-term memories are not a consideration, the justification for expecting them to attain automaticity with the number facts is the assumption that the cognitive distraction of having to work out a number fact while solving an algebraic equation, for example, will cause them to not learn algebra as well as they might.

Thus regardless of the age of children, the justification for holding them accountable to the automaticity standard is the same, the assumed need to avoid cognitive distractions while computing, else drown in mistakes in arithmetic and only dimly perceive advanced math. But what if the assumption that children cannot multitask while computing without dire consequences is false? *What if it is a myth* — a rationale for requiring children to measure up to the automaticity standard that teachers deem plausible without ever scrutinizing? With that possibility in mind, this chapter examines the justification for the automaticity standard to determine if avoiding the cognitive distraction of having to work out a number fact while computing is reason enough to force the memorization of the number facts.

Explaining to Second Graders Why They MUST Memorize the Addition and Subtraction Facts

Imagine that you are quizzing a class of second graders with flashcards on the 100 addition facts from 0 + 0 = 0 to 9 + 9 = 18 and the 100 subtraction facts from 0 − 0 = 0 to 18 − 9 = 9. What might you tell them if they ask why they have to memorize them even though they learned how to work them out in the first grade using the counting and thinking strategies listed in 1.OA.6, the Common Core performance standard for that grade? If you were to tell them that memorizing the addition and subtraction facts is how you learned to add and subtract, that would not answer their question, so what else might you tell them?

You could tell the second graders that they cannot add and subtract unless they know all 200 addition and subtraction facts, which is true, *but they can know them in three strikingly different ways*. One way is to memorize them. However, performance standard 1.OA.6 cites two additional ways to know them:

o **They can count on.** For an addition fact, they can count on either addend from the other addend. The sum is the last number in either count. For example, for 5 + 8, they can count on 5 from 8 (9-10-11-12-<u>13</u>) or 8 from 5 (6-7-8-9-10-11-12-<u>13</u>) for the answer <u>13</u>, the last number in either count. For a subtraction fact, they can count on from the subtrahend to the minuend and keep track of the counts. The *difference* is the number of counts. For example, for 14 – 9, they can count on from 9 to 14 (10-11-12-13-14) and keep track of the counts on their fingers or with tally marks for a difference of <u>5</u>, the number of counts.

o **They can use mental math**. For an addition fact like 7 + 9, they can "make a ten." 7 + 9 = 6 + 10 = <u>16</u> or 7 + 9 = 10 + 6 = <u>16</u>. For a subtraction fact like 13 – 5, they can "decompose a number leading to a ten." 13 – 5 = 13 – 3 – 2 = 10 – 2 = <u>8</u>.

So how would you want the second graders to know the addition and subtraction facts? Would you want them to know all three ways of knowing them, which would include being able to recall them instantly from memory, the same as they can recall their own names, or could you agree to their just knowing how to work them out when needed with a combination of counting them out and figuring them out mentally?

According to the automaticity standard, you should tell them that they *must* memorize the addition and subtraction facts even if they know how to work them out because they will make mistakes if they work them out while adding and subtracting. But what if they ask *"How do you know we will make mistakes if we work them out while doing that?"* Although not entirely true, you could tell them that they are *"too young"* to remember their place in the computational process if they stray from it by working out a number fact. But what if they persist and ask *"What does being young have to do with forgetting what we are doing while we are doing it?"*

Before struggling to answer the question just asked by the pesky class of second graders, consider the following: *What if cognitive distractions are built into the standard algorithms for arithmetic?* If so, what is the point in making children memorize the number facts if the only reason for making them do so is to avoid cognitive distractions while computing? <u>There is no point if it can be shown that cognitive distractions are unavoidable with the standard algorithms for paper-and-pencil arithmetic regardless of how well the number facts are known.</u>

Examination of Eight Textbook Series for Elementary School Math

In 2011, in anticipation of writing this book someday, I made repeated trips to the textbook repository of the Texas Education Agency in Austin, Texas to conduct a page-by-page review of the following textbook series for elementary school math that were in use at the time:

Saxon Mathematics, K-5, Harcourt Achieve, Inc., © 2008

Texas enVisionMath, K-5, Scott Foresman-Addison Wesley, © 2009

Texas Everyday Math, K-5, University of Chicago School Math Project, © 2008

Texas HSP Math, K-5, Harcourt School Publishers, © 2009

Texas Math, K-5, Houghton Mifflin, © 2009

Texas Mathematics, K-5, MacMillan McGraw-Hill, © 2009

Then, in 2018, in getting serious about starting on the book, I conducted a page-by-page review of *Go Math!* (K-3, Houghton Mifflin Harcourt, © 2012) that I reviewed online, and I leafed through *enVisionMath 2.0* (K, 2-3, Scott Foresman-Addison Wesley, © 2015) that I purchased. As I expected, they were essentially interchangeable with the books I had surveyed before.

My objective in examining the textbook series was to show that their grade-by-grade coverage of arithmetic in K-3 is the same regardless of the publisher. That is, *it was to reveal that the arithmetic curriculum for those grades in every textbook-based math program in every public, private, and charter school in America is the same* — the same because the math content in every textbook-based math program is determined by the same set of standards, namely, those listed in the NCTM Standards and its derivative, the CCSSM.

To examine a textbook series, I examined each textbook in the series by first determining the number of pages that dealt with specific subjects, such as arithmetic, geometry, and measurement. Then, for the pages dealing with arithmetic, I determined how many of them dealt with whole numbers, fractions, decimals, or percent. Finally, for the pages dealing with each of the foregoing headings, I determined the number of pages that dealt with topics that were germane to them. For example, for whole numbers, I determined the number of pages that dealt with numbers and numeration, the number facts, and adding, subtracting, multiplying, and dividing whole numbers.

Except for *enVisionMath 2.0*, my findings for the series examined are included in the appendix. In viewing them, you can see that regardless of the publisher, their grade-by-grade coverage of arithmetic for K-3 is the work of copycats:

Kindergarten: Count to 100. Read and write the numerals 0, 1, 2, 3 … 100. Introduction to the addition and subtraction facts (e.g., 3 + 5 = 8, 8 − 3 = 5).

Grade 1: Add and subtract two **two**-digit numbers. Rarely add *three* two-digit numbers.

Grade 2: Add and subtract two **three**-digit numbers. Rarely add *three* three-digit numbers. Introduction to the 100 multiplication facts (the times table).

Grade 3: Add and subtract two **four**-digit numbers. Rarely add *three* four-digit numbers. Introduction to multiplication.

Except for adding and subtracting slightly bigger numbers in grades 1-3, the textbook version of the arithmetic curriculum for K-3 is the same as the one in the CCSSM for those grades, which is not surprising, because the textbooks and the CCSSM stem from the same source, the NCTM Standards.

So what does my examination of textbook series for elementary school math have to do with the automaticity standard? A lot, actually, because of an oddity with the addition problems in all of the series I examined. For every addition problem with three or more addends, there were up to a hundred or more with only two addends. Wondering why, I solved a few problems with three addends and immediately discovered why.

The Automaticity Standard Is Incompatible with Column Addition

Column addition problems can be sorted into two categories based on the sums in adding the digits in a column. If, in adding the digits in a column, all of the sums are addition facts, it is a Category 1 column addition problem. Otherwise, it is a Category 2 column addition problem.

To illustrate, in adding the digits in the ones column in the Category 1 column addition problem below, $6 + 2 = 8$ and $8 + 5 = 13$, both of which are addition facts. Likewise, in adding the digits in the tens column (after writing the number 3 beneath the ones column and the number 1 above the tens column), $1 + 3 = 4$, $4 + 4 = 8$, and $8 + 1 = 9$, all of which are addition facts.

Category 1 Column Addition Problem

```
                    1             1
   36              36            36
   42              42            42
 + 15  6+2=8,8+5=13  + 15          + 15
              →           →
                     3            93
```

In contrast, in adding the digits in the ones column in the Category 2 column addition problem below, $6 + 8 = 14$ and $14 + 7 = $ *Huh!* 14 + 7 is NOT an addition fact that should automatically come to mind. Children are not expected to have memorized the answer to 14 + 7 *because they are not supposed to memorize it.* They are supposed to compute it, as shown, or work it out in the moment by counting on 7 from 14 (15-16-17-18-19-20-21) or mentally creating an equivalent but easier or known sum (e.g., $14 + 7 = 20 + 1 = 21$), as promoted in 1.OA.6, the Common Core performance standard for the first grade.

Category 2 Column Addition Problem

```
┌──────────────┐
│  Cognitive   │
│ Distraction  │
└──────┬───────┘
       │
       ↓
                          2            2
   46         1          46           46
   18  6+8   14          18           18
 + 27   →   + 7        + 27         + 27
            ────          →
            21  →  1                  91
```

However, the automaticity standard warns of negative consequences if 14 + 7 is solved by any of those methods because solving it while computing would be a cognitive distraction — a deviation from the overall task. Nonetheless, 14 + 7 must be solved in order to complete the task, so the automaticity standard must be waived for 14 + 7.

Accepting that there is no way to escape having to solve addition problems like 14 + 7 aside from the overall task in column addition, what is the point in allocating all of K-2 to memorizing the 100 addition facts? There is no point if the only reason for memorizing them is to avoid cognitive distractions while adding with paper and pencil.

Having waived the automaticity standard for 14 + 7, it should also be waived for 6 + 8 in adding the numbers in the first column. It should be waived because 6 + 8 can be worked out by <u>counting on</u> 8 from 6 (7-8-9-10-11-12-13-<u>14</u>) or 6 from 8 (9-10-11-12-13-<u>14</u>) or by <u>mentally creating an equivalent but easier or known sum</u> (e.g., 6 + 8 = 10 + 4 = <u>14</u>), *the same as for 14 + 7.*

Can we agree that we would sound stupid if we tried to explain to the aforementioned class of second graders why, in column addition, it is okay to use paper and pencil or the counting and thinking strategies in performance standard 1.OA.6 to solve an addition problem like 14 + 7 but not 6 + 8 in adding the numbers in the ones column in 46 + 18 + 27 written vertically?

Taking the Easy Way Out

Assuming agreement, a way to avoid sounding stupid by trying to explain the unexplainable would be to prevent the unexplainable from coming up, which is why Category 2 column addition problems are practically nonexistent in the textbooks for elementary school math until the fourth grade. But look at what happens in adding just two numbers:

1	**11**	**111**	**1111 1111**
28	394	4876	4739226514
+ 56	+ 298	+ 2989	+ 3494867589
84	692	7865	8234094103

If the sum of the numbers in a column with just two numbers is greater than nine, the number that is written at the top of the next column to the left is *always* the number 1 because the largest sum possible with just two numbers is 9 + 9 = 18. Therefore, it is conceivable that children are being *conditioned* in grades 1-3 to *always* write the number 1 at the top of columns when they add. If they are, they are being set up to get the wrong answer to the Category 2 column addition problem under examination. After figuring out 14 + 7 = 21, they will put the 1 in 21 above the tens column and the 2 in 21 beneath the ones column, as shown below.

$$
\begin{array}{cccc}
& & \overset{1}{} & 1\\
46 & \mathit{1} & 46 & 46\\
18 \xrightarrow{6+8} 14 & & 18 & 18\\
+27 & +7 & +27 & +27\\
& \mathit{21} & 2 & 82 \text{ \textbf{Wrong!}}
\end{array}
$$

Then, in adding the digits in the tens column, 1 + 4 = 5, 5 + 1 = 6, and 6 + 2 = 8, for an answer of 82, *which is wrong*. The correct answer is 91, as noted earlier.

Please reflect on the following three questions about the Category 2 column addition problem under consideration:

○ Can we agree that children getting 82 for an answer indicates that they do not understand base ten numeration — that they do not realize that the number 2 in 21 should have been put at the top of the tens column because it represents the number of tens in 21 because of its *place* in the numeral?

o If we agree on that, can we agree that *lots* of children getting 82 for an answer would indicate that adding mostly two numbers for three years — *all of grades 1-3* — undermines children's understanding of base ten numeration?

o If we agree on that as well, can we agree that it is imperative to find out if lots of children would get 82 for an answer?

Assuming that we are in total agreement, beginning in 1987 (*two years before the release of the NCTM Standards*) and every year thereafter for 12 years, the Category 2 column addition problem under consideration was administered to nearly 19,000 children in grades 1-8 in Texas. How that was accomplished will be explained in Chapter 8. For now, suffice it to say that I acquired the results for that problem while I was a Professor in the education department at the University of Houston in Victoria, Texas.

Results for 46+18+27 in Dallas, Houston, and S. Texas, 1987-1999

46 18 + 27	Correct 91	Incorrect 82	Incorrect Not 82	Did Not Try	N = 18,915 Total per Grd.
Grade 1	151 (5%)	620 (20%)	1,920	398	3,089
Grade 2	606 (19%)	685 (22%)	1,665	201	3,157
Grade 3	1,124 (43%)	656 (25%)	753	80	2,613
Grade 4	1,925 (72%)	294 (11%)	430	21	2,670
Grade 5	2,077 (84%)	133 (5%)	249	17	2,476
Grade 6	2,267 (81%)	209 (7%)	294	37	2,807
Grade 7	1,316 (95%)	16 (1%)	45	6	1,383
Grade 8	669 (93%)	22 (3%)	26	3	720

Forget for a moment about how poorly America's children rank in international testing. The results for 46 + 18 + 27 reveal that the majority of those tested could not add three two-digit numbers until the fourth grade. Still, the results are for Texas children more than 20-30 years ago. Now, the results might be noticeably better. To find out, you can use the same test I used, the MOVE IT Math concepts test. A copy of the test is in the appendix, along with instructions on how to administer it. Give it to some children you know, or ask the principal of an elementary school or middle school to have it administered to one or more classes.

Meanwhile, for the 18,915 children in grades 1-8 who were tested with the item, …

o How is it that only five percent of the first graders could work the problem correctly?

o How is it that 20 percent of the <u>first graders</u> put the 1 in 21 at the top of the tens column and got the wrong answer?

- How is it that a greater percent (22 percent) of <u>second graders</u> put the 1 in 21 at the top of the tens column than first graders?

- How is it that a greater percent (25 percent) of <u>third graders</u> put the 1 in 21 at the top of the tens column than second graders?

- How is it that it took until the fourth grade for more than half of the children to work the problem correctly?

- How is it that nearly 20 percent of the sixth graders could not work the problem correctly?

- How is it that it took until the seventh grade for most students to work the problem correctly, that is, *to finally reach the fifth rung of the Math Ladder for addition of whole numbers?*

This is what happens when the automaticity standard is enforced in K-3. Its effect on the arithmetic curriculum in those grades is to turn it into a puzzle whose pieces children must memorize and put together in their minds without ever seeing a picture of what the puzzle will look like put together.

Can we agree that a curriculum that conditions children to think that only the number 1 can be put at the top of columns when adding needs fixing? Given that, can we agree that knowingly and deliberately perpetuating that curriculum without fixing it would be malpractice? If yes, what is America to do? Jail teachers and publishers who knowingly and deliberately keep perpetuating it or change the curriculum? We know the answer, but changing it will be unimaginably difficult.

Changing the curriculum means changing the textbooks that reflect it, but changing them means changing the curriculum they embody, which means changing the CCSSM, which means changing the NCTM's perception of how arithmetic should be taught, which may take a while. If changing the curriculum were easy, it would have been changed more than twice since 1880.

The Teachers Were Not to Blame

The teachers of the thousands of children who attempted to solve 46 + 18 + 27 were not to blame for the dismal results for this item. The teachers did their jobs. They taught the curriculum they were given — a derivative of the NCTM Standards as expressed in the textbook series their school or school district adopted.

Unlike high school teachers and university professors, grade 1-8 teachers have very little academic freedom. For math, they do not choose the content for their grade. Nor do they have much latitude in how to teach it. Their job is to teach the curriculum they are given as it was designed to be taught. For whole number arithmetic, that means hammering in the number facts with worksheets, flash cards, and speed drills and sticking to the textbook series they were given for math.

As for the children who got 82 for an answer, they at least showed that they had been paying attention compared to those who got the wrong answer some other way or failed to try. They spotted the pattern that occurs in adding two numbers, and spotting patterns is the holy grail for mathematicians and scientists. Unfortunately, the pattern the children noticed that works every time when adding two numbers works only some of the time when adding three or more numbers.

Two Subtraction Algorithms

As explained in Chapter 2, prior to the Progressive Education era in America that began in 1880, the curriculum for elementary school math integrated all three aspects of arithmetic: basics, concepts, and applications. However, those leading the Progressive Education movement replaced it with a utilitarian curriculum that focused on just the basics: how to compute with paper and pencil. They claimed that focusing on that in elementary school best served the future needs of the nine out of ten students back then who quit school after the eighth grade.

For subtraction, focusing on the basics meant teaching the equal additions[19] (borrow-pay back) algorithm or the decomposition (fair trades) algorithm with or without marks, the crossed out numbers and numbers that are added to a subtraction problem to show how it was solved. For 73 − 28 below, if solved using the equal additions algorithm, the marks are the crossed out 2 and the added numbers 3 and 1. If solved using the decomposition algorithm, the marks are the crossed out 7 and the added numbers 6 and 1.

Equal Additions Algorithm with Marks	**Decomposition Algorithm with Marks**
$$\begin{array}{r} 7\ 3 \\ -\ 2\ 8 \\ \hline \end{array} \longrightarrow \begin{array}{r} 7\ {}^13 \\ -\ {}_3\!\!\!/2\ 8 \\ \hline 4\ 5 \end{array}$$	$$\begin{array}{r} \ 6 \\ 7\ 3 \\ -\ 2\ 8 \\ \hline \end{array} \longrightarrow \begin{array}{r} \!\!\!/7\ {}^13 \\ -\ 2\ 8 \\ \hline 4\ 5 \end{array}$$
Borrow ten ones to add to the three ones in the minuend. Pay back by adding a ten to the two tens in the subtrahend: 13 − 8 = **5**, 7 − 3 = **4**.	Trade one of the seven tens in the minuend for ten ones to add to the three ones in the minuend: 13 − 8 = **5**, 6 − 2 = **4**.

As for the two ways to subtract, which one to teach became an either-or (not both) issue during the 1930s, *as did the question of how it should be taught*, with or without marks? For the first 50 or so years of the Progressive Education movement, *subtraction was taught without marks*. For instance, for 73 − 28, subtraction was taught as follows:

$$\begin{array}{r} 7\ 3 \\ -\ 2\ 8 \\ \hline \end{array} \longrightarrow \begin{array}{r} 7\ 3 \\ -\ 2\ 8 \\ \hline 5 \end{array} \longrightarrow \begin{array}{r} 7\ 3 \\ -\ 2\ 8 \\ \hline 4\ 5 \end{array}$$

Since subtraction in the ones column was not possible, children were supposed to just know to subtract 8 from 13 (= **5**) and, if using the equal additions algorithm, 3 from 7 (= **4**) instead of 2 from 7, or, if using the decomposition algorithm, 2 from 6 (= **4**) instead of 2 from 7, and record the answers **5** and **4**, as shown.

Thus for subtraction during the early years of the Progressive Education era, children were taught to NOT show their work! The reason for that was one that you might have heard before. The marks that showed how a subtraction problem was solved were shunned as "crutches" that allowed the "inept" with subtraction to perform as if they were "adept," as if that were deplorable, *the same as counting out a number fact while computing is shunned and thought of today.*

---

[19] "Equal additions" refers to adding equal amounts to the minuend and subtrahend to keep the difference between the two the same. For example, consider 35 and 15. The difference between those numbers is 20 (35 − 15 = 20). Now add 10 to both 35 and 15. The result is 45 and 25, but the difference between the numbers is still 20 (45 − 25 = 20).

To answer the questions about which subtraction algorithm to teach and how to teach it, with or without marks, Brownell (1939) compared the effectiveness of the algorithms with classes of third graders, some taught one or the other of the algorithms <u>without marks</u> and some taught one or the other of the algorithms <u>with marks</u> and concrete materials to explain the marks.[20] Of the four methods, students taught the decomposition algorithm with marks and concrete materials outperformed those taught subtraction by the other methods, thereby becoming the preferred algorithm. Since the 1940s, the decomposition algorithm with marks has been the only way subtraction has been taught in most American schools.

Thanks to Brownell, unless you did not attend elementary school in America, you were probably taught how to subtract using the decomposition algorithm with marks. *So how does it feel to know that you were taught how to subtract with a method that would have cast you as a math dummy in earlier times?* Are you okay with that, or do you wish you had been taught how to subtract without marks/crutches?

Cognitive Distractions Are Built Into the Decomposition Algorithm

To teach the decomposition algorithm for subtraction with marks by rote, children are taught the following procedure:

A. Write the problem vertically by aligning the ones, tens, hundreds, and so on in the minuend[21] and subtrahend.

B. Subtract column by column from right to left, beginning with the ones column.

C. If unable to subtract in a column,

  a. Strike out the digit in the minuend to the immediate left of the column.

  b. Write one less than that digit above it.

  c. Write the number 1 to the right of the digit that was struck out.

As you can see, step C in the algorithm for subtraction is a <u>cognitive distraction</u>. Having to trade a ten for ten ones or a hundred for ten tens, and so on, to enable subtraction in a column is a subtask of the overall task. However, the automaticity standard warns against cognitive distractions while computing, so step C must be automated. It must be practiced until it can be performed automatically so as not to be a cognitive distraction. So even if the decomposition

---

[20] The concrete materials were sticks and bundles of ten sticks, bundles of ten of those bundles, and so on. By solving a subtraction problem with the sticks as bundled, children could literally see that the marks were just a record of how they solved the problem with the sticks.

[21] A <u>minuend</u> is a number from which another number, the <u>subtrahend</u>, is to be subtracted. The result is the <u>difference</u> between the minuend and the subtrahend. For example, for $13 - 8 = 5$, 13 is the minuend, 8 is the subtrahend, and 5 is the difference.

algorithm is taught meaningfully with marks and concrete materials to explain the marks, *it is not used meaningfully.* It is used mechanically to comply with the automaticity standard.

Most children can solve subtraction problems like 73 – 28 with the decomposition algorithm with marks that evoke step C just once, but many struggle with subtraction problems that evoke it more than once, as in solving 935 – 576, or when subtracting across zeros, as in solving 5020 – 463, the last item on the MOVE IT Math concepts test that was administered to nearly 19,000 children in grades 1-8.

Subtracting Across Zeros

Solving 5020 – 463 with the decomposition algorithm with marks evokes step C three times:

○ Cannot subtract in the ones column. Strike out the 2 in the tens column and put 1 above it and 1 in front of the zero in the ones column. Subtract: 10 – 3 = 7.

$$
\begin{array}{r}
1 \\
5\ 0\ \not{2}\,{}^{1}0 \\
-\ \ \ 4\ 6\ 3 \\
\hline
7
\end{array}
$$

○ Cannot subtract in the tens column. Strike out the 5 in the thousands column and put 4 above it and 1 in front of the zero in the hundreds column.

$$
\begin{array}{r}
\mathbf{4}\ \ \ \ 1 \\
\not{5}\,{}^{1}0\ \not{2}\,{}^{1}0 \\
-\ \ \ 4\ 6\ 3 \\
\hline
7
\end{array}
$$

○ Still cannot subtract in the tens column. Strike out the 10 in the hundreds column and put 9 above it and 1 in front of the 1 at the top of the tens column. Subtract: 11 – 6 = **5** and 9 – 4 = **5**. Bring down the 4 in the thousands place in the minuend (because there are zero thousands in the subtrahend and 4 – 0 = **4**).

$$
\begin{array}{r}
4\ \ \mathbf{9}\,{}^{1}1 \\
\not{5}\,{}^{1\!\!\!/0}\ \not{2}\,{}^{1}0 \\
-\ \ \ 4\ 6\ 3 \\
\hline
\mathbf{4\ 5\ 5\ 7}
\end{array}
$$

Results for 5020 – 463 in Dallas, Houston, and S. Texas, 1987-1999

5020 – 463	Correct 4,557	Incorrect	Did Not Try	N = 18,776 Total per Grade
Grade 1	50 (2%)	2,239 (74%)	747	3,036
Grade 2	156 (5%)	2,536 (80%)	462	3,154
Grade 3	310 (12%)	2,079 (83%)	131	2,520
Grade 4	1,095 (41%)	1,555 (58%)	30	2,680
Grade 5	1,418 (57%)	996 (40%)	62	2,476
Grade 6	1,769 (63%)	922 (33%)	116	2,807
Grade 7	1,051 (76%)	329 (24%)	30	1,383
Grade 8	609 (85%)	95 (13%)	16	720

The test results for computing 5020 – 463 show that a scandalous percentage of children cannot subtract across zero regardless of how many years they have spent on subtraction in elementary school.

○ How is it that only 12 percent of the third graders tested could solve 5020 – 463 after spending three years — *all of K-2* — on subtracting?

○ How is it that it took until the fifth grade for at least half of the children tested to solve 5020 – 463 correctly?

○ How is it that nearly one-fourth (24 percent) of the seventh graders tested could not solve 5020 – 463 even after "graduating" from elementary school?

The shockingly poor results for 5020 – 463 are not due to the decomposition algorithm. The algorithm is easily and meaningfully (but not quickly) taught if taught with arithmetic blocks or colored counters. What causes the poor results are the marks and the practice of mindlessly inserting the number 1 in front of numbers in the minuend and the confusion that causes if done more than once in a problem or when subtracting across zeros.

Inserting the number 1 in front of a number is a shortcut — a slick way to add ten to a number if you know what you are doing. Being a shortcut, it saves writing, thus time, which used to matter as shoppers waited for clerks to figure their change after paying for something. Today, with cash registers that compute the change, shaving a few seconds off the time it takes to solve a subtraction problem does not matter that much. What matters is knowing that ten is being added to a number when the number 1 is stuck in front of it, because losing sight of that can lead to mistakes, even by adults, as in subtracting mixed numbers, as shown on the next page.

$$5\frac{3}{8} \longrightarrow 5\!\!\!/\,^4{}_1\frac{3}{8}$$

$$-2\frac{7}{8} \qquad -2\frac{7}{8}$$

$$2\frac{6}{8} = 2\frac{3}{4}$$

Wrong!

$$5\frac{3}{8} \longrightarrow 5\!\!\!/\,^4\frac{^{11}\bigcirc\!\!\!{8}}{{\bigcirc\!\!\!3}\,8}$$

$$-2\frac{7}{8} \qquad -2\frac{7}{8}$$

$$2\frac{4}{8} = 2\frac{1}{2}$$

Correct

To add the number 1 to 3/8 to enable subtracting 7/8, *one adds eight eighths (8/8)*, not ten eighths (10/8), as indicated by the number 1 in front of the 3 in 3/8 in the first problem above.

Inserting the number 1 in front of a number to enable subtraction means different things based on what is being subtracted. For instance, if subtracting time, money, and measurements, inserting the number 1 in front of quantities such as minutes, pennies, and inches in the minuends radically changes the problems, as shown below for three minutes, three pennies, and three inches, because instead of ten, there are 60 minutes in one hour, five pennies in one nickel, and 12 inches in one foot.

5 hours, 3 minutes	4ᵍ hours, ¹3 minutes	4ᵍ hours, ⁶³⁽⁶⁰⁾3 minutes
− 2 hours, 7 minutes	− 2 hours, 7 minutes	− 2 hours, 7 minutes
	2 hours, 6 minutes **Wrong!**	2 hours, **56** minutes **Correct.**

5 hours, 3 minutes
− 2 hours, 7 minutes

4
5̸ hours, ¹3 minutes
− 2 hours, 7 minutes
2 hours, 6 minutes **Wrong!**

4 ⁶³
5̸ hours, ⁽⁶⁰⁾3 minutes
− 2 hours, 7 minutes
2 hours, **56** minutes **Correct.**

5 nickels, 3 pennies
− 2 nickels, 7 pennies

4
5̸ nickels, ¹3 pennies
− 2 nickels, 7 pennies
2 nickels, 6 pennies **Wrong!**

4 ⁸
5̸ nickels, ⁽⁵⁾3 pennies
− 2 nickels, 7 pennies
2 nickels, **1** penny **Correct.**

5 feet, 3 inches
− 2 feet, 7 inches

4
5̸ feet, ¹3 inches
− 2 feet, 7 inches
2 feet, 6 inches **Wrong!**

4 ¹⁵
5̸ feet, ⁽¹²⁾3 inches
− 2 feet, 7 inches
2 feet, **8** inches **Correct.**

Continuing, there are 60 seconds in a minute, 24 hours in a day, seven days in a week, 52 weeks in a year, two nickels in a dime, five nickels in a quarter, four quarters in a dollar, five $1 bills in a $5 bill, two $5 bills in a $10 bill, four $5 bills in a $20 bill, five $20 bills in a $100 bill, three feet or

36 inches in a yard, 1760 yards or 5280 feet in a mile, and lots more equivalences like that with an exchange rate other than ten in science, technology, engineering, and advanced mathematics, as well as in daily living, _so how does one not think when they compute?_

Inserting the number 1 in front of a number in the minuend of a subtraction problem is risky unless one knows what they are doing. With that in mind, can we agree that children should not be taught to do that.[22] Instead, they should be taught to write ten above the minuend in the column under consideration, as shown below for 5020 – 463, _to show that they know what they are doing._

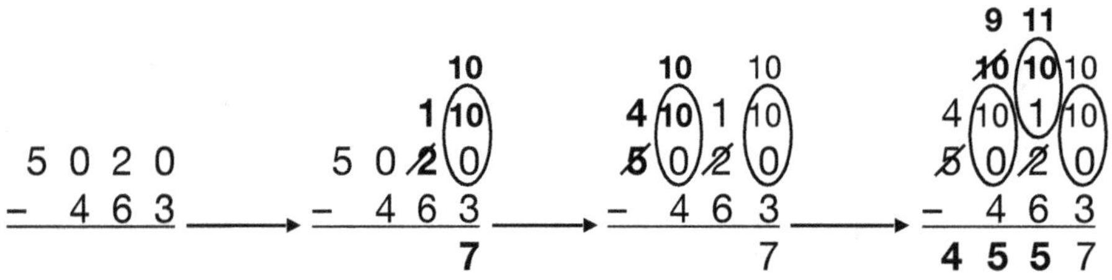

$$
\begin{array}{r}
5\ 0\ 2\ 0 \\
-\ \ 4\ 6\ 3 \\
\end{array}
\longrightarrow
\begin{array}{r}
1\ 10 \\
5\ 0\ 2\ 0 \\
-\ \ 4\ 6\ 3 \\
\hline
7
\end{array}
\longrightarrow
\begin{array}{r}
10\ \ \ \ 10 \\
4\ 10\ 1\ 10 \\
5\ 0\ 2\ 0 \\
-\ \ 4\ 6\ 3 \\
\hline
7
\end{array}
\longrightarrow
\begin{array}{r}
9\ 11 \\
10\ 10\ 10 \\
4\ 10\ 1\ 10 \\
5\ 0\ 2\ 0 \\
-\ \ 4\ 6\ 3 \\
\hline
4\ 5\ 5\ 7
\end{array}
$$

Children should be taught to _think_ their way through a subtraction problem. However, thinking while subtracting violates the automaticity standard, so the standard should be revoked for subtraction unless America's children today are much better at subtracting across zeros than they were 20-30 years ago.

The Automaticity Standard Is Irreconcilable with Multiplication and Division

Principle of Comparative Difficulty: _If an easier task is too difficult to accomplish, then a harder one most certainly is too._ — Michael Behe, 2019

The automaticity standard must also be revoked for multiplication and division because the cognitive distractions of having to solve problems "off to the side" while performing those operations are unavoidable, as you will see in solving 6 x 87 (= 522) and 522 ÷ 6 (= 87) on the next page. To solve 6 x 87, the addition problem 48 + 4 must be solved aside from the main task, and to solve 522 ÷ 6, the subtraction problem 52 – 48 must be solved aside from the main task.

---

[22] If students discover the shortcut of inserting the number 1 in front of a number in the minuend to enable subtraction or have been taught to do that by another teacher or a family member, I do not think they should be told to stop doing that. Instead, I think they should be cautioned about doing that and shown how it can lead to mistakes, as with mixed numbers, time, money, and measurement, as shown.

To solve 6 x 87, multiply 7 by 6 (= 42) and put the 2 in 42 in the ones place in the answer and the 4 in 42 above the tens column in the problem, as shown below. Then multiply 8 by 6 (= 48) and add 48 and the 4 that was put above the tens column. *Huh, again!* 48 + 4 is *NOT* an addition fact that should automatically come to mind. It is an addition *problem* — a cognitive distraction that must be solved aside from the main task in order to complete the main task.

$$\boxed{\begin{array}{c}\textbf{Cognitive}\\\textbf{Distraction}\end{array}}$$

$$
\begin{array}{ccccc}
& \overset{\mathbf{4}}{87} & & \overset{\mathbf{1}}{48\ (6 \times 8)} & \overset{4}{87} \\
\underline{\times\ 6}\ \ \underset{6\times7=42}{} \to & \underline{\times\ 6}\ \ \underset{6\times8+4}{} \to & & \underline{+\ 4} & \underline{\times\ 6} \\
& \mathbf{2} & & \mathbf{52} \longrightarrow & \mathbf{522}
\end{array}
$$

To solve 522 ÷ 6, guess/hope that 6 divides 52 at most 8 times and put the guesstimate above the 2 in 52 in the dividend, as shown below. Then multiply 6 by 8 (= 48), and, as shown, put the answer 48 directly beneath 52 in the dividend and subtract 48 from 52. *Whoa!* 52 – 48 is *NOT* a subtraction fact that should automatically come to mind. It is a subtraction *problem* — a cognitive distraction that must be solved aside from the main task in order to complete the main task.

$$\boxed{\begin{array}{c}\textbf{Cognitive}\\\textbf{Distraction}\end{array}}$$

$$
\begin{array}{cccc}
& \overset{\mathbf{8}}{6\overline{)522}} & \overset{\mathbf{4}}{\cancel{5}12} & \overset{\mathbf{87}}{6\overline{)522}} \\
6\overline{)522}\ \underset{52\div6=8?}{} \to & \underline{\mathbf{48}} & \underline{-4\ 8} & 48\downarrow \\
& & \mathbf{4} \longrightarrow & \mathbf{42} \\
& & & \underline{\mathbf{42}}\ (7 \times 6) \\
& & & \mathbf{0}\ (42-42)
\end{array}
$$

As with 14 + 7, the addition problem that occurred in adding the numbers in the ones column of 46 + 18 + 27 written vertically, the responses to 48 + 4 and 52 – 48 are not likely to instantly come to mind *because they are not supposed to*. Children do not have to memorize 48 + 4 = 52 and 52 – 48 = 4. They do not practice them until they know them as well as they know their own names. Instead, they learn how to compute them, as shown, or how to solve them during the computational process with counting and thinking strategies like the ones listed in CCSSM 1.OA.6, the Common Core performance standard for the first grade.

○ For 48 + 4, they can count on 4 from 48 (49-50-51-<u>52</u>) or mentally create an equivalent but easier problem (e.g., 48 + 4 = 50 + 2 = 52) for the sum <u>52</u> in either case.

○ For 52 – 48, they can count on from 48 to 52 (49-50-51-52) and keep track of the counts on their fingers or with tally marks (||||) for the difference <u>4</u>, the number of counts, or mentally create an equivalent but easier problem (e.g., 52 – 48 = 54 – 50 = <u>4</u>) with the same answer.

However, the automaticity standard warns that solving 48 + 4 and 52 − 48 by *any* method will have negative consequences, because solving them would entail devoting cognitive attention to subtasks of the main tasks. Nonetheless, they must be solved in order to complete the main tasks. Therefore, the purported ill effects of solving 48 + 4 and 52 − 48 must be ignored, which is why multiplication and division must be exempted from the automaticity standard. Otherwise, the answers to *every* addition problem like 48 + 4 that can arise in multiplying and *every* subtraction problem like 52 − 48 that can arise in dividing would have to be memorized so they would not have to be solved while multiplying or dividing.

An objection to revoking the automaticity standard for multiplication and division would be that having to solve addition and subtraction problems off to the side are built into the algorithms for those operations. So? Having to solve them even by rote is still paying attention to subtasks of the main tasks, and the automaticity standard warns of dire consequences for doing that, but to complete the main tasks, the warning must be disregarded, which is tantamount to dismissing the automaticity standard for multiplication and division.

Closing Remarks

The objective of this chapter was to examine the justification for the automaticity standard — the assumption that the cognitive distraction caused by working out a number fact while computing has negative consequences. In examining the assumption, it was shown that cognitive distractions are unavoidable in adding a column of numbers and in solving even the simplest of multiplication or division problems regardless of how well the number facts are known.

○ In adding a column of three or more numbers, cognitive distractions in the form of addition problems like 14 + 7 are certain to occur and have to be solved aside from the main task to complete the main task, as shown earlier in this chapter in solving 46 + 18 + 27.

○ In multiplying <u>by even a single-digit</u>, cognitive distractions in the form of addition problems like 48 + 4 are certain to occur and have to be solved aside from the main task to complete the main task, as just shown in solving 6 x 87.

○ In dividing <u>by even a single digit</u>, cognitive distractions in the form of subtraction problems like 52 − 48 are certain to occur and have to be solved aside from the main task to complete the main task, as just shown in solving 522 ÷ 6.

Therefore, claiming that cognitive distractions while computing will result in negative consequences cannot be used as a reason to make children memorize the number facts. Were that true, teaching children how to add a column of numbers, multiply, or divide would be malpractice, which is nonsense.

The automaticity standard is either a standard or not, the same as "i before e except after c" is either a rule or not in spelling. According to the Merriam-Webster dictionary, said spelling rule is *not* a rule because of the exceptions to it (e.g., "weird"). Likewise, the automaticity standard is not a standard because of the exceptions to it.

The automaticity standard is not the result of serious thinking. Most likely, it was created to deter finger counting. At best, it is advice on how to teach arithmetic. Its message in that regard is that the place to start is with the number facts, which is good advice, because arithmetic is built on them. The number facts come up in each step of the step-by-step procedures that are used in the algorithms for addition, subtraction, multiplication, and division. However, when arithmetic was first taught in America hundreds of years ago, no one knew how to teach the number facts with mental math or strategies for counting them out. They only knew that if children practiced them enough, many of them would eventually memorize them well enough to get by in arithmetic.

Note that discrediting the justification for the automaticity standard does not discredit the standard. It only discredits the *reason* it is enforced and has been for as long as anyone can remember. The automaticity standard can still be enforced for reasons other than the supposed need to avoid cognitive distractions while computing. In fact, the NCTM is doing exactly that in supporting the automaticity standard in the six Common Core performance standards for arithmetic in K-3. Why I believe it is doing that is a history lesson on the evolution of how to teach the number facts that will be addressed in the next chapter.

As for subtraction, step C in the decomposition algorithm with marks amounts to mechanically doing the same three things, as listed below, if unable to subtract in a column:

a. Strike out the number in the minuend to the immediate left of the column.

b. Write one less than that number above it.

c. Write the number 1 to the right of the number that was struck out.

Step C can be automated with sufficient practice to comply with the automaticity standard, but why have children automate it when it is known that they are likely to make mistakes if they have to evoke step C more than once in a subtraction problem or when subtracting across zeros? Why not have them learn how to think their way through a subtraction problem?

References

"Automaticity of the Basic Facts of Arithmetic." Accessed online in 2020. To access, google the title of the article.

Behe, Michael J. *Darwin Devolves*, HarperOne, 2019.

Brownell, William A. "Learning as Reorganization: An Experimental Study in Third Grade Arithmetic." Durham, NC: Duke University Press, 1939.

CCSSM. See *Common Core State Standards for Mathematics*.

Chi, Michelene T.H. "Short-term Memory Limitations in Children: Capacity or Processing Deficits?" *Memory & Cognition*, 1976.

Common Core State Standards for Mathematics, National Governors Association in collaboration with the Council of Chief State School Officers, 2010.

Curriculum and Evaluation Standards for School Mathematics, National Council of Teachers of Mathematics,1989.

enVisionMath, K-5, Scott Foresman-Addison Wesley, © 2009.

enVisionMath 2.0, K, 2-3, Scott Foresman-Addison Wesley, © 2015.

Everyday Math, K-5, University of Chicago School Math Project, © 2008.

Go Math, K-3, Houghton Mifflin Harcourt, © 2012.

NCTM Standards. See *Curriculum and Evaluation Standards for School Mathematics.*

Saxon Mathematics, K-5, Harcourt Achieve, Inc., © 2008.

Texas HSP Math, K-5, Harcourt School Publishers, © 2009.

Texas Math, K-5, Houghton Mifflin, © 2009.

Texas Mathematics, K-5, MacMillan McGraw-Hill, © 2009.

Van Eeden, Frederik. Famous Quotes & Sayings, quotestats.com.

Wilson, W. Stephen. "Elementary School Mathematics Priorities," *AASA Journal of Scholarship & Practice*, 2009.

Chapter 5: The Common Core Arithmetic Curriculum for K-3 Is Inconsistent

Children are easily taught, for they readily accept and believe lies told by their elders. — Dee Hock, founder and former CEO of VISA

The quotes below are from recent textbook series for elementary school math. Because of when they were published, the first quote is based on the NCTM Standards (1989) by the National Council of Teachers of Mathematics, the second one on the *Common Core State Standards for Mathematics* (2010) by the National Governors Association and the Council of Chief State School Officers.

The goal for mastery of the basic facts is automaticity. A student is considered to have achieved automaticity when he or she can give an answer to a basic fact in less than 3 seconds <u>without using finger counting</u>. Students who do not memorize the basic number facts will flounder as more complex operations are required. There is no real mathematical <u>fluency</u> without memorization of the most basic facts. — Joyce McLeod, **Texas HSP Math, Grade 3**, Harcourt School Publishers, 2009

<u>Fluency</u> should not require rote memorization. Instead, students should either have a fact meaningfully [?] memorized or be able to produce that fact through a highly efficient, <u>automatically executed</u>, strategy. A student has mastered a math fact if they can produce an answer within 3 seconds, through either <u>recall</u> or a <u>highly efficient strategy application</u>. — Kling & Bay-Williams (2015), **McGraw-Hill**, 2019

If you spotted the reference to three seconds in the preceding passages, you may be thinking *"What's going on?"* The automaticity standard requires the ability to recall the number facts *instantly* without conscious thought. How is it, then, that children now have three seconds to work out a number fact while computing if they cannot recall it automatically?

Before, even *hesitating* to recall a number fact while computing was unacceptable and the root of all sorts of negative consequences. Now, we are to believe that figuring out a number fact *mentally* while computing will not result in dire consequences if it can be accomplished within three seconds. *Who decided that and why three seconds?* Why not five seconds? Why even a time limit? *And why does the number fact have to be figured out mentally?* Why can it not be counted out?

To explain the three-second "softening" of the automaticity standard for the number facts if figured out mentally while computing, I need to provide some historical context. Until the mid 1970s, the automaticity standard had been widely accepted by teachers from kindergarten on because the only way commonly known to "teach" the number facts before then was to force-feed them to children via drill and practice, and the automaticity standard justified that practice. However, by 1980, two ways to actually *teach* the number facts — mental math and TouchMath — were being touted as alternatives to just having children practice them until they someday hopefully memorized them.

Mental Math

Mental math is juggling numbers in the mind with thinking strategies, like the ones listed in CCSSM 1.OA.6, the Common Core performance standard for grade 1, and mathematical principles, like the associative and commutative laws for addition and multiplication,[23] to turn tedious calculations into ones that can be done mentally. *It is not, as you might think, paper-and-pencil arithmetic on an imaginary sheet of paper in one's forehead.* It is about simplifying arithmetic problems by mentally changing the numbers in the problems without actually changing the problems (e.g., $36 + 58 = 40 + 54 = 94$ and $48 \times 9 = 48 \times 10 - 48 = 480 - 48 = 432$).

The main uses of mental math are in adding and subtracting small numbers in one's mind,[24] estimating the answers to big number problems,[25] and determining the reasonableness of answers acquired with a calculator. Additionally, mental math may be used to figure out some of the number facts with rules like the following to turn unknown number facts into known ones.

○ **Zero rules:** $N + 0 = N$, $N - 0 = N$, and $N \times 0 = 0$. For example, for $N = 5$, $5 + 0 = 5$, $5 - 0 = 5$, and $5 \times 0 = 0$. **Note:** $N \div 0$ is not equal to anything, except perhaps infinity to indicate that the smaller the divisor, the larger the quotient.[26]

○ **One rules:** $N \times 1 = N$, $N \div 1 = N$, and $N \div N = N/N = 1$. For example, for $N = 8$, $8 \times 1 = 8$, $8 \div 1 = 8$, and $8 \div 8 = 8/8 = 1$.

○ **Nine rule for addition:** $9 + N = 10 + (N - 1)$. For example, for $N = 6$, $9 + 6 = 10 + (6 - 1) = 10 + 5 = 15$.

○ **Nine rule for multiplication:** $9 \times N = (10 \times N) - N$. For example, for $N = 7$, $9 \times 7 = (10 \times 7) - 7 = 70 - 7 = 63$.

○ **Make-a-ten rule for addition and subtraction:** For example, for addition, $6 + 8 = 10 + 4 = 14$, and $9 + 5 + 2 = 10 + 4 + 2 = 10 + 6 = 16$. For subtraction, $8 - 3 = 10 - 5 = 5$, and $13 - 7 = 16 - 10 = 6$.

Other rules are the double plus or minus one rules based on knowing the doubles for the addition facts from $1 + 1 = 2$ to $9 + 9 = 18$:

---

[23] The **associative laws** for addition and multiplication are $(a + b) + c = a + (b + c)$ and $(a \times b) \times c = a \times (b \times c)$, respectively. For example, $(2 + 3) + 5 = 2 + (3 + 5) = 10$ and $(2 \times 3) \times 5 = 2 \times (3 \times 5) = 30$. The **commutative laws** for addition and multiplication are $a + b = b + a$ and $a \times b = b \times a$, respectively. For example, $5 + 20 = 20 + 5 = 25$ and $5 \times 20 = 20 \times 5 = 100$.

[24] For example, to figure out $348 + 26$ mentally, think $348 + 26 = 350 + 24 = 374$. To figure out their difference mentally, think $348 - 26 = 352 - 30 = 322$.

[25] For example, to estimate the product of 97 and 38 (= 3686), think $100 \times 40 = 4000$, which would be an upper bound for the answer.

[26] For example, note how dividing the number 1 by increasingly smaller numbers results in increasingly larger quotients: $1 \div 1 = 1$, $1 \div 0.1 = 10$, $1 \div 0.01 = 100$, $1 \div 0.001 = 1000$, and so on with the quotient approaching infinity as the divisor approaches zero.

- **Double plus one rule** for addition facts like 4 + 5 where the second number is one more than the first number: $N + (N + 1) = (N + N) + 1 = 2N + 1$. For example, for $N = 4$, $4 + 5 = 2 \times 4 + 1 = 8 + 1 = 9$.

- **Double minus one rule** for addition facts like 7 + 6 where the second number is one less than the first number: $N + (N - 1) = (N + N) - 1 = 2N - 1$. For example, for $N = 7$, $7 + 6 = 2 \times 7 - 1 = 14 - 1 = 13$.

- **Commutative rules for addition and multiplication**: $M + N = N + M$ and $M \times N = N \times M$, respectively. For example, $7 + 4 = 4 + 7 = 11$ and $7 \times 4 = 4 \times 7 = 28$.

- **Fact family[27] relationships**:

  Every **addition fact** is related to its "twin" and two subtraction facts. For example, the fact family for $5 + 7 = 12$ is its twin, $7 + 5 = 12$, and $12 - 7 = 5$ and $12 - 5 = 7$.

  Every **subtraction fact** is related to its "twin" and two addition facts. For example, the fact family for $10 - 7 = 3$ is its twin, $10 - 3 = 7$, and $3 + 7 = 10$ and $7 + 3 = 10$.

  Every **multiplication fact** is related to its "twin" and two division facts. For example, the fact family for $5 \times 7 = 35$ is its twin, $7 \times 5 = 35$, and $35 \div 5 = 7$ and $35 \div 7 = 5$.

  Every **division fact** is related to its "twin" and two multiplication facts. For example, the fact family for $48 \div 8 = 6$ is its twin, $48 \div 6 = 8$, and $6 \times 8 = 48$ and $8 \times 6 = 48$.

Unfortunately, as noted passionately by Ashcraft (1985) in defense of having children *memorize* the number facts,[28] *the mental math rules listed do not yield all 390 number facts.*

What are the other rules? What rule, for instance, yields the answer 13 to the problem 8 + 5? What rule or procedure generates the 56 for 8 x 7? The 7 for 4 + 3? The 36 for 9 x 4?

For instance, there are no double plus or minus two or three rules for addends that differ by those amounts because they are not easily learned by young children. So in asking *"What rule, for instance, yields the answer 13 to the problem 8 + 5?"* Ashcraft is pointing out that there is no double minus three rule that could be used to recast $8 + 5$ as $8 + (8 - 3) = (8 + 8) - 3 = 16 - 3 = 13$.

Ashcraft also points out the absurdity of expecting children to know that $10 - 7 = 3$ because it is in the same fact family as $7 + 3 = 10$:

To suggest that "some subtraction combinations may be <u>efficiently reconstructed</u> from their addition counterparts (e.g., 10 – 7 is 3 because 7 + 3 is 10)" is completely evasive. What rule generates 7 + 3 = 10? This is not a minor, nagging incompleteness. ... <u>It is an absolutely central flaw</u>.

So mental math is not the panacea for teaching the number facts as the National Council of Teachers of Mathematics (NCTM) would have America believe.

---

[27] Constructing fact families to generate the number facts is a strategy advanced in *Everyday Math*, a textbook series for elementary school math that was developed by the University of Chicago School Math Project.

[28] As explained in chapter 7, Ashcraft is incorrect in believing that children can memorize the number facts and remember them indefinitely if they just practice them enough in elementary school.

TouchMath

TouchMath is a prekindergarten through grade 3 and up math program that was created by Janet Bullock, an elementary school teacher, who began disseminating the program in the 1970s via workshops for teachers near her home in Colorado Springs, Colorado. Today, the program is distributed in kits for those grades by Innovative Learning Concepts, a company of Ms. Bullock's making. Information about the program and sample activity sheets may be downloaded free @ touchmath,com.

I learned of TouchMath in 1980 during my first year at Metropolitan State College (now a university) in Denver, Colorado as an Associate Professor in the math department where I taught math and how to teach math. I had just returned from a two-year working holiday in Perth, Western Australia where I taught how to teach math at Churchlands College (now a university) and was unaware of TouchMath until some of my students in my math for teachers class informed me of it and urged me to investigate it.

I resisted my students' insistence that I learn about TouchMath. I had dedicated my career to improving how math is taught and had the credentials and experience to show for it, yet I was being pressured to be tutored on how to teach arithmetic by a former grade school teacher with scant knowledge of mathematics. Still, being curious and just an hour's drive from Colorado Springs, I stifled my hubris and called Ms. Bullock and asked if she would teach me the basics of the program she had developed. Her gracious response to my request more than 40 years ago and the friendship and mutual respect that evolved from it inexorably led me to write this book.

In its entirety, TouchMath is non-controversial. It consists of the conventional algorithms or procedures for adding, subtracting, multiplying, and dividing whole numbers, fractions, and decimals and is aligned with the NCTM Standards and the Common Core arithmetic curriculum for K-6. Within it, though, is a methodology for teaching the number facts that to this day is as controversial as it is effective.

In TouchMath, each numeral from 1 through 9 is assigned touchpoints — dots and double dots — that depict the number of elements it represents. A dot gets touched and counted once. A double dot — a dot inside a circle — gets touched and counted twice. Zero gets no dots and is not touched or counted.

The Touchpoints Are Abstractions of Things that Can Be Counted

One of the first tasks in teaching arithmetic is to teach the meaning of the numerals 1 through 9, which is accomplished by relating them to the amounts they represent. The numeral 3, for instance, stands for the concept of "threeness" — the numeric property of all sets with three elements. To teach that, children are shown the numeral 3 in connection with three things or pictures of three things, like three apples, three letters, and three ducks, as shown on the next page.

Such instruction is all well and good, but since it is a math lesson, why not generalize on it? Let a dot stand for anything that can be counted. Then three dots stand for any three things that can be counted, like three apples, three letters, or three ducks. Now put the three dots on the numeral 3 to show what the numeral stands for. *I cannot think of why that is not a good idea.* Can you?

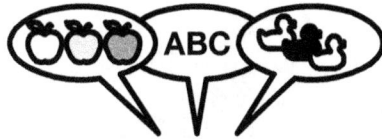

The child's ability to give number words meaning depends on the availability of perceptual items. — Paul Cobb, 1987

Counting Out the Number Facts

By counting forward, backward, and skip counting[29] on the touchpoints that are shown or imagined on the numerals 1 through 9, <u>all 390 number facts can be counted out</u>, as follows:

Counting Out an Addition Fact

For an addition fact, count the touchpoints on both numbers. The <u>sum</u> is the last number in the count. For example, for 9 + 3, count 1, 2-3, 4-5, 6-7, 8-9; 10-11-<u>12</u>. The sum is <u>12</u>, the last number in the count.

$$9 + 3 = 12$$

---

[29] To skip count by a number, count by the number times 1, times 2, times 3, and so on. For example, to skip count by 5 for ten counts, count 5-10-15-20-25-30-35-40-45-50, which is equivalent to counting 5 x 1, 5 x 2, 5 x 3 ... 5 x 10. As you can see, skip counting skips a lot of numbers, which is why it is called skip counting.

Alternatively, <u>count on</u>, as prescribed by CCSSM 1.OA.6, the Common Core performance standard for the first grade. For 9 + 3, count on 9 from 3 on the touchpoints of the nine (4, 5-6, 7-8, 9-10, 11-<u>12</u>), or count on 3 from 9 on the touchpoints of the three (or on the *"pointy things on the three,"* as in the last frame of the Peanuts cartoon below: *"ten-eleven-<u>twelve</u>."*) In either case, the answer is <u>12</u>, the last number in either count.

$$9 + 3 = 12$$

Note: The order in which the touchpoints and double touchpoints on a numeral are counted is irrelevant. They may be counted from top to bottom, bottom to top, left to right, right to left, or however. What matters is that 1) all of the touchpoints and double touchpoints are counted and 2) that each touchpoint is counted just once and each double touchpoint exactly twice.

Counting Out a Subtraction Fact

For a subtraction fact, <u>count back</u> from the minuend on the touchpoints of the subtrahend. The <u>difference</u> is the last number in the count. For example, for 13 – 7, count back 7 from 13 by counting back on the touchpoints of the seven: 12, 11-10, 9-8, 7-<u>6</u>. The difference is <u>6</u>, the last number in the count.

$$13 - 7 = 6$$

Alternatively, <u>count on</u> from the subtrahend to the minuend and keep track of the counts. The <u>difference</u> is the number of counts. For 13 – 7, count on from 7 to 13 (8-9-10-11-12-13) and keep track of the counts on your fingers or with tally marks (||||| |). As before, 13 – 7 = <u>6</u>, except with respect to the number of counts.

Counting Out a Multiplication Fact

For a multiplication fact, <u>skip count</u> by one of the numbers on the touchpoints of the other number. The <u>product</u> is the last number in either count. For example, for 5 x 8, skip count by 5 on the touchpoints of the eight (5-10, 15-20, 25-30, 35-<u>40</u>) or by 8 on the touchpoints of the five (8-16-24-32-40). The product is <u>40</u>, the last number in either count.

$$5 \times 8 = 40$$

$$5 \times 8 = 40$$

Counting Out a Division Fact

For a division fact, <u>skip count</u> by the divisor to the dividend and keep track of the counts. The <u>quotient</u> is the number of counts. For example, for 21 ÷ 3 (interpreted as *"How many 3s in 21?"*), skip count by 3 to 21 <u>without touchpoints</u> (3-6-9-12-15-18-21) and keep track of the number of counts on your fingers or with tally marks (||||| ||). The quotient is 7, the number of counts or 3s in 21.

$$21 \div 3 = 7$$

Note: The skip counting for the division facts is **FORWARD** — *the same as for the multiplication facts* — even though division is the opposite of multiplication. Thus there is no need to teach children how to skip count backwards, *nor is there any mathematical benefit in doing so.*

Skip Counting Revolutionizes Teaching the x and ÷ Facts

Skip Counting enables actually teaching the multiplication and division facts instead of just having children practice them year after year in the hope that they will eventually memorize them. Instead of having to memorize all 190 of them, they only have to learn the skip counting sequences for

2 through 10[30] for ten counts each (2-4-6- ... 20, 3-6-9- ... 30, 4-8-12- ... 40, and so on to 10-20-30- ... 100), *which are easily taught with songs,* the same as the letters of the alphabet are taught with the alphabet song: *A, B, C, D, E, F, G ... W, X, Y, Z. Now I know my ABCs.*

In MOVE IT Math, the skip counting sequences for 2 through 10 are taught with the songs and dances on *Multiplication Motivation*,[31] a CD distributed by Melody House @ melodyhousemusic.com. Research confirms the common sense notion that the more of the five senses that are involved in a lesson, the better and more lasting the outcome. Lessons that involve whole-body movements, like dancing to the skip counting songs while singing them, are known to *"boost learning, increase brain development, enhance retention, and improve classroom behavior"* (Math & Movement, 2020).

Kindergarteners learn the songs and dances for 2, 5, and 10, first graders those for 3 and 4, and second graders those for 6, 7, 8, and 9. By distributing the memorization of the skip counting sequences over three years, the stressfulness in memorizing them is virtually nil compared to tears aplenty in having to memorize all 190 multiplication and division facts all at once in the third grade, as mandated by the Common Core arithmetic curriculum for that grade.

Once a MOVE IT Math class can skip count by a number, they are shown how to count out the multiplication and division facts for that number. For instance, once a kindergarten class has learned how to skip count by 5, they are ready to count out the multiplication and division facts for 5 once they know how to *read* (interpret) a multiplication or division fact.

To read 7 x 5 or 5 x 7, kindergarteners are taught to think of them as asking *"How much is seven 5s or 5 added to itself seven times?"* To answer the question, they are taught to skip count by 5 on the touchpoints of the seven (5, 10-15, 20-25, 30-<u>35</u>). The <u>product</u> is <u>35</u>, *the last number in the count.*

$$7 \times 5 = 35$$
$$5 \times 7 = 35$$

To read 35 ÷ 5, kindergarteners are taught to think of it as asking *"How many 5s in 35?"* and to answer the question by skip counting by the divisor 5 to the dividend 35 *without*

---

[30] Strictly speaking, the multiples of 10 are not multiplication facts, but they are taught along with the multiples of 2 through 9 because multiples of 10 come up often when working with time, money, and measurement. For instance, 3:40 on an analog clock equals 10-20-30-40 minutes past 3:00, four dimes equal 10-20-30-40 cents, and four centimeters equal 10-20-30-40 millimeters.

[31] Thanks to Suzanne Rogers, a third grade teacher in New Braunfels, Texas, who discovered the CD. Children learn the skip counting sequences for 2 through 10 for ten counts each by singing and dancing to the following songs on the CD: The Twos Stroll, The Circle of Threes, The Fours Hop, The Alive Fives, The Sixes Surprise, The Waltzing Sevens, The Boogie Woogie Eights, The Nines Blues, and The Tens Promenade. As of the writing of this book, the songs are on YouTube.

touchpoints (5-10-15-20-25-30-35) while keeping track of the number of counts on their fingers or with tally marks (||||| ||). The <u>quotient</u> is 7, the number of counts or 5s in 35.

$$35 \div 5 = 7$$

Whoa! If you were not startled by what you just read, you must not have been paying attention, so please read the preceding three paragraphs again. *Teaching multiplication and division beginning in kindergarten?* Absolutely!

With TouchMath, kindergartners learn the multiplication and division facts so easily and enjoyably with the skip counting songs and dances that no child is left behind in arithmetic. Virtually *every* child learns them at about the same time as their classmates, thereby skirting the loss of self-confidence and self-esteem that occurs when children realize that they are falling behind their peers because they keep forgetting some of the multiplication or division facts.

The Common Core Curriculum for the x and ÷ Facts

Teaching the multiplication facts in the Common Core arithmetic curriculum is delayed until grade 3, and, according to CCSSM 3.OA.7, one of the two performance standards for that grade, all 100 of them from 0 x 0 = 0 to 9 x 9 = 81 must be memorized by the end of the grade. The methodology for teaching them is repetition with a nod to mental math to figure them out in spite of having no mental math rules besides multiplication being commutative (e.g., 4 x 7 = 7 x 4 = 28) and the nine rule for the multiplication facts (e.g., 9 x 6 = 10 x 6 – 6 = 60 – 6 = 54).

For the division facts, third graders are taught the gazintas. For 40 ÷ 5, for instance, they are taught to interpret that as asking *"How many times does 5 gazinta (go into) 40?"* However, to answer the question, they are taught to answer the related multiplication problem *"What times 5 is 40?"* If 8 x 5 = 40 pops into their heads, they are told that means that 40 ÷ 5 = 8. If it does not pop into their heads, they are taught to guess the answer to the question and check to see if they guessed correctly (the same as in long division). If they guess 6, 6 x 5 = 30, so they must guess again. If they guess 7, 7 x 5 = 35, so they must guess again. If they guess 8, 8 x 5 = 40. Yay! They finally guessed correctly, so 40 ÷ 5 = 8.

Tapping Pencils

In 1987 — *two years before the release of the NCTM Standards* — the NCTM published "Beware of Tapping Pencils" by Flexer and Rosenberger in the *Arithmetic Teacher.* Said article must have been widely read because it became the go-to reference for rejecting TouchMath. Since the article's publication, I have yet to read an article in a professional journal as misleading as that one.[32] The following is the first paragraph from the article:

----

[32] For a small fee, you can download the article from the NCTM @ nctm.org.

Advocates of [TouchMath], a new, "magical" method for teaching addition, subtraction, multiplication, and division are attracting the attention of many elementary school teachers. The method involves teaching children to tap reference points on numerals to count out sums, differences, and products. <u>Teachers like it because it's so easy. Students catch on quickly and get the right answers. Even students who might have had trouble with arithmetic before can tap out correct answers.</u> What's the problem, then?

Indeed! If teachers like it because it is easy to teach, and children learn it quickly and get right answers, *What is the problem?* According to the authors of the article, *"the problem with the new 'magic' is that a mechanical technique for getting answers is displacing learning with understanding,"* which is laughable. Adherence to the automaticity standard, which has overshadowed the teaching of arithmetic for as long as anyone can remember, requires that the algorithms for all four operations (+, −, x, ÷) be performed "automatically" without conscious thought. The authors' assessment of "the problem" in the article indicated that they did not know what they were talking about.

TouchMath is *not* a "magical" method for teaching addition, subtraction, multiplication, and division. *What is magical about TouchMath is its number facts program* — its system of touching and counting on the numerals 1, 2, 3 … 9 to count out all 390 number facts. *Period!* That is what is magical about TouchMath, just its number facts program.

TouchMath's presentation of the algorithms for addition, subtraction, multiplication, and division is the standard presentation except for its use of visual cues that specify the usual steps in the standard algorithms for the four operations, as shown below for the addition algorithm.

| Double-Digit Addition with Visual Cues | Double-Digit Addition with Regrouping | Three-Digit Addition with Visual Cues |

Written as if it were based on research, the tapping pencils article is a shocking example of deception based on opinions and dubious assumptions. Nonetheless, except for K-2 and special education teachers, the article was shamefully successful in creating a negative response to TouchMath and, more broadly, in linking it to the age-old intolerance of counting on one's fingers while computing with paper and pencil.

The NCTM's response to TouchMath was immediate and reactive. It forcefully rejected it with as much conviction and histrionics as a toddler rejecting green peas. Its voiced complaint was that counting out a number fact while computing lacked "fluency" without ever explaining what it meant by that in meaningful terms. Nonetheless, in spite of the NCTM's unrelenting campaign to stamp out TouchMath for the past 50 years, *it has not extinguished the program.* Nor will it ever because of the program's proven effectiveness with the addition and subtraction facts in K-2.

Problems with TouchMath in Grade 3

K-2 teachers tend to be open to using TouchMath because the Common Core arithmetic curriculum for those grades only deals with the addition and subtraction facts, which are easy to teach with TouchMath. (Students *"catch on quickly and get the right answers,"* even those *"who might have had trouble with arithmetic."*) Once children can count forward to 18, count on from 1 through 9, and count backward from 18, which they can learn how to do in kindergarten in a matter of weeks, they can *"quickly and accurately"* count out all 200 addition and subtraction facts. *All is well until the third grade.*

Grade 3 teachers resist TouchMath because the Common Core arithmetic curriculum requires the memorization of the multiplication facts by the end of grade 3. Thus grade 3 teachers tend to view counting on the numerals as a gimmick to avoid memorizing the number facts and therefore discourage their students from counting them out on the numerals. In the teachers' defense, their students were not taught how to skip count in K-2 except for the numbers 5, 10, and 100 (2.NBT.2), so they have not been prepared to count out the multiplication facts except for the fives, but how to do that by skip counting by 5 (5-10-15- ... -50) is not addressed in the CCSSM.

This tension between K-2 teachers who use TouchMath and grade 3 teachers who forbid it is toxic for children. Children who learned TouchMath in K-2 are shunned in grades 3-6 for "finger counting." If you were ever admonished for counting out a number fact on your fingers (or on sticks drawn in the margins of your worksheets and then erased to hide that you had counted) when you were in elementary school, you may have decided back then that guessing the answer to a number fact, even if wrong, was preferable to being "corrected" for counting it out.

Forgotten Number Facts Are Bonafide Arithmetic Problems

If a child encounters a number fact like 8 + 5 while computing and the number 13 does not pop into the child's head, *8 + 5 is a bonafide arithmetic problem for the child* — one with consequences if it comes up while taking a test. Only adult arrogance can claim otherwise. A teacher disparaging a child for not knowing what the child is *"supposed to know by now"* is comparable to a would-be physician scolding an adult patient for having a childhood ailment like an earache.

As for allowing three seconds to figure out a number fact with mental math while computing, limiting the amount of time down to the second for children to figure out a number fact sends the wrong message about arithmetic. Adults do not race against a timer while performing a calculation, whether with paper and pencil or a calculator. They use arithmetic *deliberatively*, as in preparing a budget or estimating the cost of items in a shopping basket. How long it should take a child to figure out a number fact should be however long it takes them to work it out *correctly*, not quickly (within three seconds) and possibly getting it wrong.

What Is the Problem with *"quickly and accurately"*?

*Anyone who has spent time teaching in the elementary grades realizes how many students are unsuccessful at rote memorization and how often they revert to counting on their fingers. We would agree that third or fourth graders who are counting on their fingers certainly have not reached a level of **fluency**, even though they may do it pretty quickly and accurately!* — Linda Gojak, **President, NCTM** (2012-14)

Again, *What is the problem with* *"quickly and accurately"?* Nothing, of course, except that counting out a number fact while computing violates the NCTM's fluency requirement for paper-and-pencil arithmetic. As pointed out in Chapter 3, "fluency" is defined in the Common Core arithmetic curriculum as *"skill in carrying out procedures <u>flexibly</u>, accurately, <u>efficiently</u> and <u>appropriately</u>,"* but without ever explaining what is meant by the underlined words. However, their meaning is apparent. They mean that counting out a number fact while computing is unacceptable and grounds for admonishment.

Of the three words that define fluency, the one that stands out is the word "appropriately." Said word implies an *attitude*, and the NCTM's attitude toward counting out a number fact while computing is that it is like a food fight during dinner, *even if "quick and accurate,"* and the NCTM is the self-appointed "grownup" telling children how they should behave at the arithmetic table.

I imagine that most high school math teachers support *"quick and accurate"* for working out a number fact, regardless of how it is worked out, because they cannot afford to take time from teaching algebra, for instance, to make sure that all of their students can instantly recall all 390 number facts. Speaking for myself, when I taught algebra in high school, if a student forgot a number fact while solving an equation, I would not have cared if they had stood up in class and counted on their fingers and toes to work it out. *What I cared about was that they solved the equation correctly.*

No Fishing Allowed in the Common Core Arithmetic Curriculum

Math facts, themselves, are a small part of mathematics, and they are best learned through the use of numbers in different ways and situations. — Jo Boaler, 2015

Think about a story you have probably heard before — the one about giving a needy person a fish versus giving them a fishing pole. According to the story, if given a fish, the person will be dependent on the person who gave it to them to give them another one when it is gone, but if given a fishing pole and taught how to fish, *they do not have to depend on a benefactor when they want a fish.* They can catch one themselves. Now think of the number facts as a school of fish.

Forcing children to memorize the number facts is like giving them a fish for each one memorized. So when they forget one, which is inevitable as you will learn in the chapter on forgetting, they need to be given the fish again with more drill and practice. In contrast, if they are given the mental math and TouchMath "fishing poles" for working them out while computing, *they can catch all the number-fact fish they want whenever they want one.*

Working out a number fact is like looking up a word in a dictionary. If, like me, you ever looked up the same word in a dictionary more than once, I bet I can tell you when you quit looking it up. *When you remembered what it meant!* So when do children quit "fishing" for a number fact? *When they remember it!*

In MOVE IT Math, children are given both fishing poles for the number-fact fish along with the MOVE IT Math No Guessing rule to account for a <u>gradual</u>, <u>stress-free memorization</u> of most of the number facts <u>through use</u> instead of endless practice:

If you know it, write it down. If you don't, or aren't sure, <u>work it out</u>. NO GUESSING!

The Common Core arithmetic curriculum for K-3 focuses on the wrong things. In putting a premium on memorization and rapid responses, it is designed for bookkeepers and tax accountants before they had access to calculators and computers. Instead, it should be aimed at making arithmetic understandable and accessible to *all* children instead of the few who "get it," no matter how it is taught.

Simply put, *the number facts are too fundamental to success in ALL of mathematics to be entrusted to memory alone.* Children need to know that they can readily access them whenever they need to know one in case they forget one. Knowing that they can quickly and accurately work out any that they might forget is a major confidence builder, *as is never having to guess while adding, subtracting, multiplying, and dividing*.

> *Self-confidence built on success is the most important objective of the mathematics curriculum.* — National Research Council, 1989

The Standard That Isn't

> *If you are confused, it means you are paying attention.* — Josh Gates, 2020

Imagine a class of second graders asking their teacher why, if they forget a number fact while adding and subtracting, they are allowed to figure it out mentally but not by counting it out? What should their teacher tell them? To side with the NCTM, their teacher should tell them that they cannot count out a number fact while computing because that would violate the organization's fluency requirement for paper-and-pencil arithmetic. If asked *"How so?"* they should tell them that the NCTM claims that counting out a number fact while computing lacks <u>flexibility</u>, is <u>inefficient</u>, and not <u>appropriate</u>, which tells them nothing. To be truthful, they should tell them that the NCTM insists on no counting out a number fact while computing because they said so.

Imagine physicians who reject some life-saving procedure that is easy to administer and effective with *every* patient because they are hung up on another procedure even though it requires years of treatment and is effective with only some of their patients. The NCTM is those physicians. Ever since TouchMath became known 50 or so years ago, the NCTM has shunned its system for counting out the number facts while computing and advocated mental math in place of it.

The Common Core arithmetic curriculum for K-3 requires children to memorize all 390 number facts by the end of grade 3, which implies that every child can do that and never forget them if they practice them enough in elementary school, which the NCTM knows is wishful thinking; otherwise, they would not anticipate forgetting by allowing three seconds to mentally figure out the ones they forget. The reality that plagues forcing children to memorize the number facts is that many children keep forgetting some of them, even ones that they once recalled easily.

> *Accurate and automatic production of the basic number combinations is a major objective of elementary mathematics education. Typically, it is not an objective that is easily and quickly attained. Indeed, teachers regularly lament about how much difficulty children have in mastering the basic number facts.* — Arthur Baroody, 1985

The truth of the matter is that many children require *years* to memorize the number facts, *until grade 3 or 4 to memorize just the addition and subtraction facts* (Ashcraft, 1982; Carpenter &

Moser, 1984; Lankford, 1974; Woods et al., 1975). However, <u>by the end of grade 2</u>, *every* child can learn how to count out all 390 number facts by <u>counting forward</u> or <u>counting on</u> for the addition facts, <u>counting back</u> or <u>counting on</u> for the subtraction facts, and <u>skip counting</u> for the multiplication and division facts. The result is an arithmetic curriculum in which all of the puzzle pieces for whole number arithmetic are put together in *kindergarten* to reveal the picture on the box the puzzle comes in.

The Common Core Arithmetic Curriculum for K-3 Is Double Talk

A hallmark of mathematics is that the basic assumptions and rules that govern how numbers and shapes are dealt with are *consistent*, which means that they do not contradict one another. If they did, everything derived from them would be both true and false, including $2 + 2 \neq 4$. Nonetheless, as explained in the preceding chapter, the six Common Core performance standards for K-3 compel adherence to the automaticity standard, which mandates the memorization of the number facts with the caveat that unless recalled instantly while computing, children will lose their place in the algorithm they are using and make mistakes and fail to grasp the meaning of the math at hand. Yet the same six performance standards promote thinking strategies for figuring them out while computing, which is inconsistent with recalling them instantly.

Initially, the purpose of the automaticity standard was probably to force the memorization of the number facts because they are foundational to arithmetic, and nobody knew how to teach them until the 1970s except with drill and practice. Now, however, children can be taught how to count them out with TouchMath and figure out some of them with mental math. So why is the NCTM still supporting the automaticity standard in the Common Core arithmetic curriculum for K-3? Why is it doing that while simultaneously violating the standard by allowing three seconds to solve a number fact mentally while computing?

Closing Remarks

What is now evident is that holding children accountable to the automaticity standard is indefensible. Allowing three seconds to work out a number fact while computing is an admission that children do not have to memorize the number facts in order to add, subtract, multiply, and divide. However, to comply with the NCTM's fluency requirement, they must be figured out mentally. They cannot be counted out, and were you to ask *"Why not?"* you would be told that counting them out is *"inflexible, inefficient, and inappropriate."* Never mind that it is quick and accurate and easily learned by even 5-year-olds.

<u>Mental math is *not* a viable alternative to TouchMath for working out a number fact while computing</u>. Mental math only helps with some of the addition and subtraction facts and hardly any of the multiplication and division facts, which are the most difficult facts to remember for any length of time. In contrast, TouchMath's system of touching and counting on the numerals 1 through 9 generates all 390 number facts. Still, mental math should be taught, but to note how the operations are related to one another, not as a false equivalency with TouchMath for generating the number facts.

References

Ashcraft, Mark H. "The Development of Mental Arithmetic: A Chronometric Approach," *Developmental Review*, Vol. 2, pp. 213-236, 1982.

Ashcraft, Mark H. "Is It Farfetched that Some of Us Remember Our Arithmetic Facts?" *Journal for Research in Mathematics Education*, Vol. 16, No. 2, pp. 99-105, 1985.

Baroody, Arthur J. "Mastery of Basic Number Combinations: Internalization of Relationships or Facts?" *Journal for Research in Mathematics Education*, Vol. 16, No. 2, pp. 83-98. 1985.

Boaler, Jo, Cathy Williams and Amanda Confer. "Fluency Without Fear: Research Evidence on the Best Ways to Learn Math Facts," *youcubed*, Stanford University, 2015.

Carpenter, Thomas P. & James M. Moser. "The Acquisition of Addition and Subtraction Concepts in Grades One through Three," *Journal for Research in Mathematics Education,* Vol. 15: 179-202,1984.

CCSSM. See *Common Core State Standards for Mathematics.*

Cobb, Paul. "An Analysis of Three Models of Early Number Development," *Journal for Research in Mathematics Education*, Vol. 18, No. 3, pp. 163-179, 1987.

Common Core State Standards for Mathematics, National Governors Association in collaboration with the Council of Chief State School Officers, 2010.

Curriculum and Evaluation Standards for School Mathematics, National Council of Teachers of Mathematics,1989.

Flexer, Roberta J. and Naomi Rosenberger. "Beware of Tapping Pencils," *The Arithmetic Teacher*, National Council of Teachers of Mathematics, Vol. 34, No. 5, 1987.

Gates, Josh. "The Hunt for the Golden Owl," Expedition Unknown, Discovery Channel.

Gojak, Linda M. NCTM President (2012-14), *Summing Up*, National Council of Teachers of Mathematics, November 1, 2012.

Hock, Dee. Founder and former CEO of VISA, quotefancy.com/dee-hock-quotes.

Kling & Bay-Williams (2015). "A Research-Based Approach to Math Fact Fluency," McGraw Hill, 2019.

Lankford, F.G., Jr. "What Can a Teacher Learn About a Pupil's Thinking through Oral Interviews?" *The Arithmetic Teacher*, National Council of Teachers of Mathematics, Vol. 21, pp. 26-32, 1974.

Math & Movement. "Exercise Body and Mind with Math & Movement's Play-Based Learning Techniques," mathandmovement.com.

McLeod, Joyce. *Texas HSP Math*, Grade 3, Harcourt School Publishers, 2009.

MOVE IT Math™, moveitmath.com.

Multiplication Motivation (CD), Melody House @ melodyhousemusic.com.

National Council of Teachers of Mathematics (**NCTM**), 1906 Association Drive, Reston, VA 20191-1502, nctm.org.

National Research Council, *Everybody Counts: A Report to the Nation on the Future of Mathematics Education*. National Academy Press, Washington, D.C.,1989.

NCTM Standards. See *Curriculum and Evaluation Standards for School Mathematics.*

Peanuts. See Schultz.

Schultz, Charles M. Peanuts, featuring "Good ol' Charlie Brown," United Features Syndicate, Inc., 1981.

TouchMath, Innovative Learning Concepts (ILC). Free catalog available at touchmath.com.

Woods, S.S., Lauren B. Resnick, and G.J. Groen. "An Experimental Test of Five Process Models for Subtraction," *Journal of Educational Psychology*, Vol. 67, pp. 17-21, 1975.

Chapter 6: Solving Arithmetic Word Problems

The best way to create interest in a subject is to render it worth knowing, which means to make the knowledge gained usable in one's thinking beyond the situation in which the learning has occurred. — Jerome Bruner, 1960

After Ms. Bullock, the TouchMath lady, showed me how she taught the number facts by counting them out, I was sold on the methodology. My only complaint with it was that I did not think of it. The clincher for me was how skip counting not only simplifies teaching the number facts for multiplication and division but reveals the *nature* of the operations, that in spite of there being four operations, *arithmetic just combines and separates amounts*.

Addition and Multiplication Are Combining Actions

Multiplication is a combining action, as is addition, but whereas addition combines *any* amounts, multiplication only combines *equal* amounts, and skip counting demonstrates that. For example, counting out 5 x 8 by skip counting by 5 on the touchpoints of the eight yields 5-10-15-20-25-30-35-40, which is a <u>running total</u> of the sum of eight 5s (5+5+5+5+5+5+5+5 = 40), which shows that multiplication is <u>repeated addition</u>.

Subtraction and Division Are Separating Actions

Division is a separating action, as is subtraction, but whereas subtraction separates a quantity into *any* amounts,[33] division separates an amount into *equal* amounts, and skip counting by the divisor to the dividend and keeping track of the counts determines the number of equal amounts. For example, counting out 40 ÷ 5 by skip counting by 5 to 40 (5-10-15-20-25-30-35-40) and keeping track of the counts on your fingers or with tally marks (||||| |||) shows that 5 can be subtracted from 40 eight times (40 − 5 − 5 − 5 − 5 − 5 − 5 − 5 − 5 = 0), which shows that division is <u>repeated subtraction</u>.

Multiplication adds equal amounts, and division subtracts equal amounts. Thus in spite of their being four operations in arithmetic, *we just add or subtract.* That is, we just combine or separate, after which it is a matter of "style." *How did we combine or separate?* <u>If we combined *any* amounts, we **added**, but If *equal* amounts, we **multiplied**. If we separated a quantity into *any* amounts, we **subtracted**, but if into *equal* amounts, we **divided**.</u> Envisioning the four operations as combining or separating actions with any or equal amounts makes arithmetic *real and relevant*. It connects the operations to daily activities.

The connection between the four operations and combining or separating with any or equal amounts is made by assigning "homework" — homework in the sense that it involves children's homes and communities. Children are tasked to make a list of common, everyday events in their homes and elsewhere that involve combining or separating any or equal amounts. When done, they pool their findings with those of their classmates and, except for duplicates, list what they found for each operation on a chart for that purpose, as shown on the next page with examples of such events.

---

[33] For example, 15 − 8 = 7 indicates that the minuend 15 has been separated into two amounts, 8 and 7, the subtrahend 8 and the *difference* 7 between the subtrahend and the minuend, respectively.

Combining and Separating Events with Any and Equal Amounts

Combining = Amounts

Combining Any Amounts

Separating Any Amounts

Separating into = Amounts

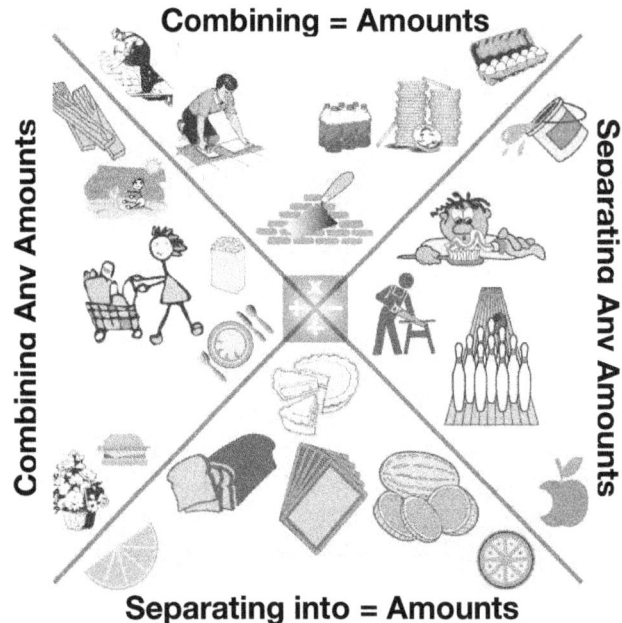

Combining/Separating Chart

Combining Any Amounts	Separating Any Amounts	Combining = Amounts	Separating into = Amounts
Loading a dishwasher	Emptying a sink of dishes and silverware	Buying dozens of eggs	Putting eggs in cartons
Putting groceries in a shopping cart	Cutting off the end of a board	Putting nickels in a piggy bank	Trading quarters for dollars
Putting coins in a piggy bank	Unpacking groceries	Trading dollars for quarters	Folding socks
Saving money	Letting some air out of a tire	Building a brick wall	Tearing down a brick wall
Taking packages to a post office to ship	Whittling a stick	Putting postcards in a mail box	Turning inches into feet
Collecting rocks	Taking a bite out of an apple	Buying gloves or shoes	Dealing hands from a deck of cards
Decorating a Christmas tree	Taking the ornaments off a Christmas tree	Selling tickets to attend a concert	Unraveling a scarf
Putting photos in a picture album	Getting a haircut	Knitting a scarf	Eating an ear of corn three rows at a time
Sweeping the floor	Clipping one's toenails	Skip counting by 5	Shoveling dirt
Raking leaves	Filing one's fingernails	Pumping up a tire	Bailing out a boat
Building a rock wall	Sanding wood	Consuming a bowl of soup by the spoonful	Emptying a bowl of soup by the spoonful
Making a sandwich	Digging a hole	Blowing up a balloon	Scooping ice cream
Putting clothes in a washing machine	Pruning a tree	Picking 4-leaf clovers	Picking teams
Packing a suitcase	Robbing a piggybank	Filling saltshakers	Filling pint-sized bottles with water (from a bucket)
Putting a puzzle together	Abbreviating a word	Filling pint-sized bottles with water (from a hose)	
	Shelling a peanut		

To add events to the combining/separating chart, the teacher and children mime the events to decide on their placement in the chart. The objective in having them act out the events is for the children to physically experience the combining or separating with any or equal amounts that is occurring in the events. Ultimately, the objective is for them to sense the same actions in arithmetic word problems as they act them out in their minds. Once they can do that, they are ready to learn how to solve arithmetic word problems *and their real-life equivalents when adults*.

Solving Arithmetic Word Problems

To solve an arithmetic word problem, a student must …

1. Understand the problem.

2. Decide to add, subtract, multiply or divide.

3. Compute or calculate the answer.

If a student understands an arithmetic word problem, they know which operation to perform or key to tap on a calculator, so problem solved. However, if they do not understand it, they are stuck, but if they know that the four operations are combining or separating actions with any or equal amounts, *and that they are opposites of one another*, they can be taught the What's Happening heuristic that will guide them through the first two steps in solving the problem.

What's Happening Heuristic

A heuristic is an algorithm of sorts. Like an algorithm, it is a step-by-step procedure, but one for solving non-computational problems. The What's Happening heuristic is a step-by-step procedure for solving arithmetic word problems and the like in daily living. Notably, it consists of just two steps, namely, asking the two questions below in the order given:

1. *What's happening?* **Combining or separating?**

2. *How's it happening?* **With any or equal amounts?**

Answering these questions pinpoints the operation key to tap on a calculator to solve an arithmetic word problem because every such problem is about combining or separating any or equal amounts, *or it is not an arithmetic word problem!*

If the answer to the first question is *"combining"* and that for the second one is *"any amounts,"* it is an **addition problem**. If *"equal amounts,"* it is a **multiplication problem**. If the answer to the first question is *"separating"* and that for the second one is *"into any amounts,"* it is a **subtraction problem**. If *"into equal amounts,"* it is a **division problem**. Thus the What's Happening heuristic guides children through the understanding and decision-making stages in solving an arithmetic word problem. The diagram on the next page to enlarge and display in a classroom encapsulates that and shows the connection between combining/separating events and the operations of addition, subtraction, multiplication, and division.

What's Happening?

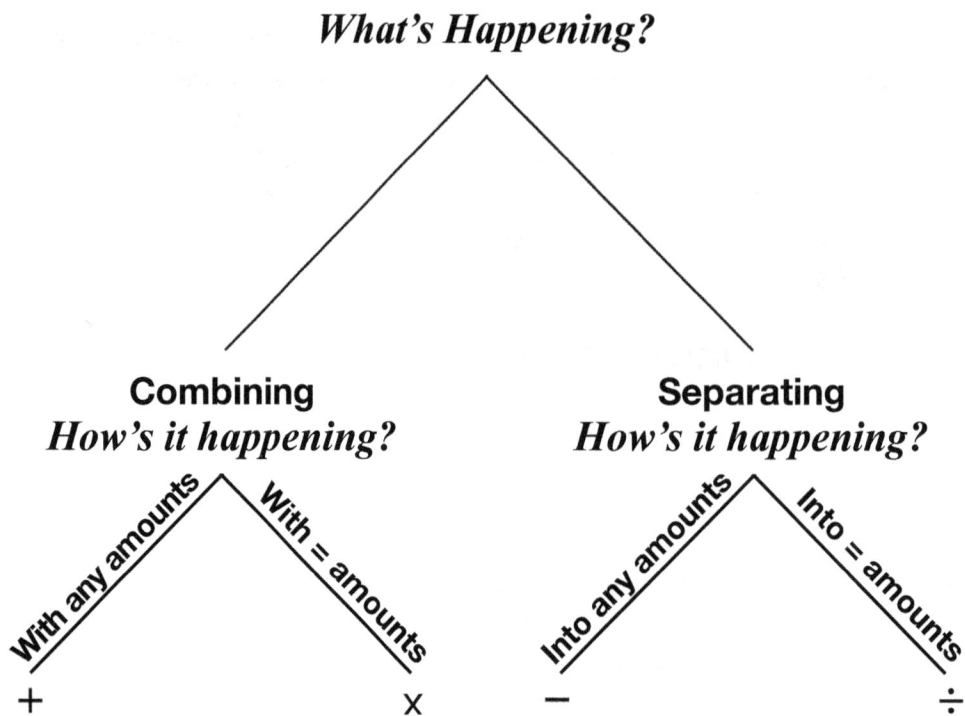

For most arithmetic word problems, the answers to the What's Happening questions are self-evident because they are *opposites* of one another. However, if the answer to the first question is unclear for an arithmetic word problem, STOP! There is no point in trying to solve the problem until the combining or separating question has been answered, but knowing that the answer must be one or the other helps. It narrows the decision making to just one of those two choices. To illustrate, note how the heuristic directs the problem-solving process in solving the following McDonald's hamburger problem.

Stacking Billions of McDonald's Hamburgers to Reach the Moon

I got my first hand-held calculator for Christmas about 50 years ago. When I got it, McDonald's was bragging on its marquees that it had sold more than 15 billion hamburgers. So to check out my new toy after I had opened all of my presents, I used it to figure out the height, in miles, of a stack of that many McDonald's hamburgers to see if it would reach from the earth to the moon, a distance of about 238,855 miles (384,400 km).

To figure the height of the stack of hamburgers, I assumed that each hamburger was two inches thick from top to bottom. Then, in picturing the stack in my mind, I could "see" that figuring its height was a combining problem with equal amounts — that of stacking 15 billion 2-inch thick hamburgers on top of one another. Thus the height of the stack of hamburgers was 2 x 15,000,000,000 = 30,000,000,000 inches, which I figured out without my calculator by doubling the 15 in 15 billion. Then, knowing the height of the stack in inches, my plan was to convert the height of it in inches into feet, the height of it in feet into miles, and the height of it in miles into trips to the moon.

To convert the height of the stack of hamburgers that was in inches into feet, I imagined the inches being separated into 12-inch sections because there are 12 inches in a foot. Thus I knew to divide 30 billion by 12 because I was separating the inches into equal amounts, namely, 12-inch sections. I could have worked that out by skip counting by 12 to 30 billion (12-24-36- … 30,000,000,000) and noting the number of counts with tally marks (||||| ||||| ||||| … |||||), but at the rate of one count per second for 40 hours a week and about 4,000 tally marks per page, that would have taken me 334 years and used up 625,000 pages of paper![34] So instead, I worked it out on my calculator.

- o   I entered 30,000,000,000, tapped the division key, entered 12, and tapped the equals key to get the height of the stack of hamburgers in feet: <u>2,500,000,000 feet</u>.

Before continuing, please note what I did with my calculator to get to this point in my McDonald's hamburger problem: 1) I entered a number, 2) tapped the division key, 3) entered another number, and 4) tapped the equals key. *A crow could be trained to do that and would get the correct answer to a division problem every time.* Really! (Butterworth, 2022; Clemmitt, 2010). However, a crow could not be trained to *decide* to tap the division key instead of the addition, subtraction, or multiplication key. In other words, a crow could not be trained to use the What's Happening heuristic.

Returning to my McDonald's hamburger problem, knowing the height of the stack of hamburgers in feet, I needed to turn the feet into miles. To do so, I imagined the feet being separated into mile-high stacks 5,280 feet high. Thus I knew to divide 2,500,000,000 feet by 5,280 because I was separating the feet into equal amounts, namely, stacks 5,280 feet high.

- o   So I entered 2,500,000,000 in my calculator, tapped the division key, entered 5280, tapped the equals key, and found that 2,500,000,000 feet equals <u>473,485 miles</u>.

Then, to convert that many miles into trips to the moon, I imagined the miles being separated into stacks 238,855 miles high because that is the approximate distance from the earth to the moon. Thus I knew to divide 473,485 miles by 238,855 because I was separating 473,485 into equal amounts, namely, into stacks 238,855 miles high.

- o   So I entered 473,485 in my calculator, tapped the division key, entered 238,855, tapped the equals key, and found that 473,485 miles equals <u>1.98 trips to the moon</u>.

I was surprised to learn that a stack of 15 billion McDonald's hamburgers would reach all the way to the moon and within a mere 4,225 miles[35] of all the way back. Before I did the arithmetic, I thought the stack of that many hamburgers *might* reach the moon, but I was not sure of that until I worked out the height of the stack with my calculator.

---

[34] To skip count by 12 to 30 billion at the rate of one count per second would take 30,000,000,000 ÷ 12 = <u>2,500,000,000 seconds</u>, which equal 2,500,000,000 ÷ 60 = <u>41,666,667 minutes</u> (because there are 60 seconds in a minute), which equal 41,666,667 ÷ 60 = <u>694,444 hours</u> (because there are 60 minutes in an hour), which equal 694,444 ÷ 40 = <u>17,361 weeks at 40 hours per week</u>, which equal 17,361 ÷ 52 = <u>334 years</u> (because there are 52 weeks in a year).

Assuming 4,000 tally marks per page, recording 2,500,000,000 tally marks (from 30 billion divided by 12) would use up 2,500,000,000 ÷ 4000 = 625,000 pages of paper.

[35] The distance from the earth to the moon and back is 238,855 x 2 = 477,710 miles, thus the difference between that many miles and the distance traveled is 477,710 − 473,485 = 4,225 miles.

What amazed me was that I had found that out in about ten minutes: five minutes playing with my new toy to make sure it was giving me the correct answers to arithmetic problems I could work in my head and five minutes looking up the distance from the earth to the moon the old-fashioned way in a hand-held encyclopedia.

After solving my McDonald's hamburger problem with my calculator and being able to visualize the enormity of McDonald's achievement in selling 15 billion hamburgers, I was motivated to make up more McDonald's hamburger problems to solve with my calculator. To do that, I needed to know more about a McDonald's hamburger than my estimate of its height, so I talked to the manager of a nearby McDonald's and learned that they used one gallon of mustard for every 2,400 hamburgers, one gallon of ketchup for every 300 hamburgers, and one gallon of pickles for every 350 hamburgers. With that information, I calculated that to make 15 billion hamburgers, McDonald's had used enough mustard to fill 240 railroad tank cars, enough ketchup to paint the District of Columbia 17 times, and enough pickles to build a wall of pickle jars 12 feet high around Long Island, New York.

To share my cognitive adventure with my calculator, I wrote a short account of my mind-boggling calculations for 15 billion McDonald's hamburgers and submitted it to the National Council of Teachers of Mathematics (NCTM) who published it in the *Arithmetic Teacher* (Shoecraft, 1975). To conclude said adventure, I sent a copy of the article to McDonald's corporate address. In return, I received a coupon for a free hamburger, drink, and fries at any McDonald's.

The point of my McDonald's hamburger story was to show the connection between counting out the multiplication and division facts by skip counting and knowing when to tap the multiplication or division key on a calculator. The connection is straightforward. <u>Multiplication combines equal amounts, and division separates a quantity into equal amounts</u>, and counting out the multiplication and division facts by skip counting reinforces that about the operations. In contrast, memorizing the multiplication facts and learning the division facts as "backwards" multiplication problems (e.g., 42 ÷ 6 = 7 because 6 x 7 = 42) disclose nothing about what the operations do to numbers or how they appear in daily events as in the combining/separating chart.

Your Turn

Filling in the combining/separating chart connects the operations to activities that children experience in and out of school. That being the case, filling it in makes solving arithmetic word problems relevant to children because the entries in the chart can be made into arithmetic word problems, as shown below for four underlined entries in the chart.

1. A ticket to an outdoor event cost $20. The event was rained out. If 2,375 tickets to the event were sold, how much money had to be refunded?

2. If $250 is deposited in a savings account with a balance of $1,350, how much money is in the account?

3. A boy made a compact with his parents to put 50 cents from his allowance in a piggybank every week for a year. After a year, he had $26 in his piggybank. To reward his steadfastness in doing what he had agreed to do, his parents said he could take $15 of

the money he had saved to spend however he wished. How much money will he have in his piggybank after he does that?

4. How many egg cartons are needed to put 10,000 eggs in egg cartons?

To experience how the What's Happening heuristic guides one through the understanding and decision-making stages in solving arithmetic word problems, please use it to decide which operation key to tap on a calculator to solve each of the made-up problems. For each one, first decide if it is a combining or separating event. Then decide if the combining or separating is with any or equal amounts. *No need to actually solve the problem.* When finished, check your answers by seeing if they agree with the placement of the underlined events in the combining/separating chart.

Does It Matter How the x and ÷ Facts Are Taught?

It absolutely matters, because what matters is that children, when adults, know which operation key on a calculator to tap to solve the arithmetic problems they will encounter in daily living, as when paying bills, planning a trip, or buying some product on time. That is why children are taught how to solve arithmetic word problems, but a teacher doing so by showing children how *they* would solve an arithmetic word problem is just pulling a rabbit out of a hat. For many children, what may have made sense to them when their teacher solved the problem did not make sense to them when they tried to solve a similar problem on their own.

Being able to automatically recall a multiplication fact from memory may be a wondrous accomplishment in the moment, but it has no bearing on deciding to multiply when solving an arithmetic word problem. Likewise, knowing that the division facts are related multiplication facts has nothing to do with deciding to divide when solving an arithmetic word problem. So solving problems that involve multiplication and division requires knowing more than just the number facts for those operations. What must be known is that multiplication combines equal amounts and that division separates a quantity into equal amounts.

When children count out a multiplication fact by skip counting by one of its multipliers on the touchpoints of the other multiplier, *they are actually multiplying*. They are combining an equal amount multiple times. For example, counting out 3 x 7 by skip counting by 3 for seven counts (3-6-9-12-15-18-21) or by 7 for three counts (7-14-21) actually combines the number 3 seven times or the number 7 three times. Either way, 3 x 7 = 7 x 3 = 21.

Similarly, when children count out a division fact by skip counting by its divisor to its dividend while keeping track of the number of counts, *they are actually dividing*. They are separating the dividend into equal amounts the size of the divisor. For example, counting out 21 ÷ 3 by skip counting by 3 to 21 (3-6-9-12-15-18-21) and keeping track of the counts on one's fingers or with tally marks (||||| ||) actually separates the number 21 into seven 3s, which answers the question *"How many 3s in 21?"* Thus 21 ÷ 3 = 7.

Skip counting to count out the multiplication and division facts connects the operations to the kinds of arithmetic word problems and problems in daily living that can be solved with multiplication and division. Every time a child counts out a multiplication or division fact by skip counting, they are acting out how those operations appear in common, everyday events that combine equal amounts (like putting chairs in rows) and that separate a quantity into equal amounts (like sharing the cost of a meal with friends). Knowing that about the operations from the moment they are introduced in *kindergarten* is surely better preparation for knowing when to multiply or divide when an adult than memorizing the multiplication facts and knowing that the division facts are backwards multiplication facts.

More Needs to Be Said about Skip Counting

In addition to skip counting by 2 through 9 to generate the number facts for multiplication and division, skip counting by the numerators and denominators of fractions generates lists of equivalent fractions that reveal those with common denominators that may be added and subtracted. For instance, skip counting by the numerators and denominators of ⅔ and ⅘ yields the following lists of equivalent fractions that show that ⅔ = 10/15 = 20/30 and that ⅘ = 12/15 = 24/30.

$$\mathbf{\frac{2}{3}} = \frac{4}{6} = \frac{6}{9} = \frac{8}{12} = \mathbf{\frac{10}{15}} = \frac{12}{18} = \frac{14}{21} = \frac{16}{24} = \frac{18}{27} = \mathbf{\frac{20}{30}} = \ldots$$

$$\mathbf{\frac{4}{5}} = \frac{8}{10} = \mathbf{\frac{12}{15}} = \frac{16}{20} = \frac{20}{25} = \mathbf{\frac{24}{30}} = \frac{28}{35} = \frac{32}{40} = \frac{36}{45} = \frac{40}{50} = \ldots$$

Then, using ⅔ = 10/15 and ⅘ = 12/15 — the fractions equivalent to ⅔ and ⅘ with the least common denominator (LCD) — ⅔ and ⅘ may be added and subtracted as shown below:

⅔ + ⅘ = 10/15 + 12/15 = 22/15

⅘ − ⅔ = 12/15 − 10/15 = 2/15

With skip counting, making addition and subtraction of fractions understandable just got easy were it not for the NCTM's intolerance of TouchMath and students visibly counting while computing.

No Skip Counting by 2, 3, 4, 6, 7, 8, 9 in the Common Core Arithmetic Curriculum

The Common Core arithmetic curriculum for elementary school math includes a skip counting standard for 5, 10, and 100, as shown by 2.NBT.2, a Common Core *content* standard for grade 2: *Count within 1000; skip-count by 5s, 10s, and 100s.* Knowing how to skip count by 5, 10, and 100 aids in determining the value of an assortment of coins (e.g., 5-10-15-20 cents for four nickels), telling time (e.g., 10-20-30-40 minutes past the hour), and measuring in metric units (e.g., 100-200-300- ... 1,000 meters in a kilometer). However, the NCTM is silent on skip counting by 2, 3, 4, 6, 7, 8, 9.

Were the NCTM to condone teaching children how to skip count by those numbers, it would be condoning counting out the multiplication and division facts, which the NCTM disfavors in its ongoing rejection of TouchMath. Instead, the NCTM *insists* that the multiplication and division facts for those numbers must be memorized or figured out with mental math if forgotten.

The only aids for figuring out the multiplication and division facts with mental math are the nine rule for multiplication (e.g., 9 x 7 = 10 x 7 – 7 = 70 – 7 = 63), the commutative property of multiplication (e.g., 4 x 8 = 8 x 4 = 32), and "fact families — the realization that every multiplication fact is related to its "twin" and two division facts (e.g., 6 x 8 = 48 is in the same "fact family" as its "twin," 8 x 6 = 48, and 48 ÷ 6 = 8 and 48 ÷ 8 = 6) and that every division fact is related to its "twin" and two multiplication facts (e.g., 56 ÷ 8 = 7 is in the same "fact family" as its "twin," 56 ÷ 7 = 8, and 8 x 7 = 56 and 7 x 8 = 56).

Closing Remarks

In physics, we learn that every action has an equal but opposite reaction, which is true for combining and separating events, as well. <u>Every combining action has a related separating action and vice versa</u>. For instance, consider loading a dishwasher. Doing so is listed in the combining/separating chart as a "combining any" action. However, the things being loaded into the dishwasher had to come from somewhere, like a sink, so a related action is removing them from that somewhere, which is a "separating any" action.

Additionally, some combining actions can also be interpreted as separating actions and vice versa depending on how they are imagined. For instance, note that the last entries in the "combining equal" and "separating equal" columns in the combining/separating chart are the same except for what is in parentheses. If "filling pint-sized bottles with water" is thought to be from a hose, the action would be on filling the bottles with water, which would be combining equal amounts of water. However, if the bottles are being filled with water from a bucket a cup at a time, the action would be separating equal amounts of water from the bucket.

The fact that some combining actions are also separating actions and vice versa may be one of the reasons children struggle with arithmetic word problems. If the picture that forms in a child's mind after reading a word problem differs from the picture that was in the mind of the person who wrote the problem, the child may use the wrong operation to solve it even if their interpretation of the problem was correct from their viewpoint. Thus time spent on having children mime, visualize, and think of different interpretations of the actions in the combining/separating chart will increase their awareness and depth of understanding of what is happening in such events.

How to teach children — *all children* — how to use the What's Happening heuristic to solve arithmetic word problems is surprisingly easy once the children are aware of the many combining and separating events they experience in and out of school. I learned how easily they could learn to use it while observing a lesson taught by a third-grade teacher in Houston, Texas.

To begin, the teacher split her class into small groups and gave each group a bag containing about 50 arithmetic word problems that were on slips of paper, one problem per slip, like the messages inside Chinese fortune cookies.[36] She then issued the following instructions to the groups and circulated among them to make sure they were doing as instructed.

---

[36] To acquire the several hundred arithmetic problems the teacher had bagged, she copied pages of arithmetic word problems from math textbooks and cut the problems apart with scissors.

1. Take turns taking a problem out of the bag and reading the problem. Then, with everyone in your group, mime the combining or separating action in the problem.

2. With everyone in your group, use the What's Happening heuristic on the wall to decide which operation key on a calculator would be tapped to solve the problem.

3. When everyone in your group agrees on which key would be tapped to solve the problem, pass the bag to the person on your left. *Do not solve the problem.* You can solve one like it later.

4. Continue taking problems out of the bag, reading them, acting them out, and deciding on which operation key on a calculator would be tapped to solve them.

I was amazed by the effectiveness of the lesson. Within the hour, the teacher's students "got it." They worked through every word problem they took out of the bags without ever asking their teacher for help.

MOVE IT Math reveals elementary school teachers' creativity. Its methodology liberates them to actually *teach* arithmetic instead of making children memorize it. *Any adult can make children memorize arithmetic,* but a teacher who knows how to teach for understanding makes arithmetic resonate in children's minds. They tell them enough to get them thinking about what they know so they can see the connection between that and whatever new material their teacher wants them to know. However, for their teacher to be able to do that requires enough understanding of arithmetic on their part to guide children through that process, which is why elementary school teachers need instruction in the first course in MOVE IT Math: All Children Can Learn Arithmetic.

With the What's Happening lesson just described in mind, I can now explain how MOVE IT Math evolved into the child-centered, teacher-friendly elementary school math program it became. My job in creating the program was to get the math right and work out how to make it meaningful to elementary school teachers. When meaningful to them, their job was to come up with ways to make the math they had learned meaningful to their students. I was humbled by their work ethic and creativity in thinking of engaging ways to do that.

So here is how to develop a much-needed K-6 hands-on algebra curriculum. Get the algebra right with the many hands-on materials already available for teaching algebra to children, as listed in the appendix beginning on page 279, and make it meaningful to elementary school teachers. Then take notes on how they parcel it out among themselves to make it meaningful to children.

Creating a K-6 hands-on algebra curriculum would be a simple matter and fun to do. Any elementary school could make their own with the help of the middle school and high school math teachers in their district to get the algebra right with the aforementioned hands-on materials in the appendix and make it meaningful to elementary school teachers.

Getting back to the What's Happening lesson, was it a math lesson or a reading lesson? *It was both*, and an exceptionally good one for reading because of its focus on the children <u>comprehending</u> what they were reading to one another. Because of that, the lesson could be conducted during the time allotted in the curriculum for reading.

What I am getting at is how teachers can extend the time they would normally spend on teaching math if they occasionally teach both math and reading during reading time. The cumulative effect of using just one hour a week from reading time to teach both math and reading would add about 36 hours per school year to the time spent on teaching math. That is like extending the school year for teaching math by almost two months without having children go to summer school. <u>That is teaching smarter, not harder</u>.

References

Bruner, Jerome. *The Process of Education, A Landmark in Educational Theory*, Harvard University Press, Cambridge, Massachusetts and London, England, 1960.

Butterworth, Brian. *Can Fish Count? What Animals Reveal about Our Uniquely Mathematical Minds,* Hachette Book Group, 2022.

Clemmitt, Marcia. "Animal Intelligence: Do Animals Think?" CQR - CQ Press Library, 2010.

CCSSM. See *Common Core State Standards for Mathematics.*

Common Core State Standards for Mathematics, National Governors Association in collaboration with the Council of Chief State School Officers, 2010.

National Council of Teachers of Mathematics (**NCTM**), 1906 Association Drive, Reston, VA 20191-1502, nctm.org.

Shoecraft, Paul. "15 Billion Hamburgers Is a Lot of Multiplication and Division," *The Arithmetic Teacher*, National Council of Teachers of Mathematics,1975.

TouchMath, Innovative Learning Concepts (ILC). Free catalog available at touchmath.com.

Chapter 7: Your Brain Says *"Use it or lose it."*

Knowledge one has acquired without sufficient structure to tie it together is knowledge that is likely to be forgotten. An unconnected set of facts has a pitiably short half-life in memory.
— Jerome Bruner, 1960

The automaticity standard asserts that every child must memorize the number facts until they can recall them "automatically," *which means instantly, without thinking*, the same as they can recall their own names. Now add to that assertion *"and be able to do that for the rest of their lives."* Does that challenge your thinking about the automaticity standard if you ascribe to it?

Although unstated, the automaticity standard assumes that once children memorize the number facts, they will not forget them. *Not ever!* That means that if you, the reader, once memorized them, you should still be able to instantly recall them. See if you can recall at least 40 or more multiplication facts in 60 seconds by timing yourself on the speed drill on the multiplication facts in the appendix on page 319. If you can, good for you. If not, take solace in knowing that you are normal.

Forgetting facts once known occurs when the reason for remembering them is gone. To verify that, try to remember facts that you once knew as well as you know your own name, facts like your address and home phone number when you were in elementary school if it has changed or the zip codes of your last three addresses if you have moved around. I have asked thousands of elementary school teachers questions like that to see if they could still remember them and have been surprised by the result: *About one in five could still remember them!*

Maybe you can remember facts that you have not had to remember for a long time. If so, then you and the teachers I asked who could still remember such facts are wonderfully odd. The four in five, including me, who could not remember them, are, by definition, *normal*. Forgetting facts that used to come to mind automatically is to be expected when remembering them is no longer required.

Ebbinghaus Forgetting Curve

Research on forgetting factual information has been ongoing since 1885 ever since Hermann Ebbinghaus (1850-1909) published his forgetting curve (Burton, 2023) that is shown on the next page. Said curve showed how quickly he forgot a list of three-letter nonsense words like "BUP" and "TOV" after he quit practicing them (Murre & Dros, 2015). After 20 minutes, his forgetting curve showed that he had forgotten more than 40 percent of the three-letter "words" he had memorized. After a day, more than half of them. After a month, more than 70 percent of them.

The number facts are factual information about numbers, which is why they are called "facts." As such, the Ebbinghaus forgetting curve predicts that children will forget some of them once they no longer practice them. You know this if you ever crammed for a test when you were in high school or college. I bet you did not quit cramming a week before the test. Instead, I bet that you crammed right up to the night before the test because you knew you would start forgetting what you had memorized once you stopped reviewing it.

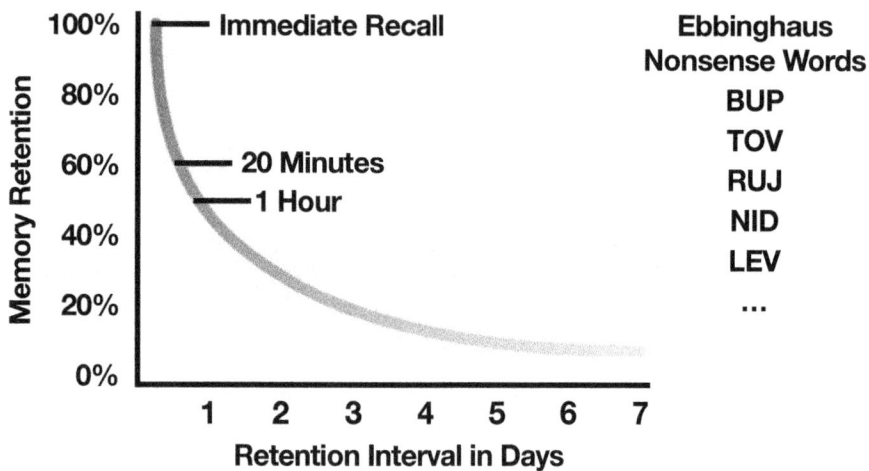

In follow-up studies, Ebbinghaus established the concept of spaced repetition to mitigate forgetting. He showed that repeating what had been memorized, first within days, then within larger spans of time, even years, significantly reduced forgetting, especially if what had been memorized was a derivative of something already known (Wikipedia.org/Ebbinghaus). Thus Ebbinghaus' findings about forgetting and how it can be stymied by periodic review and "overlearning" (practicing a skill long after it was mastered) implies that it is possible to stifle forgetting *indefinitely* with enough spaced review, but falsely so.

Believing that reviewing the number facts all through K-8 will result in every child remembering them for the rest of their lives is the reason for all of the worksheets, flash cards, and speed drills on the number facts in those grades. However, we know that no amount of review on the number facts has ever been enough to permanently fix them in every child's mind. We know that because of the need in grades 3-8 to review the number facts all over again at the start of the school year because of the forgetting that occurs during the summer break, *and there is nothing their teachers could have done to prevent that*. Teachers cannot make students remember *anything* when their students are not in school.

According to neuroscience, no amount of spaced review or overlearning of material that has been memorized is enough to permanently fix it in memory, as will be established later in this chapter, but the "use it or lose it" maxim is sufficient to establish that now. Forgetting factual information that has been memorized occurs no matter how well it is known if not practiced regularly, and evidence of that cannot be denied. Why do world-class musicians and athletes practice their skills throughout their careers? *We know the answer.* No matter how accomplished, they practice them in order to stay at the peak of their performance levels.

> *If I miss one day's practice, I notice it. If I miss two days' practice, the critics notice it. If I miss three days' practice, the public notices it.* — Paderewski (1860-1941), a famous Polish composer and pianist, 1911

So what do highly skilled musicians and athletes tell us about children being able to automatically recall the number facts for as long as they live? Again, *we know the answer*. They just have to practice them for the rest of their lives. Fat chance of that happening, and why should it if children can be taught how to *"quickly and accurately"* figure out any number fact they might forget by counting or reasoning it out?

Accepting the "use it or lose it" maxim for what we remember, what is wrong with teaching the number facts solely with flash cards, worksheets, and speed drills? Outside of school, that sort of practice is not encountered unless being tutored. In contrast, children routinely count forward, count on, count back, and skip count when not in school. They count on or back to figure out how many minutes before leaving for school or how many days until their birthday or a special holiday. They skip count by five and ten to determine the value of nickels and dimes in their pocket or purse. They practice these counting skills in and out of school without even thinking about what they are doing, which is why learning how to count out the number facts sticks.

Still, many adults remember the number facts long after they quit practicing them in elementary and middle school. How can that be if forgetting facts that have been memorized is certain to occur if not recalled periodically? Aside from the wonderfully odd people who just remember stuff, *they use the number facts*, and we can predict some of those who do: those who teach math from grade 3 on, those who perform calculations on a regular basis, even if with a calculator (e.g., bookkeepers, accountants, and real estate agents), and those who work in the STEM disciplines: science, technology, engineering, and mathematics.

Failure in Math Begins with the Number Facts

The Common Core arithmetic curriculum for K-3 requires children to memorize all 100 addition facts by the end of the second grade (CCSSM 2.OA.2) and all 100 multiplication facts by the end of the third grade (CCSSM 3.OA.7). The standard practice is to hammer them into children's heads with worksheets, flash cards, and speed drills. Paradoxically, a major problem with this practice is that by the end of the school year, *it seems to have worked!* By then, many children perform reasonably well on the number facts and appear to know them. Thus elementary school teachers have reason to believe in the efficacy of "hammering" to teach the number facts.

What every elementary school teacher needs to do is talk to the seventh grade math teachers in their school district. They need to ask them if they were satisfied with how well their students knew the number facts when they began the seventh grade. That never happens, but if it did, they would be told *"No,"* especially for the multiplication facts. To fact-check me on that, *talk to a seventh grade math teacher.*

I have talked to lots of seventh grade math teachers about how well their students knew the number facts when they began the seventh grade. Every one of them complained about how many of their students had forgotten enough of them to merit hammering them into their heads with worksheets, flash cards, and speed drills. *Again!*

Ancespitoria

As noted in Chapter 3, the rungs for the number facts on the math ladder are slippery. Children keep losing their footing on the rungs because they keep forgetting some of the number facts if they have not practiced them for a while, like during summer vacation, a school holiday, or even a weekend. To appreciate why, welcome to Ancespitoria, a made-up place with a made-up name with different symbols for the numerals 0 through 9:

+	x	=	0	1	2	3	4	5	6	7	8	9
+	x	=	●	■	✿	⬤	⊗	★	▣	⊠	◪	⊙

To you, number sentences like 2 + 3 = 5 and 4 x 7 = 28, read *"two plus three equals five"* and *"four times seven equals 28,"* respectively, are meaningful. You do not puzzle over them because you are familiar with them. To children, though, they are gibberish when first encountered. As you can literally *see* in the tables below, the equivalent of 2 + 3 = 5 and 4 x 7 = 28 in Ancespitoria is ✿ + ◉ = ★, read *"flower plus donut equals star,"* and ⊗ x ⊠ = ✿◪, read *"x in a circle times x in a square equals flower square donut."*

Now you see it.

+	0	1	2	3	4	5	6	7	8	9
0	0	1	2	3	4	5	6	7	8	9
1	1	2	3	4	5	6	7	8	9	10
2	2	3	4	5	6	7	8	9	10	11
3	3	4	5	6	7	8	9	10	11	12
4	4	5	6	7	8	9	10	11	12	13
5	5	6	7	8	9	10	11	12	13	14
6	6	7	8	9	10	11	12	13	14	15
7	7	8	9	10	11	12	13	14	15	16
8	8	9	10	11	12	13	14	15	16	17
9	9	10	11	12	13	14	15	16	17	18

X	0	1	2	3	4	5	6	7	8	9
0	0	0	0	0	0	0	0	0	0	0
1	0	1	2	3	4	5	6	7	8	9
2	0	2	4	6	8	10	12	14	16	18
3	0	3	6	9	12	15	18	21	24	27
4	0	4	8	12	16	20	24	28	32	36
5	0	5	10	15	20	25	30	35	40	45
6	0	6	12	18	24	30	36	42	48	54
7	0	7	14	21	28	35	42	49	56	63
8	0	8	16	24	32	40	48	56	64	72
9	0	9	18	27	36	45	54	63	72	81

Now you don't!

+	●	■	✿	◉	⊗	★	⊡	⊠	◪	⊙
●	●	■	✿	◉	⊗	★	⊡	⊠	◪	⊙
■	■	✿	◉	⊗	★	⊡	⊠	◪	⊙	■●
✿	✿	◉	⊗	★	⊡	⊠	◪	⊙	■●	■■
◉	◉	⊗	★	⊡	⊠	◪	⊙	■●	■■	■✿
⊗	⊗	★	⊡	⊠	◪	⊙	■●	■■	■✿	■◉
★	★	⊡	⊠	◪	⊙	■●	■■	■✿	■◉	■⊗
⊡	⊡	⊠	◪	⊙	■●	■■	■✿	■◉	■⊗	■★
⊠	⊠	◪	⊙	■●	■■	■✿	■◉	■⊗	■★	■⊡
◪	◪	⊙	■●	■■	■✿	■◉	■⊗	■★	■⊡	■⊠
⊙	⊙	■●	■■	■✿	■◉	■⊗	■★	■⊡	■⊠	■◪

X	●	■	✿	◉	⊗	★	⊡	⊠	◪	⊙
●	●	●	●	●	●	●	●	●	●	●
■	●	■	✿	◉	⊗	★	⊡	⊠	◪	⊙
✿	●	✿	⊗	⊡	◪	■●	■✿	■⊗	■⊡	■◪
◉	●	◉	⊡	⊙	■✿	■★	■◪	✿■	✿⊗	✿⊠
⊗	●	⊗	◪	■✿	■⊡	✿●	✿⊗	✿◪	◉✿	◉⊡
★	●	★	■●	■★	✿●	✿★	◉●	◉★	⊗●	⊗★
⊡	●	⊡	■✿	■◪	✿⊗	◉●	◉⊡	⊗✿	⊗◪	★⊗
⊠	●	⊠	■⊗	✿■	✿◪	◉★	⊗✿	⊗⊙	★⊡	⊡◉
◪	●	◪	■⊡	✿⊗	◉✿	⊗●	⊗◪	★⊡	⊡⊗	⊠✿
⊙	●	⊙	■◪	✿⊠	◉⊡	⊗★	★⊗	⊡◉	⊠✿	◪■

If you think of the 200 Ancespitorian addition and multiplication facts as zip codes, that is a lot of zip codes to remember, even if some of them are "easy" to recollect, like zero plus any number is that number (e.g., ● + ■ = ■, read *"circle plus square equals square"*), zero times any number is zero (e.g., ● x ■ = ●, read *"circle times square equals circle"*), and one times any number is that number (e.g., ■ x ⊡ = ⊡, read *"square times square bullseye equals square bullseye"*). Additionally, half of them are the reverse of one another because addition and multiplication are commutative (e.g., ⊗ + ▣ = ▣ + ⊗ = ■● and ⭘ x ⊠ = ⊠ x ⭘ = ✿■, and never mind how they are read), but that is still a lot to remember.

Nonetheless, in spite of all that to remember, the performance standards for the Common Core arithmetic curriculum stipulate that *all* children must memorize the equivalent of all 100 Ancespitorian addition facts and all 100 Ancespitorian multiplication facts by the end of grade 3 *and never forget them*. Find me an elementary school teacher that can make that happen, and I will find you a unicorn. According to a study by DeMaioribus (2011), that simply *never* happens for most students.

DeMaioribus gave a timed test on the multiplication facts to college students to determine the relationship between automaticity with the number facts and being successful in math. For automaticity defined as *"the ability to recall [the number] facts with speed and accuracy at an unconscious level,"* she found that automaticity with the number facts had a positive bearing on achievement and attitude for math, but that turned out to be an aside. Her main finding was *"the lack of basic multiplication fact automaticity in more than 90 percent of the college students tested!"* So what is the point in spending years trying to hammer the multiplication facts into children's heads with worksheets, flash cards, and speed drills when we know that the vast majority of children will eventually forget them?

Imagine a car company initiating a zero-tolerance policy about a manufacturing standard. Imagine further that 90 percent of the cars it manufactured ended up being rejected because they failed to meet the standard. In this scenario, management can either brag about the perfection of the few cars it makes that are not rejected, or it can change the way it makes cars so that fewer of them are rejected. Which would you advise if you were managing this company?

Research confirms that students who have achieved automaticity with the number facts tend to do better in math than those who have not achieved it (Olson, 2021; Robinson, 1999; Williams, 2014). So? Confirming that is merely claiming the obvious, like claiming that boys who are 6 feet, 6 inches tall or taller tend to play basketball better than shorter boys. So how does a basketball coach build a winning team with a bunch of short boys? *They teach them how to play as if they were 6 feet, 6 inches tall or taller.*

So how do elementary school teachers build winning math teams with their students knowing that 90 percent of them will probably be "too short" with the multiplication facts to play math well after they graduate from high school or college? How do they make every child "tall enough" to overcome a predictable deficiency with the multiplication facts when they no longer have to practice them? *They teach them how to count them out with skip counting* — a skill that once learned is resistant to forgetting because it is a *felt* skill. It is rhythmic, like a song, and instead of 100 separate multiplication facts to remember, there are just eight "songs" to remember, the songs for the 2s (2-4-6- … 20) through the 9s (9-18-27- … 90).

Neuroscience: How We Learn and Why We Forget

There is no such thing as memorizing. We can think, we can repeat, we can recall and we can imagine, but we aren't built to memorize. Rather our brains are designed to think and automatically hold onto what's important. — Gabriel Wyner, 2014

Neuroscience is the study of the structure and function of the brain and nervous system. It is a multidisciplinary branch of biology that draws from a variety of fields, including psychology, molecular biology, and mathematical modeling. Its quest is to identify the fundamental properties of <u>neurons</u> (nerve cells), the only cells in the body that do not touch other cells, and <u>synapses</u>, minute gaps between neurons that allow them to communicate with each other.

The Role of Neurons and Synapses in Learning and Forgetting

Brain research is a new frontier, and studies show that neuroscience has had little impact on the institution of education. — Blakemore & Frith, 2005

A major objective of neuroscience is to identify the role of neurons and synapses in learning and forgetting. When neurons communicate, the synapse that connects them is strengthened. However, if the sending neuron does not repeatedly transmit the same information to the receiving neuron, the synapse that connects them will weaken over time and eventually be removed or "pruned" (Modern Neuroscience, 2019). Thus <u>learning is the creation and strengthening of synapses, and forgetting is the pruning that occurs when a synapse is rarely used and is removed</u>, *all of which occurs subconsciously at the brain's discretion.*

Learning occurs at the synapses when they are changed to increase the connectivity between neurons. That occurs when a sending neuron that has been stimulated transmits an electrical or electrochemical signal across a synapse to a receiving neuron for a response to the stimulus. If no synapse exists, one will be created to enable the transmission. Otherwise, the synapse that is already there will be used and strengthened — made more responsive to new input of the same type. <u>This increase in connectivity in the synapse is what has been learned</u>. — Learning and Memory, 2019

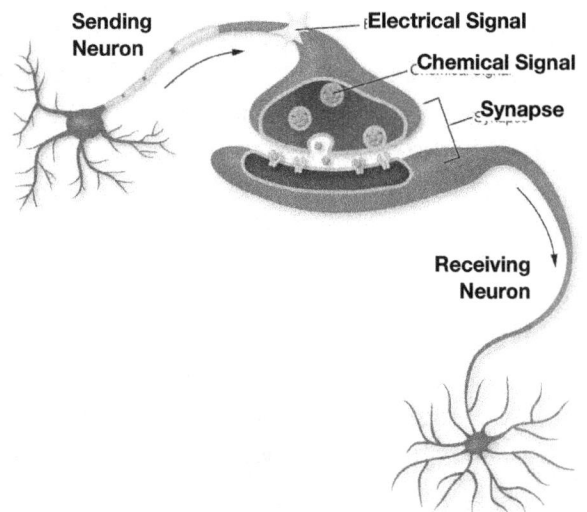

At birth, a child's brain has about 100 billion neurons, about all the neurons it will ever have, and *trillions* of synapses; however, most of the neurons have yet to be connected by the synapses, and connecting the neurons and strengthening or weakening the synapses that connect them is dependent on the child's experiences.

In the first decade of life, a child's brain forms <u>trillions</u> of connections or synapses. ... Each individual neuron may be connected to as many as 15,000 other neurons, forming a network of neural pathways that is immensely complex. ... As the neurons mature, more and more synapses are made. At birth, the number of synapses per neuron is 2,500, but by age two or three, it's about 15,000 synapses per neuron. ... If they are not used repeatedly, or often enough, they are eliminated. In this way, <u>experience plays a crucial role in "wiring" a young child's brain</u>. — Judith Graham, 2020

Whether or not a synapse is pruned is influenced by the experiences a developing child has with the world around them. Constant stimulation causes synapses to grow and become permanent. But if a child receives little stimulation, the brain will keep fewer of those connections. — Healthline.com

The Automaticity Standard Is Dead, Long Live the Automaticity Standard!

The automaticity standard assumes that once children memorize the number facts, they will remember them indefinitely, *but they do not choose what they remember!* Their brains do, based on how often a synapse is used and the length of time since it was last used. The longer the time, the more likely the synapse will be pruned.

Once the brain forms a synapse, it can either be strengthened or weakened. This depends on how often the synapse is used. In other words, the process follows the "use it or lose it" principle: Synapses that are more active are strengthened, and synapses that are less active are weakened and ultimately pruned. — Healthline.com

Neuroscience reveals that the automaticity standard is unrealistic. Without frequent practice, the number facts that are not used very often get pruned. So what are children supposed to do if they forget a number fact while computing? Short answer:

Work it out! No guessing allowed.

Learning mathematics is not about just knowing math. It is about *knowing* that you know it based on understanding and reasoning, not *hoping* that you remember it. Thus requiring arithmetic and the number facts, in particular, to be memorized is a questionable practice and always has been.

No More Struggling with Arithmetic

As stated in the introduction, this book explains why so many children struggle with arithmetic and identifies two changes that must be made to end the struggling. The reason so many children struggle with arithmetic was identified in Chapter 3 and is restated below:

Children struggle with arithmetic because of the **automaticity standard**. To meet the standard, they must "automate" (memorize) arithmetic. They must be able to recall from memory all 390 number facts without hesitation and know how to add, subtract, multiply, and divide whole numbers, fractions, and decimals by rote.

Four chapters later, I can now identify both changes that must be made to end the struggling:

o Discard the automaticity standard. Claiming that arithmetic and the number facts must be memorized to avoid cognitive distractions while computing is indefensible because said distractions are unavoidable in arithmetic. No matter how well a student knows the number facts, they are still going to encounter cognitive distractions when adding a column of numbers and when multiplying and dividing even with single-digit multipliers and divisors.

o No more memorization of the number facts and no more just adding and subtracting small numbers by rote in K-3. Instead, children should be taught how to count out the number facts and should be adding, subtracting, multiplying, and dividing whole numbers of any size with understanding beginning in kindergarten.

My First TouchMath Lesson

Within days of my learning about TouchMath, as reported in Chapter 5, I taught my first TouchMath lesson. My student was my oldest daughter, Jessica, a first grader at the time. She knew how to count to 18, the prerequisite for counting out the addition facts in TouchMath, so I taught her the TouchMath numerals and how to count out the addition facts on the touchpoints on the numerals. To my amazement, it took her just one hour to learn how to do that and show me that she could count out any of the 100 addition facts without making a mistake.

My other daughter, Bradlee, a year away from entering kindergarten, had been listening to my lesson with her sister and wanted to know how she, too, could count on the TouchMath numerals to do "pluses." Again, to my amazement, she learned how to count out all 64 of the addition facts with sums up to 10 that very day, 10 being as high as she could count at the time.

One word comes to mind as I reflect on my daughters' response to learning how to count out the addition facts: *ownership*. For Jessica, in particular, the addition facts were no longer somebody else's math that was being crammed into her head with flash cards, worksheets, and speed drills. They were *her* addition facts. She knew them and *knew* that she knew them. She knew that given 6 + 8, for instance, that if 14 did not jump into her head, she could *"quickly and accurately"* come up with it on her own.

About 20 years after my first TouchMath lesson, Bradlee surprised me by reminding me of that day when I showed her how to do "pluses." She was in college at the time and was excitedly telling me about an astronomy course she was taking. In learning about the kinds of problems she was solving in the course, I was impressed with her intuitive use of calculus in solving some of them and told her so. Her response was not what I expected. *She thanked me for never making her memorize the number facts*. When I asked if she still counted out the number facts, she said *"just the 'hard' ones"* and then surprised me again by adding *"but sometimes just for fun!"*

What I witnessed in teaching my daughters how to count out the addition facts was career changing. In addition to teaching Jessica how to count out all 100 addition facts in about an hour, I could have taught her how to count out all 100 subtraction facts in another hour. To count out 13 – 5, for instance, I could have instructed her to <u>count on</u> from the subtrahend 5 to the minuend 13 (6-7-8-9-10-11-12-13) and keep track of the number of counts on her fingers or with tally marks (||||| |||) for a *difference* of 8, the number of counts. Thus 13 – 5 = 8. Moreover, as admitted by the naysayers of counting out the number facts, she would then have been able to *"quickly and accurately"* count out any of the 200 addition and subtraction facts and, according to neuroscience, not forget how to do that because of the counting she naturally did when not in school.

I was astounded by how easily my daughters had learned how to count out the addition facts, and my mind was buzzing with the implications of that. Had I taught Jessica how to count out both the addition facts and subtraction facts in my first TouchMath lesson, I could have fully accomplished in about <u>two hours</u> what takes <u>three years</u> to only partially accomplish in the Common Core arithmetic curriculum for K-2 because of forgetting once those facts are no longer practiced. Thus if children are allowed to count out the number facts while computing, *the barrier raised by having to memorize the number facts that holds them back in math in the early grades tumbles down <u>and stays down</u>!*

Closing Remarks

In the long run, it is not the memorization of mathematical skills that is particularly important — without constant use, skills fade rapidly — but the confidence that one knows how to find and use mathematical tools whenever they become necessary. — National Research Council, 1989

So America has a choice. It can support the Common Core arithmetic curriculum for K-2 that echos the NCTM's penchant for "fluency" while computing with paper and pencil. It can choose to limit the coverage of arithmetic in those grades to just memorizing the addition and subtraction facts, the same as always. Alternatively, it can choose to allow children to count out the addition and subtraction facts in a few months *in kindergarten* and be done with them, as affirmed by my daughter Bradlee who *never* memorized them but distinguished herself in math and science in high school and college.

The problem with America's choice is that <u>*America does not realize that it has a choice*</u>, which is why I am writing this book. The NCTM has been relentless in squelching the choice in its war with TouchMath, and it succeeded as reflected in the Common Core arithmetic curriculum for K-3 and the CCSSM stamp on the covers of virtually every textbook and workbook for elementary school math in use in America.

Moreover, the NCTM was not alone in squelching America's choice. As explained below, Pearson Education, the world's leading publisher of educational materials, suppressed my effort to make America aware of the choice in 2005 with a book I wrote that was based on a book I coauthored with Terry Clukey, an elementary school teacher.

I met Terry in 1978 at his school in Greenville, Maine, a small town in the northern part of the state next to Moosehead Lake. At the time, I was an Associate Professor in the education department at the University of Maine in Farmington and had just addressed the faculty at Terry's school on the need to reform the elementary school math curriculum.

Terry was fired up when he accosted me after my presentation at his school. He all but drug me to his classroom to show me a number facts program that he had created. What he showed me was a big cardboard box full of hundreds of worksheets on the number facts for grades 1-6 that he had prepared and had tossed in the box. So? Nothing new here, I thought, but I listened to Terry's explanation of how the worksheets were used.

The worksheets were administered on a weekly basis, one per day, on the same batch of number facts for the week. For example, the worksheets for the first week in the program for grade 1 concentrated on the addition facts with sums less than ten. To that end, the 55 addition facts that are less than ten were apportioned equitably on the five worksheets for the week. Thus in that week, the children would practice recalling or working out (with no restrictions on how they might work them out) all 55 addition facts with sums less than ten, with each of those facts being practiced at least twice.

What made the program popular was how it was administered. The worksheets were passed out face down and were to remain that way until the children were told to turn them over and begin, which started a timer. The children were to recall or work out as many number facts as they could <u>in the order given</u> in just one minute. When the minute was up, they were told to stop and determine their score as their teacher read the answers to the problems.[37]

Their score was the number of problems worked correctly <u>up to the first one missed or skipped</u>. If they worked all 30 problems correctly, their score was 30, but if they missed or skipped the second one, for example, their score was one, even if they worked all of the rest of the problems correctly. The reason for the severe penalty for missing or skipping a number fact was to emphasize the need to know *all* of the number facts and not to guess if not sure of one, in which case they were to work it out.

After Terry explained how his number facts program worked, he asked me to make a book out of his box full of worksheets and find a publisher for it. At the time, I knew that many elementary school teachers spent an inordinate amount of time on practicing the number facts, so I agreed to Terry's request in order to rein in that time in order to free up time for other aspects of math, mainly, hands-on algebra.

My job was to 1) organize the worksheets Terry had made on the number facts for grades 1-6 into six-week programs consisting of 30 worksheets per grade, 2) create an eight-week program for grades 7-8 consisting of 40 worksheets on the number facts, reducing fractions, and fraction-decimal-percent equivalences (e.g., ½ = 0.5 = 50%), and 3) write the instructions on how to implement the program. Terry and I are not famous, *but our book is famous.*

Working evenings and weekends, I completed my job for the book in about six weeks and submitted it to Addison-Wesley who snapped it up at first sight and published it in 1980 with the moniker *The Mad Minute, a Race to Master the Number Facts*, for grades 1-8. *Said book became a best seller and has outsold all other number facts programs since its publication.* As of 2020, more than 250,000 copies of it had been sold worldwide, and copies of it were still available at Amazon for about $40. That said, everything written up to now about the book Terry and I wrote is background information for what follows.

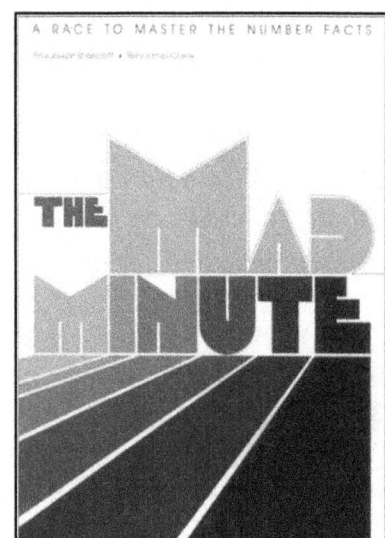

In 1998, Dale Seymour Publications, a subsidiary of Addison-Wesley, called and asked if I would write another book like *The Mad Minute* because it was still a best seller. When first asked, I said no, because by then I had learned how to actually teach the number facts with TouchMath instead of just having children practice them endlessly; however, after more telephone conversations with Addison-Wesley — and with Terry's permission — I relented and agreed to write the *Mad Minute Primer*, a book similar to *The Mad Minute*, except just for K-3 and with touchpoints on the numerals with instructions on how to count out all 390 number facts.

---

[37] Research on the effect of speed drills on the number facts on elementary school children yields cautionary results. Boaler (2012) asserts that they cause math anxiety, and Parker-Stanford (2015) claims that they teach children to hate math. Negative outcomes like these occur with the speed drills in *The Mad Minute* if the children's scores on the drills are linked to their grade in math, which has occurred in spite of that not being recommended in the introduction in the book on how to use the program.

I procrastinated on writing the *Mad Minute Primer*, and by the time I decided to get started on it, Pearson Education had purchased Addison-Wesley and its subsidiaries, thereby adding to Pearson's stature as the world's leading publisher of educational materials. In making the purchase, Pearson acquired the rights to the *Mad Minute Primer*, which it published in 2005. However, immediately upon doing so, *it killed it*, even though it had created a new cover for *The Mad Minute* to go with the cover it had created for the *Mad Minute Primer*.

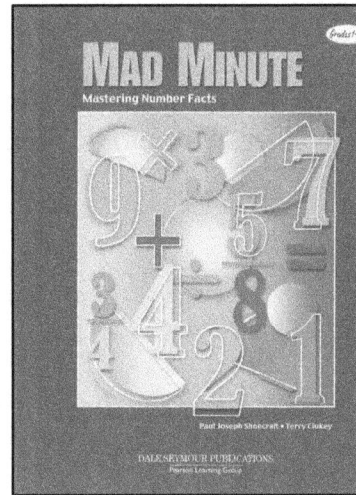

Pearson never acknowledged that the Mad Minute Primer existed. Although it continued to sell *The Mad Minute*, and did so with the revised cover, it never alluded to the *Mad Minute Primer* in its catalogs or on its website. The only way to buy the book was to call Pearson and order it by name over the phone, but to do that, you had to know the book existed, and Pearson never disclosed that outside of Pearson. The only way to know the book existed was to goggle *The Mad Minute* and by chance see a copy of the *Mad Minute Primer* for sale on eBay. How it got there is a fluke.

I doubt if Pearson sold more than 100 copies of the *Mad Minute Primer*. Within a year or two of it being published, if someone called Pearson and asked for it, they were told that it was *"out of print,"* the same as if they had asked for two other books I had written that Pearson killed after it acquired them from Addison-Wesley. The books were volumes 1 and 2 of *Math Games and Activities* — the books that Anne Shaw used with her first graders that got her fired for teaching *"too much"* math, as recounted in Chapter 3.

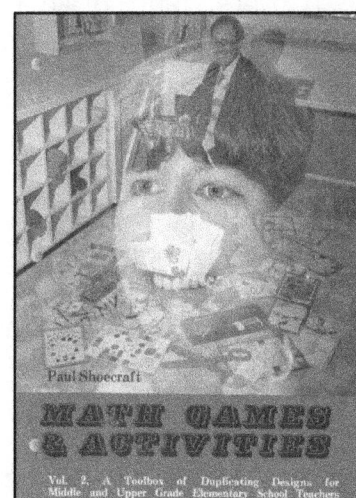

I was not surprised when Pearson killed the *Mad Minute Primer*. Actually, I was amazed that they had even bothered to publish it, but I now realize that doing so enabled them to control its dissemination, which was to kill it along with everything else I had published with Addison-Wesley and its subsidiaries.

If you turn to pages 320 and 321 in the appendix, you can see why Pearson killed the book. The *Mad Minute Primer* made a mockery of the textbooks and testing materials for elementary school math that Pearson sells. As shown on those pages, *all 100 addition facts, all 100 subtraction facts, and all of the multiplication and division facts for the numbers 2, 5, and 10 can be taught in* <u>kindergarten</u> *and not be forgotten because of the counting children do when not in school.*

References

Blakemore, S.J. & U. Frith. "The Learning Brain: Lessons for Education: A Précis," *Developmental Science*, 2005.

Boaler, Jo. "Timed Tests and the Development of Math Anxiety," *Education Week*, 2012.

Burton, Colin. Ebbinghaus Forgetting Curve, Thinkific Plus, August 15, 2023.

Bruner, Jerome. *The Process of Education, A Landmark in Educational Theory*, Harvard University Press, Cambridge, Massachusetts and London, England, 1960.

CCSSM. See *Common Core State Standards for Mathematics*.

Common Core State Standards for Mathematics, National Governors Association in collaboration with the Council of Chief State School Officers, 2010.

DeMaioribus, Carmel E. (2011). "Automaticity of Basic Math Facts: The Key to Math Success." Retrieved from the University of Minnesota Digital Conservancy, https://hdl.handle.net/11299/187488.

Ebbinghaus, Hermann. See Wikipedia.org/Ebbinghaus and Murre and Dros.

Graham, Judith. "Children and Brain Development: What We Know About How Children Learn," Bulletin #4356, University of Maine, c. 2020.

Healthline.com. "How does Synaptic Pruning Work?"

Learning and Memory, quizlet.com, 2019.

Modern Neuroscience, Wikipedia.org.

Murre, Jaap M. and Joeri Dros. "Replication and Analysis of Ebbinghaus' Forgetting Curve," journals.plos.org, 2015.

National Research Council, *Everybody Counts: A Report to the Nation on the Future of Mathematics Education*. National Academy Press, Washington, D.C.,1989.

Olson, Jessica. "The Importance of Math Fact Automaticity," Hamline University, DigitalCommons@Hamline, 2021.

Paderewski, Wikipedia.org.

Parker-Stanford, Clifton B. "Speed Drills Teach Kids to Hate Math," *Futurity*, 2015.

Robinson, Sandra S. "Acquiring Automaticity of Basic Math Facts and the Effect that This Has on Overall Mathematic Achievement" (1999). Theses and Dissertations. 1878. https://rdw.rowan.edu/etd/1878.

Shoecraft, Paul and Terry Clukey, *The Mad Minute*, Grades 1-8, Addison-Wesley, 1981.

Shoecraft, Paul, *Math Games and Activities*, Volumes 1 & 2, Dale Seymour Publications, 1984.

Shoecraft, Paul, *Mad Minute Primer*, K-3, Dale Seymour Publications/Pearson Learning Group, 2005.

TouchMath, Innovative Learning Concepts. Free catalog available at touchmath.com.

Williams, Nancy Blue. "Automaticity and the Learning of Mathematics," PhD dissertation, University of Georgia, 2014.

Wyner, Gabriel. *Fluent Forever: How to Learn Any Language and Never Forget It*, 2014.

Chapter 8: Multimodality Math aka MOVE IT Math

How do I know what I think until I feel what I do. — Jerome Bruner, 1960

My last teaching job was with the University of Houston (UH) system in Texas from 1986 to 2003 as a Professor in the education department. Until 1999, I worked at UHV, a branch campus of the system in Victoria, Texas, after which I transferred to the main campus in Houston where I remained until I moved on in 2003.

At UHV, I taught two courses on how to teach elementary school math: Multimodality Math 1 and 2. Both courses were derivatives of the curriculum that I developed for 'Rithmetic in Residence,[38] the summer math camp that I created and managed for two summers in 1974 and 1975 while an Assistant Professor in the math department at Arizona State University in Tempe.

Multimodality Math

○ **Multimodality Math 1: All Children Can Learn Arithmetic**. For pre-service and practicing elementary school teachers, including PreK teachers. Required for elementary education majors. Credits: 3. Prerequisites: None.

Teaching addition, subtraction, multiplication, and division of whole numbers, fractions, and decimals with games, guided discovery activities, and concrete materials. Extending the language of arithmetic to include the x, y, z language of algebra.

Need for elementary school math reform • The five concepts and ways of thinking that are fundamental to understanding arithmetic • Sensory representations of addition, subtraction, multiplication, and division • Equals as balanced on a teeter totter • Fair trades with arithmetic blocks and counters as the basis for understanding base ten arithmetic, decimals, and equivalent fractions • Counting and mental strategies for figuring out the number facts • Low-stress algorithms for whole number arithmetic • Understanding arithmetic word problems as combining and separating events • What's Happening heuristic for solving arithmetic word problems • Discovering the rules for adding and subtracting fractions with fraction cakes.[39] • Hands-on algebra with math balances.

○ **Multimodality Math 2: Enrichment & Acceleration**. For pre-service and practicing elementary school teachers, including PreK teachers. Required for elementary education majors. Credits: 3. Prerequisites: Multimodality Math 1: All Children Can Learn Arithmetic.

Teaching measurement (standard and metric), informal geometry, and hands-on algebra with games, guided discovery activities, and concrete materials. Introduction to problem solving strategies.

Fraction-decimal-percent equivalences with games • Feet/meters with yardsticks/meter sticks • Quarts/liters with containers marked in fluid ounces/milliliters • Pounds/kilograms with pan balances and weights calibrated in ounces/grams • Line, rotational, and slide

---

[38] For information about 'Rithmetic in Residence, go to moveitmath.com and click on the train at the top of the home page.

[39] Fractional parts of circles divided into halves, thirds, fourths, fifths, sixths, eighths, ninths, tenths, twelfths, fifteenths, and sixteenths.

symmetry • Equality (=), congruence (≅), and similarity (~) • Ratios vs. fractions • Proportionality of corresponding parts of similar figures • Discovering pi (π) with measuring tapes and calculators • Area, perimeter, square roots, and the Pythagorean theorem (and its generalization to similar shapes on the sides of right triangles) with square geoboards • Adding and subtracting positive and negative numbers with two-colored counters • Adding and subtracting polynomials with Algeblocks • Multiplying binomials and factoring trinomials with Algebra Tiles • Problem solving strategies, including Peek Around (ask for help).

Of the two Multimodality Math courses, the first one stands out. All Children Can Learn Arithmetic is the most mathematically sophisticated presentation on how to teach arithmetic ever devised, except, perhaps, for how arithmetic was presented prior to the progressive education movement that began in 1880. I could be wrong about that, but I do not think so. Regardless, at this point in time, it is the only such program that can claim success with virtually *all* children. Of that claim, I am certain.

As for the other Multimodality Math course, Enrichment & Acceleration, it shows how to teach the topics other than arithmetic that are introduced in elementary school math with concrete materials as referents for the math. Noteworthy is that the lessons used with teachers are the very lessons that the teachers can use with their students. By solving problems with the materials, the teachers and their students learn that math is, indeed, *a language* — a written one for recording symbolically how they solved problems with the materials.

The Mathematics Education Initiative

In addition to my teaching duties at UHV, I directed the Mathematics Education Initiative (MEI), an independent unit of the education department that was created in response to feedback I got from practicing elementary school teachers when I began teaching the first course in Multimodality Math off campus in 1987. *The teachers got it!* Many of them told me that they understood arithmetic for the first time and were eager to apply what they had learned so their students would likewise "get it."

The mission of the MEI was to foster the implementation of the first course in Multimodality Math, All Children Can Learn Arithmetic, *in every elementary school in Texas*. To that end, I met with school administrators from around the state to establish school/MEI partnerships with elementary schools whose teachers voted to implement what they would learn in the course after I explained what that would entail. In return, the MEI secured grants that covered the cost of enrolling the teachers in the course for three hours of university graduate credit and supplying them with the instructional materials that they learned how to use in the course.

Implementing Multimodality Math 1: All Children Can Learn Arithmetic

When you have solved the problem of controlling the attention of the child, you have solved the entire problem of its education. — Maria Montessori (1870-1952)

To teach arithmetic after receiving instruction in Multimodality Math 1, every K-6 classroom was supplied with the following, *regardless of the grade:*

o Math games for reviewing the number facts, understanding fractions, and more made out of cardstock from templates in *Math Games & Activities*, Volumes 1 & 2 (Shoecraft, 1984).

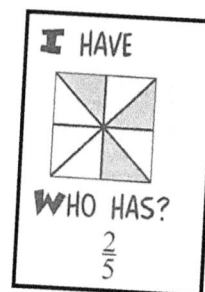

I Have, Who Has Fractions

o Math balances to illustrate that the equal sign in an equation means *"is the same as,"* not *"write the answer,"* like on a calculator. Two per classroom for an instructional center for four children; 10-20 per school for teachers to share for whole-class lessons.

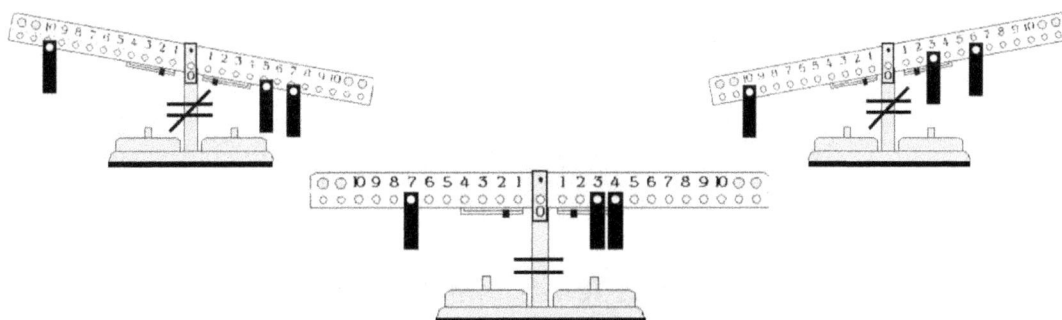

o Class set of MOVE IT Math base two and base three arithmetic blocks for children to experience hands-on what it means for a numeration system with place value to be *based* (built) on a particular number. In base two, two blocks the same make one of the next bigger block and vice versa. In base three, three blocks the same make one of the next bigger block and vice versa.

Base Two Arithmetic Blocks   **Base Three Arithmetic Blocks**

o Class set of MOVE IT Math base ten arithmetic blocks for children to experience hands-on what it means for a numeration system with place value to be *based* (built) on the number ten. In base ten, ten blocks the same make one of the next bigger block and vice versa. Ten unit-blocks make one ten-block, ten ten-blocks make one hundred-block (*not 100 unit-blocks*), and ten hundred-blocks make one thousand-block (*not 1,000 unit-blocks*).

Base Ten Arithmetic Blocks

o Class set of 1,000 colored counters like Multilinks or Unifix Cubes for illustrating the numbers 1-9 and experiencing hands-on adding and subtracting in different bases other than two, three, and ten.

Multilinks

Unifix Cubes

○ Fraction Cakes from halves to sixteenths[40] made from templates that may be downloaded *free* as explained in the footnote at the bottom of this page. One set of 12 fraction cakes per student printed on cardstock, one fraction cake per sheet, preferably a different color for each fraction. Indispensable for children to understand fraction notation, equivalent fractions, and the counter-intuitive rules for adding, subtracting, multiplying, and dividing fractions.

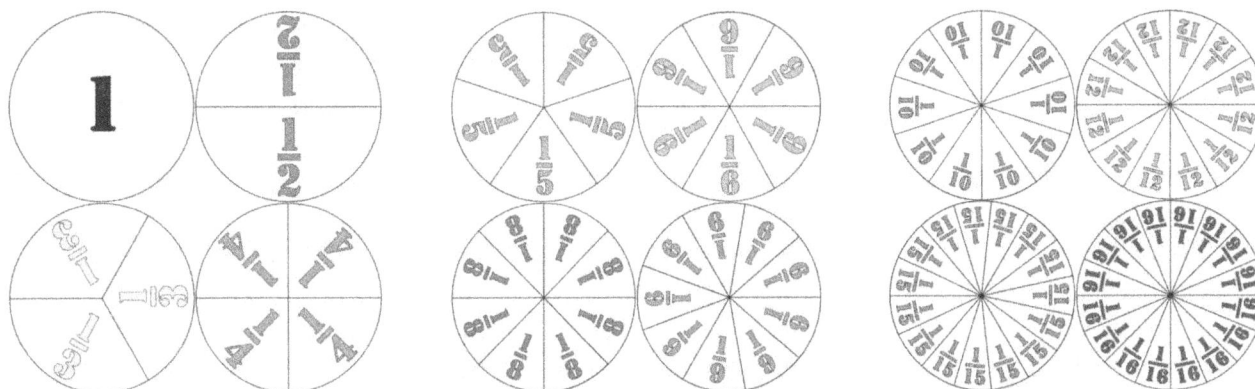

More about the Base Two and Base Three Arithmetic Blocks

Teaching addition and subtraction in base ten in K-2 is developmentally inappropriate! Research conducted decades ago by Jean Piaget (1896-1980) and associates showed that many school-age children are "non-conservers of number," meaning that their thinking about quantities is not logical because they cannot keep the number of items in quantities at the forefront of their minds. They base their belief about the size of quantities more on how they look than on their actual amounts. The classic experiment to determine if a child cannot conserve number is an experiment you can easily conduct yourself if you have access to a child ranging in age from 5 to 7, the closer to 5 the better.

The Experiment

Assuming a willing young child, place two arrangements of the same kind of candy in front of the child, one with ten pieces of candy close together, the other with ten pieces of candy spread apart, as shown below.

Then, assuming that the child can count to ten, ask them how many pieces of candy are in each arrangement. Next, assuming that they counted correctly and told you that there were ten pieces of candy in each arrangement, ask them which arrangement they would like to have.

---

40 The templates for making the fraction cakes may be downloaded **free** at moveitmath.com. Click on "Store" near the top of the home page. Then click on "Real Fractions" under Categories on the page that comes up. Then double-click on each of the two e-books on the page that comes up to download the e-books. The first one contains the templates for making the fraction cakes. The second one contains guided discovery activities for the fraction cakes that enable children to figure out for themselves the non-intuitive rules for adding, subtracting, multiplying, and dividing fractions.

If the child cannot conserve number, they will pick the one with the pieces spread out because, to them, it *looks* like it has more. Conversely, if the child can "conserve number," that is, keep (conserve) the number ten in mind as they compare the two arrangements, they will indicate that it does not matter which arrangement they pick because the number of pieces of candy in either arrangement is the same.

All children are non-conservers of number at some point in their cognitive development and then grow out of it, *but not all at the same time*. So in most K-2 classes, there will be children who can conserve number and some who cannot, and being a non-conserver of number matters.

Numerous studies have shown that non-conservers of number tend to fall behind in math, which is not surprising. For non-conservers of number, much of the math they experience during their early years in school is meaningless. For example, if shown ten beans glued to a tongue depressor to illustrate that ten ones make a ten, they perceive the amount of beans on the stick as *less* than when they were apart. Although they will eventually outgrow their lack of logic in comparing quantities, catching up is hard to do, and many never do. Fortunately, there is a workaround for school-age children who are non-conservers of number.

Try the Experiment Again.

According to follow-up studies on Piaget's findings about some school-age children being non-conservers of number, researchers found that *all* school-age children *can* conserve number if the quantities being compared are small enough. They *can* think logically about small quantities and not be mislead about their size based on how they are arranged. For instance, given two arrangements with five pieces of candy in each arrangement, most children *can* keep the number of pieces of candy in each arrangement in mind and not be misled by their appearance.

Thus to teach *every* child, *including non-conservers of number*, how to add and subtract in base ten, the workaround is to begin instruction on how to add and subtract in base two and base three because, as explained shortly, only small amounts are "allowed" in those bases. However, *instruction must be hands-on*, preferably with the MOVE IT Math base two and base three arithmetic blocks because of how they are made,[41] otherwise with colored counters with values assigned to certain colors.[42]

> *Infants as young as six to nine months of age show an incipient sense of number, being able to distinguish a set of two objects from a set of three [objects] even when the spatial configurations of the sets have been deliberately rearranged.* — Howard Gardner, 1991

---

[41] To make your own arithmetic blocks, glue sugar cubes together as shown on the next page.

[42] For instance, for base two, arbitrarily designate that two reds (units) make one blue, two blues make one green, two greens make one yellow, and so on. To make your own counters, color lima beans with food coloring or Easter egg dye.

Once children understand base ten numeration as just another numeration system with place value based on a certain number, *no more having them compute in different bases*. In particular, *no teaching them how to multiply and divide in different bases*, which was done in the New Math era in the 1960s, which attributed to the demise of the New Math. Computing in different bases became a focal point for America to reject the New Math.

Adding and Subtracting with the Base Two and Base Three Arithmetic Blocks

At each stage of development, the child has a characteristic way of viewing the world and explaining it to himself. The task of teaching a subject to a child at any particular age is one of representing the structure of that subject in terms of the child's way of viewing things. — Jerome Bruner, 1960

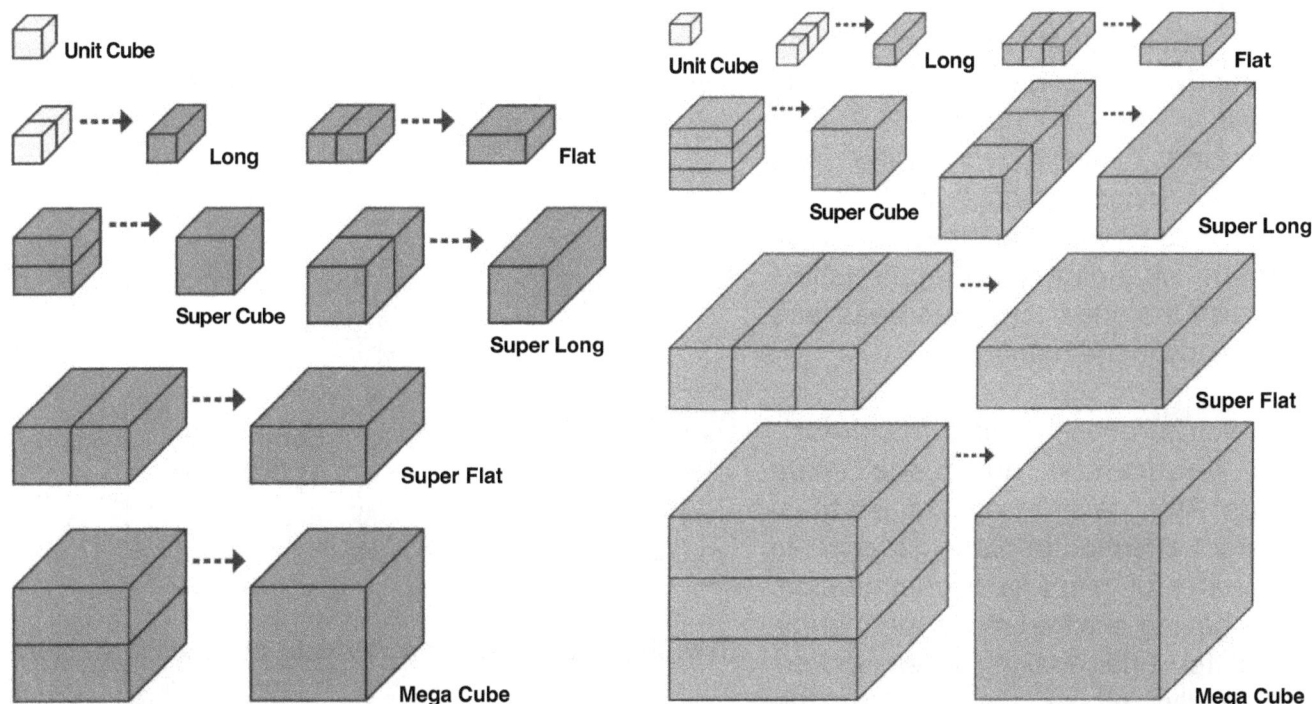

Base Two Blocks: Two the same make the next bigger block

Base Three Blocks: Three the same make the next bigger block

To teach addition and subtraction with the MOVE IT Math base two and base three blocks, a teacher imagines with their students a place called Fair Lands where the blocks are stored in two warehouses. One of the warehouses is said to be in Two Land because the base two blocks are stored there. The other warehouse is said to be in Three Land because the base three blocks are stored there. Noteworthy is how the blocks are stored in the warehouses.

Two blocks the same in the warehouse in Two Land and three blocks the same in the warehouse in Three Land *must* be traded for one of the next bigger block in the warehouses. It is against the Law of the Land in those warehouses to have more blocks the same than just stated. Knowing that and how to keep track of the number of blocks in the warehouses as blocks are bought and sold, <u>as shown on pages 325-327,</u> is all children need to know to learn how to add and subtract in base two and base three and ultimately, base ten.

More about the MOVE IT Math Base Ten Arithmetic Blocks

Simple, clear purpose and principles give rise to complex intelligent behavior. Complex rules and regulations give rise to simple stupid behavior. — Dee Hock, founder and former CEO of VISA

Consider the mantra *"Ten ones make a ten, ten tens make a hundred, ten hundreds make a thousand,"* which you may have heard more than once when you were in elementary school. For the MOVE IT Math base ten arithmetic blocks, the mantra would be *"Ten unit cubes make a long, ten longs make a flat, ten flats make a super cube,"* which shows that there is nothing inherently meaningful in the words "ones," "tens," "hundreds," and "thousands," not if words like "unit cubes," "longs," "flats," and "super cubes" can be substituted for them without changing the relationships between the ones, tens, hundreds, and thousands in the mantra.

The words "ones," "tens," "hundreds," and "thousands" are *place value* words. Authors and teachers use them to direct attention to particular digits in a number or *place* in a computation. Thus having children recite the mantra about ten ones making a ten and so on totally misses the point if the

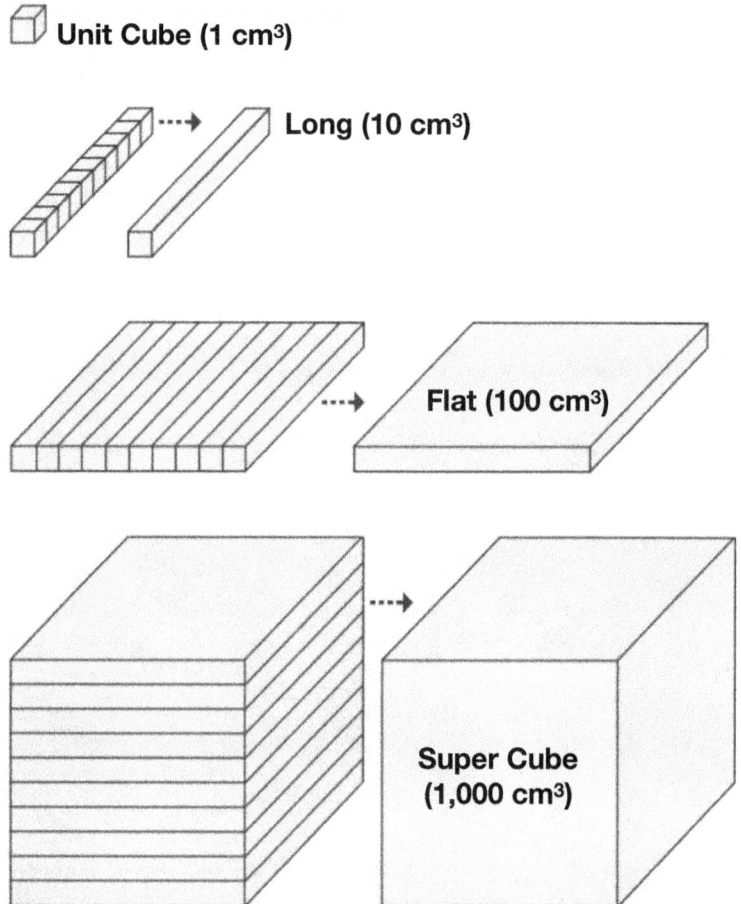

Unit Cube (1 cm³)

Long (10 cm³)

Flat (100 cm³)

Super Cube (1,000 cm³)

Base Ten Blocks: Ten the same make the next bigger block

objective is to explain base ten numeration. To do that, the mantra needs to be *"Ten the same make one of the next bigger thing,"* which captures the dynamic nature of base ten numeration, as illustrated on the odometer of a car when a string of 9s "rolls over."

Think about what happens on a car's odometer that registers 2,**999** miles when the next mile is driven. All of the 9s turn into zeros, and the digit to the left of them is increased by one. That is, 2**999** + **1** = **3000**. What would have become "ten the same" in the units place became one more ten, which, with nine tens, became one more hundred, which, with nine hundreds, became one more thousand, which bumped up the number 2 in the thousands place to 3.

The MOVE IT Math base ten blocks picture what happens to a string of 9s on a car's odometer when the 9s roll over. The scoring on the blocks shows that ten unit-blocks make one ten-block, that ten ten-blocks make one hundred-block, and that ten hundred-blocks make one thousand-block. The scoring shows that *"ten the same make one of the next bigger thing."*

In contrast, as shown below, the scoring on the so-called base ten blocks that are used today shows that a ten-block is ten unit-blocks, which is okay, but then shows that a hundred-block is 100 unit-blocks and that a thousand-block is 1,000 unit-blocks, *which is not okay* — not if the objective is to explain base ten numeration.

Showing that a hundred-block is 100 unit-blocks and that a thousand-block is 1,000 unit-blocks illustrates *place value*. It shows that three-digit numbers are about *hundreds* of units and that four-digit numbers are about *thousands* of units. For instance, it exhibits why 345 (represented with three hundred-blocks, four ten-blocks, and five unit-blocks) is read "three <u>hundred</u> forty five" and why 5,020 (represented with five thousand-blocks and two ten-blocks) is read "five <u>thousand</u> twenty."

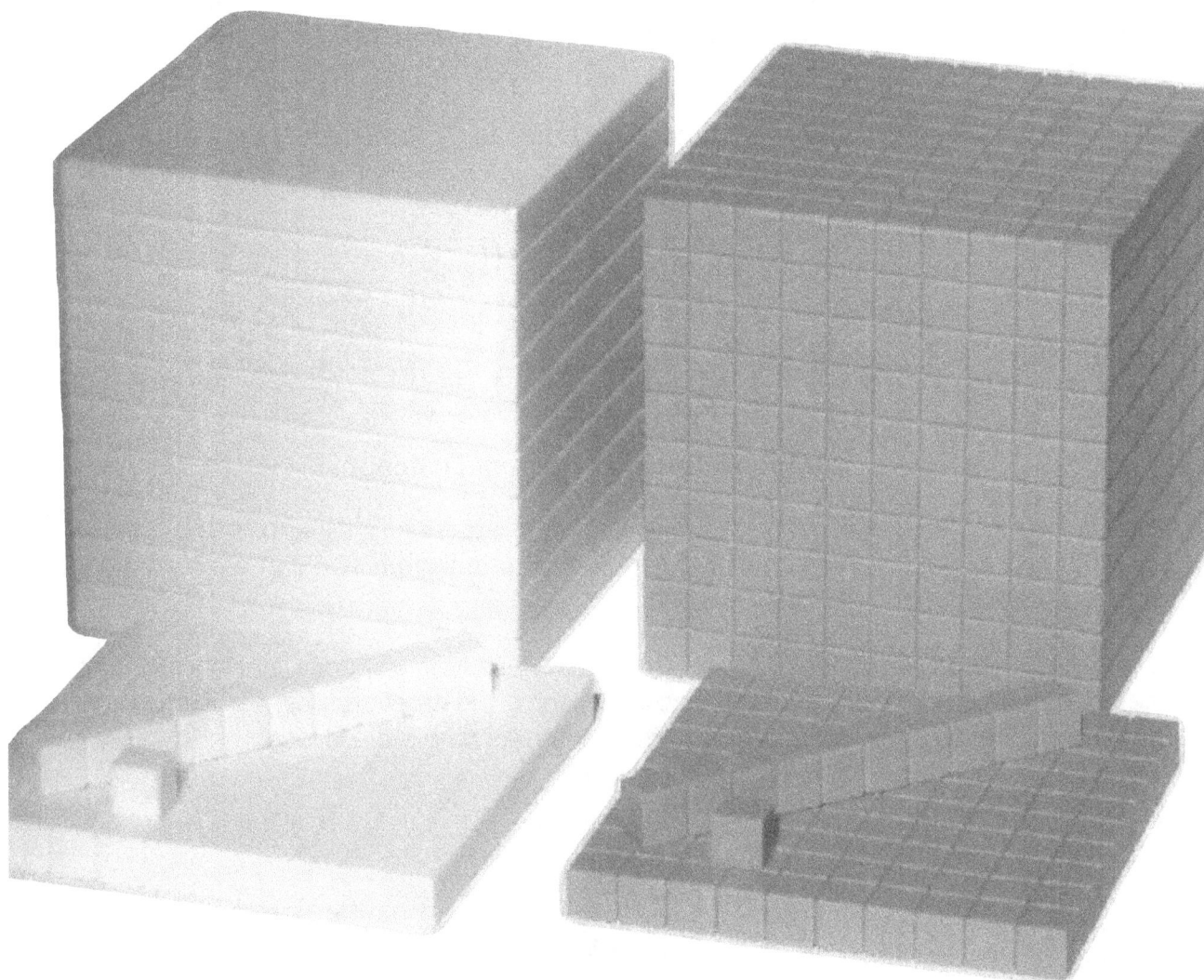

MOVE IT Math Base Ten Blocks   **Base Ten Place Value Blocks**

What is *not* okay is to use the base ten place value blocks to teach children how to add and subtract without addressing how the blocks are scored. For children to use them to add and subtract with *understanding*, they must "see" the hundred-block as made of ten ten-blocks, *not 100 unit-blocks*, and they must "see" the thousand-block as made of ten hundred-blocks, *not 1,000 unit-blocks*, which means they must be able to ignore most of the scoring on the blocks, *which many cannot do!*

If asked how many unit-blocks make a thousand-block with the base ten place value blocks, *many children will tell you 600!* They see 100 squares on each of the block's six sides. They do not see the thousand-block as made of ten hundred-blocks stacked on top of each other, nor do they see the hundred-block as ten ten-blocks next to one another, *which they must if the blocks are being used to teach addition and subtraction of whole numbers.* So if used for that purpose, it is imperative that children build the hundred-block with ten-blocks and the thousand-block with hundred-blocks multiple times until they literally *see* that ten blocks the same make one of the next bigger block.

Implementing Multimodality Math 2: Enrichment & Acceleration

To teach geometry, measurement, and algebraic concepts (e.g., positive and negative numbers) after receiving instruction in Multimodality Math 2, every K-6 classroom was supplied with the following, depending on the grade:

O Math games for reviewing the number facts, learning fraction-decimal-percent equivalences, and more made out of cardstock from templates in *Math Games & Activities*, Volumes 1 & 2 (Shoecraft, 1984).

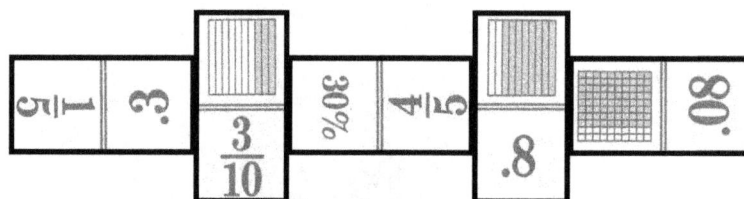

Fraction-Decimal-Percent Dominos

O Yardsticks and meter sticks for measuring lengths in standard units and metric units. Two of each per classroom for an instructional center for four children, 10-20 of each per school for teachers to share for whole-class lessons.

O Containers marked in ounces and milliliters for measuring capacity in standard units and metric units. Two sets of each per classroom for an instructional center for four children, 10-20 sets of each per school for teachers to share for whole-class lessons.

O Pan balances and weights calibrated in grams and ounces for measuring weight[43] in metric units and standard units. Two pan balances and sets of weights in grams and ounces per classroom for instructional centers for four children, 10-20 pan balances and sets of weights in grams and ounces per school for teachers to share for whole-class lessons.

---

[43] The weight of an object is a measure of the force of the earth's gravity on the object, whereas the mass of an object is a measure of the amount of matter in the object.

○ Plastic hand mirrors for investigating line symmetry. Four per classroom for an instructional center for four children, 20-40 per school for teachers to share for whole-class lessons.

○ Measuring tapes and calculators for general use and discovering and estimating pi (π). Four of each per classroom for an instructional center for eight children, 10-20 of each per school for teachers to share for whole-class lessons.

○ Square geoboards for area, perimeter, square roots and discovering the Pythagorean theorem and its generalization to similar shapes on the sides of right triangles. Four per classroom for an instructional center for four children, 20-40 per school for teachers to share for whole-class lessons.

○ Two-colored counters for adding and subtracting positive and negative numbers. Two sets per classroom for an instructional center for four children, 10-20 sets per school for teachers to share for whole-class lessons.

○ Algebra blocks for adding and subtracting polynomials. Two sets per classroom for an instructional center for four children, 10-20 sets per school for teachers to share for whole-class lessons.

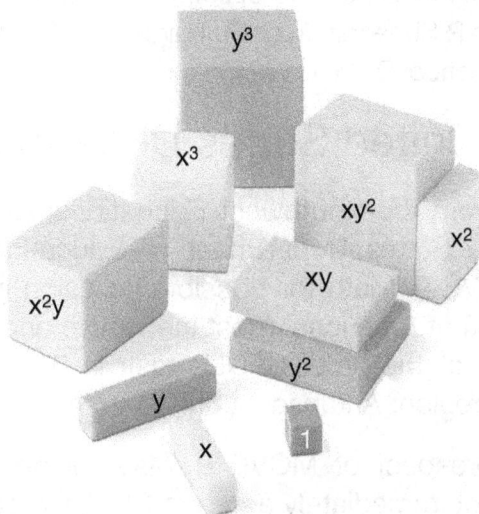

o Algebra Tiles for multiplying binomials and factoring trinomials. Two sets per classroom for an instructional center for four children, 10-20 sets per school for teachers to share for whole-class lessons.

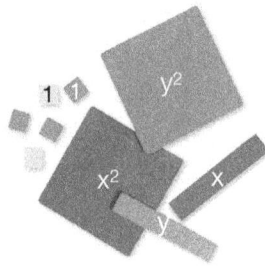

Endorsed by Texas VIPS

By 1989 — *the year the NCTM Standards were released* — a cadre of about 20 elementary school teachers who were implementing both courses of Multimodality Math in their classrooms were assisting me in teaching the courses to their peers. As word got out about the success teachers were having with the program, newspapers began to report on the success. By 1991, VIPs were visiting classes that were using the program and praising what they had witnessed. By then, Multimodality Math had been renamed MOVE IT Math to make the name of the program easier to say and remember. MOVE IT is an acronym for Math Opportunities, Valuable Experiences, Innovative Teaching.

Ann Richards, Governor of Texas, 1991-1995: *"MOVE IT Math is a quantum leap ahead of traditional math instruction now taught in Texas classrooms. The question will be how much of it we can do and how fast we can put this program on track."* — *The Victoria Advocate*, March 23, 1991

Rod Paige, Secretary of the United States Department of Education (2001-2005) under President George W. Bush: *"I have witnessed children using Move-It-Math, and I am amazed at them working fractions so successfully at such an early age [kindergarten]. The Move-It-Math program dimensions will give students the opportunity to learn from each other, to work with manipulatives, and to become mathematical experts at an early age."* — Letter to P. Shoecraft from Dr. Paige while Superintendent of the Houston Independent School District, 1995

Acclaimed by the United States Department of Education

In 1993, I received a call from the Southwest Educational Development Laboratory (SEDL), a regional unit of the United States Department of Education (USDOE). They called to ask permission to evaluate MOVE IT Math for possible inclusion in *Promising Practices*, a USDOE publication that would list 50 of Americas' most innovative and effective K-12 math and science programs that were in use at the time. SEDL's task in that regard was to identify five such programs in their five-state region: Arkansas, Louisiana, New Mexico, Oklahoma, and Texas.

Although pleased at the prospect of MOVE IT Math being announced to all of America in *Promising Practices*, I did not immediately agree to SEDL's request. SEDL was located in Austin within hearing distance of the untruths that the Texas Education Agency (TEA) was spreading about MOVE IT Math, so I was not sure if SEDL was sincere in wanting to evaluate the program. To sense their relationship with the TEA, I told them that the agency loathed it. I laughed and eased up when they told me that they took that as a good sign.

In agreeing to SEDL's request to evaluate MOVE IT Math, I directed them to the Comal Independent School District in New Braunfels, Texas. Said school district was the first to implement both courses of Multimodality Math/MOVE IT Math and had been doing so since 1987. SEDL then arranged with Comal ISD to observe K-6 classes and to interview teachers and administrators in order to judge the fit of the program with the following broad curricular guidelines in the NCTM Standards:

○ Pose tasks based on sound and significant mathematics.

○ Build on students' prior experience and knowledge.

○ Develop mathematics thinking skills that convince students of the validity of particular representations, solutions, conjectures, and answers.

○ Engage students' intellect; pose questions and tasks that elicit, engage, and challenge each student's thinking.

○ Develop students' mathematical knowledge and skills.

○ Stimulate students to make connections and develop a coherent framework for mathematical ideas.

○ Call for problem formulation, problem solving, and mathematical reasoning.

○ Promote the development of all students' dispositions to do mathematics.

○ Develop an instructional model based on the range of ways students learn mathematics.

Considering that both courses of MOVE IT Math were developed long before the publication of the NCTM Standards, you might think that the program failed to meet the curricular guidelines just listed. *Surprise!* As shown on the following pages from *Promising Practices* (1994, 1995), *it met every one of them*. Moreover, it was the *only* K-6 math program that did so. Other programs that met them were just for specific grades.

MATHEMATICS EDUCATION INITIATIVE
MOVE IT MATH
University of Houston – Victoria
Victoria, Texas

K-6 Hands-on, Discovery Mathematics for ALL Children

TOPIC: Elementary [School] Math

USER(S): K-6 Educators, Curriculum Specialists, School Administrators, College Faculty, Education Consultants, Parents

TARGET POPULATION: All Students

EMPHASIS ON: Understanding

Instructional Materials	Teaching Strategies	Assessment Tools
• Teaching Lessons/Units	• Hands-On Learning	• N/A
• Curriculum Guides	• Student-Centered Learning	
• Technology-Based Materials	• Cooperative/Group Learning	
• Supplemental Learning/ Teaching Materials	• Technology-Based Strategies	

GENERAL DESCRIPTION

MOVE IT Math is a K-6 university-supported professional development program. It advocates mathematics instruction based on the use of manipulatives. It consists of two 24-hour inservices: All Children Can Learn Arithmetic and Enrichment & Acceleration.

Innovative features: MOVE IT Math includes the following:

1. Immersion in manipulatives until the mathematics being modeled is internalized;
2. introduction to algebra as early as kindergarten;
3. use of children's literature and science to give meaning and purpose to mathematics;
4. attention to mathematics as a language for explaining certain [quantitative] events;
5. emphasis on understanding rather than memorization;
6. students discovery of the "rules" of mathematics through pattern examination;
7. flexibility in its exposition and acceptance of alternative ways to solve problems; and
8. teaching "essential elements," not covering textbook pages.

Goals: MOVE IT Math seeks a balance between skills, concepts, and problem solving in order to: 1. elevate scores on standardized exams; 2. meet the challenge of changing demographics; 3. improve student attitudes toward mathematics and teacher attitudes toward teaching mathematics; 4. mainstream "at risk" students for mathematics; 5. aid "at risk" students in meeting grade-level expectations; and 6. prepare students to participate in a mathematically literate global economy. An objective of the program is that all children will be ready for a quality algebra class in the 8th grade.

Effectiveness: Data collected in 1989 indicated that approximately 80% of the children classified as at-risk (i.e., one year below grade level in reading and/or mathematics) were no longer classified as such after being in the program for approximately six months. All children in one elementary school tested out of the "chapter" program for mathematics in 1992-93 after implementing the program. Also, teachers report increased understanding, enthusiasm, and interest in teaching mathematics. In some schools, student attendance and discipline problems have shown marked improvement.

NCISE Standards met [for science education]:

☐ Accessible to all students.

☐ Build on students' prior experience and knowledge.

☐ Use an instructional model based on the scientific process.

☐ Relate to personal and social needs.

☐ Select science concepts that are developmentally appropriate, with illustrative examples drawn from the content of multiple disciplines of science.

☐ Develop scientific thinking skills such as drawing conclusions based on evidence, using inference, creating models.

☐ Develop scientific habits of mind such as curiosity, skepticism, honesty, living with ambiguity.

☐ Use assessments to chart teaching and learning.

☐ Shift the role of teacher from importer of knowledge to designer and facilitator of learning.

☐ Seek to find relevant and significant applications of science content and concepts to students' personal and community life.

NCTM Standards met [for math education]:

■ Pose tasks based on sound and significant mathematics.

■ Build on students' prior experience and knowledge.

■ Develop mathematics thinking skills that convince students of the validity of representations, solutions, conjectures, and answers.

■ Engage students' intellect; pose questions and tasks that elicit, engage, and challenge each student's thinking.

■ Develop students' mathematical knowledge and skills.

■ Stimulate students to make connections and develop a coherent framework for math ideas.

■ Call for problem formulation, problem solving, and mathematical reasoning.

■ Promote the development of all students' dispositions to do mathematics.

■ Develop an instructional model based on the range of ways students learn mathematics.

RESOURCES / MATERIALS NEEDED FOR IMPLEMENTATION

1. Manipulatives – in particular, mathematics balance, multibase [arithmetic] blocks and/or counters for trading activities, fraction cakes [from halves to sixteenths] (to support Level 1 inservice).

2. Library of teacher resource materials covering the seven strands from the NCTM Mathematics Standards, estimated cost of $5000, which the district must agree to purchase.

3. Access to a computer in order to access the data base for selecting lessons from the library and to personalize the curriculum to address local student needs.

Equipment and Support Needed:

• Manipulatives • Orientation • Consultants / Trainers • Computer Equipment

• Staff Development • Teacher Collaboration

• Special Hands-on Equipment (exchange blocks, fraction cakes)

• Training Packet • Support Groups • Workshop / Inservice • Administrative Support

Note: Participants begin with an overview on the need for mathematics reform in the early grades. Workshops are conducted by two trainers who must have used MOVE IT Math in their classrooms for a minimum of one year. Workshops are limited to 35 participants and are supported with a 350-page instructional packet for teachers and an implementation guide for administrators. Teachers receive scope and sequence support materials and the district sponsoring the inservice receives a curriculum database keyed to essential elements. Teacher resource books and supporting videos are being produced for Level 1 MOVE IT Math. Additional teacher resource books and videos will be produced for Level 2.

FUNDED BY: Eisenhower Mathematics and Science Education Program. Typically, funds provide manipulatives and resource materials used to implement the program with a school or school district paying tuition for three hours of university graduate credit and the University of Houston-Victoria paying for instruction.

CONTACT:
Dr. Paul Shoecraft, Director MEI
Lynne Shoecraft, Assistant Director MEI
University of Houston-Victoria
2506 Red River, Victoria, TX 77901

SITE(S):
Comal Independent School District
1421 Highway 81 East
New Braunfels, TX 78130

Disseminating MOVE IT Math in Texas

By 1999, the Mathematics Education Initiative (MEI) had received more than $3 million in federal grants, state appropriations, and private and corporate donations with which to fund MEI partnerships with school districts in the following towns and cities in Texas:

> Aransas, Bay City, Beeville, Bloomington, Brownfield, Columbia-Brazoria, Columbus, Cuero, **Dallas**, Devine, Eagle Lake, El Campo, Frisco, Gladewater, Goliad, Gonzales, Gregory-Portland, Hallettsville (including Rice Consolidated and Sacred Heart), **Houston**, Kerrville, Lohn, New Braunfels, Nordheim, Palacios, Plano, Port Lavaca, Port O'Connor, Richmond/Rosenberg, Santa Fe, Schulenburg, Victoria, Vysehrad, **Waco**, Waelder, Waller, Wallis, Weimer, West Columbia, Wharton, Woodsboro, and Yoakum.

The partnerships were created as follows. A school district interested in implementing MOVE IT Math would schedule a time for me to present an overview of the program to its elementary school teachers, after which the teachers would vote if they wanted to enroll in the first MOVE IT Math course: All Children Can Learn Arithmetic. If at least 70 percent of the teachers wanted to do so (which was always the case), the school district partnered with the MEI to implement the course with the following conditions:

○ Multimodality Math 1/All Children Can Learn Arithmetic would be taught in the teachers' school.

○ Every elementary school teacher in the school, even those for sports and music, would enroll in the course for three hours of graduate credit from UHV.

○ The cost of enrolling in the course would be paid by the school district or with a grant.

○ The cost of the instructional materials for teachers to implement what they learned in the course would be paid by the MEI with a grant.

Prior to the start of the course, every K-6 classroom teacher would administer the MOVE IT Math concepts test and report the results on the form provided for that purpose, both of which are included full size in the appendix with permission to duplicate.

From 1987 to 1999, the MOVE IT Math concepts test was administered to more than 20,000 children in the towns and cities listed. Said test includes 46 + 18 + 27 and 5020 − 463 that were addressed in Chapter 4.

As for the MOVE IT Math concepts test, it is not an achievement test. It is a diagnostic tool. Items like 46 + 18 + 27 and 7 = 7 on the test were not designed to trick children. Their purpose was to reveal possible weaknesses in the K-8 arithmetic curriculum, in particular, those associated with children's understanding of the equal sign (=) and base ten numeration.

Administering the MOVE IT Math Concepts Test

Teachers were advised to give the MOVE IT Math concepts test without any helpful hints before passing it out and to allow 10-15 minutes for its completion. Teachers of young children were told that they could read the first three items to the children. To record the results, teachers were told to use the form provided and to tally the types of responses and total the tally marks, as shown below for a true/false item on the test: true (15), false (7).

True	False
///// ///// ///// 15	///// // 7

Results for "Equals" in Dallas, Houston, and S. Texas, 1987-1999

"Equals"	True/False	True	False	N = 20,967 Total True/False
Grade 1	5 = 2 + 3 7 = 7 4 + 6 = 8 + 2	1,928 1,660 1,296	1,496 (44%) 1,639 (50%) 1,914 (60%)	3,424 3,299 3,210
Grade 2	5 = 2 + 3 7 = 7 4 + 6 = 8 + 2	1,817 1,501 1,307	1,602 (47%) 1,915 (56%) 2,111 (62%)	3,419 3,416 3,418
Grade 3	5 = 2 + 3 7 = 7 4 + 6 = 8 + 2	1,946 1,367 1,209	930 (32%) 1,508 (52%) 1,668 (58%)	2,876 2,875 2,877
Grade 4	5 = 2 + 3 7 = 7 4 + 6 = 8 + 2	2,603 1,883 1,526	567 (18%) 1,304 (41%) 1,653 (52%)	3,170 3,187 3,179
Grade 5	5 = 2 + 3 7 = 7 4 + 6 = 8 + 2	2,450 1,771 1,451	346 (12%) 1,023 (37%) 1,345 (48%)	2,796 2,794 2,796
Grade 6	5 = 2 + 3 7 = 7 4 + 6 = 8 + 2	2,756 2,393 2,285	204 (7%) 638 (21%) 746 (25%)	2,960 3,031 3,031
Grade 7	5 = 2 + 3 7 = 7 4 + 6 = 8 + 2	1,394 1,213 1,205	52 (4%) 272 (18%) 282 (19%)	1,446 1,485 1,487
Grade 8	5 = 2 + 3 7 = 7 4 + 6 = 8 + 2	703 703 667	43 (6%) 43 (6%) 79 (11%)	746 746 746

According to studies about the meaning children ascribe to the equal sign (e.g. Baroody & Ginsburg, 1983; Shoecraft, 1989), many children think it is a command. They think it means "write the answer," because that is what it means on a calculator. However, that is not what it means in math. There, as in arithmetic and algebra, it means "is the same as." It asserts that the expressions on either side of an equal sign are identical in some fashion even if they look different, which is almost always the case. Thus the correct answers for items 1-3 that test for this understanding are true, true, true. If that is obvious to you, you may be surprised by so many children believing they were false.

- **Item 1**: If a child thinks $5 = 2 + 3$ is false, they think it is written backwards. If asked to "correct" it, they will write $2 + 3 = 5$.

- **Item 2**: If a child thinks $7 = 7$ is false, they think the equal sign must signify a problem. If asked to "correct" $7 = 7$, they will write $7 + 0 = 7$ or $7 \times 1 = 7$. *Thus children who think $7 = 7$ is false will think equivalent fractions like $1/2 = 5/10$ are false.*

- **Item 3**: If a child thinks $4 + 6 = 8 + 2$ is false, they think it is two problems or one big one written incorrectly. If asked to "correct" it, they will write $4 + 6 = 10$ and $8 + 2 = 10$ or $4 + 6 + 8 + 2 = 20$. *Thus children who think this item is false are going to be confused by an algebraic equation like $a + b = b + a$ that states that addition is commutative.*

If the MOVE IT Math concepts test were administered today, the results for items 1-3 on the meaning of "equals" should be better than my results for those items. Prior to the release of the *Common Core State Standards for Mathematics* (CCSSM) in 2010, virtually all of the textbooks for math in grades 1-8 showed the answer to a number sentence to the right of the equal sign (e.g., $36 + 64 = 100$). Thus no surprise that many of the children tested with the MOVE IT Math concept test in 1987-1999 had spotted that and had concluded that the equal sign meant "write the answer to the right of the equal sign." However, by the time the Common Core math curriculum was published, this misconception of the equal sign and what was causing it was becoming known, as indicated by the following CCSSM content standard for grade 1:

- **1.OA.7**

    Understand the meaning of the equal sign, and determine if equations involving addition and subtraction are true or false. For example, which of the following equations are true and which are false? $6 = 6$, $7 = 8 - 1$, $5 + 2 = 2 + 5$, $4 + 1 = 5 + 2$.[44]

Current textbooks for elementary school math, all of which are in compliance with the CCSSM, show answers on either side of the equal sign, which should result in more correct answers for items 1-3 on the MOVE IT Math concepts test. Note, however, that correct answers for those items would not reveal if children understood that the equal sign signifies "is the same as." Correct answers could signify that the equal sign was interpreted as "write the answer on either side of it," as in their textbooks.

---

[44] These questions about equality are virtually identical to the first three questions in the MOVE IT Math concepts test.

Equations Are Stories

Equations are not just equal signs with problems on one side and answers on the other side. *They are stories* — stories about relationships. For instance, C = πd is a story about how the circumference C and diameter d of a circle are related.

I did not fully appreciate the story told by C = πd until I was a senior in high school and my math teacher, Mr. Clark, asked the class to imagine a perfectly spherical earth with a steel band wrapped tightly around its equator with a circumference of exactly 25,000 miles. His request had nothing to do with what we were studying at the time, but he occasionally sidetracked us with off-the-wall requests like that to make us think, so we dutifully imagined as requested. He then asked us to think of cutting the circular band and adding three feet to its circumference. After giving us a few seconds to picture doing that, he asked if adding three feet to the circumference of the steel band would enlarge it enough to where we could crawl under it if it still encircled the equator.

I was not alone in saying *"No way!"* in response to Mr. Clark's question. How could adding a mere three feet to the circumference of a circle so big even be noticeable? If what is true in math could be decided by majority vote, my classmates and I would have overruled Mr. Clark when he told us we were wrong. However, rather than argue with us, he wrote the following on the chalkboard behind his desk.

Given C = πd and C = 25,000 miles + 3 feet, solve for d.

C = πd

πd = C [Equality is reflexive. If a = b, then b = a.]

πd/π = C/π [dividing both sides of the equation by π]

d = C/π [πd/π = d]

d = (25,000 miles + 3 feet)/π [substitution]

d = 25,000 miles/π + 3 feet/π [dividing both terms by π]

d = 25,000 miles/π + 0.96 feet [3 ÷ 3.14 = 0.96]

Mr. Clark showed us that if the steel band around the equator were touching the earth anywhere along the equator, the distance between it and the earth at the polar opposite of where the band was touching the earth would be 0.96 feet, or nearly a foot. As trim as I and my classmates were at the time, we could have easily crawled under the band with that much clearance.

Dummy me and my classmates! We should have realized that the size of the circle did not matter — that however much was added to its circumference, almost a third of that would be added to its diameter. That is the story told by $C = \pi d$.

The most famous story told by an equation is the one Albert Einstein (1879-1955) told in 1905 with $e = mc^2$ that tells how energy, e, and mass, m, are related.[45] The second most famous story told by an equation is the one told by Leonard Euler (1707-1783) in 1748 with $e^{i\pi} + 1 = 0$.

Known as Euler's Identity aka the God Equation, $e^{i\pi} + 1 = 0$ defies understanding, yet it unites in a single equation five of the most fundamental numbers in math: 0, 1, i, π, and e. Thus it unites arithmetic, algebra, geometry, and analysis: <u>arithmetic</u> (binary) with the numbers 0 and 1, <u>algebra</u> with the letter "i" defined as the solution to $i^2 + 1 = 0$, <u>geometry</u> with π equal to the ratio of the circumference, C, of a circle to its diameter, d, and <u>analysis</u> with e equal to the "limit" of $(1 + 1/n)^n$ for n > 0 as n approaches infinity. (The word "limit" may be interpreted as "gets closer to" as n gets bigger and bigger.)[46]

Results for Item 4 on Base Ten Numeration

Of the four addition problems for Item 4 on the MOVE IT Math concepts test, the only one that matters is the last one, 46 + 18 + 27, written vertically. The three problems leading up to it are warmup exercises. They engage students with a familiar task (adding two numbers) as a lead-in to challenging them with a related but somewhat unfamiliar task (adding three numbers).

The results for 46 + 18 + 27 were reported in Chapter 4. Those who solved it correctly put the 2 in 21 (from adding 6, 8, and 7 in the ones column) at the top of the tens column, which indicated an understanding of base ten notation. Those who missed it put the 1 in 21 there instead. They did that because they had been conditioned to do so. In adding mostly two numbers in grades 2 and 3, the only number they had put at the top of the tens or hundreds column was the number 1, which was the case until the publication of the Common Core math curriculum in 2010 and the following CCSSM content standard for grade 2:

○ **2.NBT.6**

   Add up to four two-digit numbers using strategies based on place value and properties of operations.

To be in compliance with the standard, some textbook series for elementary school math now include a special section on adding three or more numbers. Thus if the MOVE IT Math concepts test were given today, the results for item 4 should be better than my results for it decades ago, but I would not count on that.

---

[45] The constant c in $e = mc^2$ is the speed of light, approximately 186,000 (1.86×10^5) miles per second. Thus $c^2 = (1.86 \times 10^5)(1.86 \times 10^5) = 1.86^2 \times 10^{10} = 3.4596 \times 10^{10} = 34{,}596{,}000{,}000$, an astonishingly large number, which implies that the energy released in fusing even a grain of sand can blow things up.

[46] For n = 1, $e = (1 + 1/1)^1 = 2^1 = 2$. For n = 2, $e = (1 + 1/2)^2 = (1.5)^2 = 2.25$. For n = 10, $e = (1 + 1/10)^{10} = (1.1)^{10} = 2.5937424601$, and so on, indefinitely. Like π, e is an irrational number and can only be approximated as a decimal, namely, 2.72 to the nearest hundredth.

Results for Item 5 in Dallas, Houston, and S. Texas, 1987-1999

1 ft. 8 in. + 7 in.	Correct 1 ft. 15 in. or 2 ft. 3 in.	Incorrect 2 ft. 5 in.	Incorrect Not 2 ft. 5 in.	Did Not Try	N = 19,707 Total per Gr.
Grade 1	116 (4%)	176 (6%)	1,287	1,457	3,036
Grade 2	254 (8%)	437 (13%)	1,769	929	3,389
Grade 3	389 (14%)	705 (26%)	1,347	294	2,735
Grade 4	482 (17%)	1,184 (42%)	998	162	2,826
Grade 5	573 (22%)	1,284 (49%)	669	74	2,600
Grade 6	1,184 (42%)	1,001 (36%)	426	196	2,807
Grade 7	618 (45%)	559 (40%)	167	39	1,383
Grade 8	696 (75%)	132 (14%)	83	20	931

This item measures the conditioning effect of only computing in base ten. It determines how meaningfully children worked item 4 in case they got it correct. A surprising number of those who did so missed item 5. They added 7 and 8, got 15, and ended up with 2 feet, 5 inches, which is wrong. They did not account for the *context* of the problem — that there are 12 inches in a foot, not ten — so 15 inches should have been recorded as 1 foot, 3 inches.

Can we agree that any child who has ever measured something with a ruler should be able to add 7 inches to 1 foot 8 inches and at least get 1 foot 15 inches, if not 2 feet 3 inches? Nonetheless, it was not until the *eighth grade* that the majority of the students tested could do that. It took until then for most of them to finally *think* — to acknowledge that there are 12 inches in a foot, not ten.

The most common wrong answer was 2 feet 5 inches, resulting from 8 + 7 = 15 and putting the 5 in 15 in the inch column of the answer. However, the number 15 occurs in a context, and the context determines its meaning, *the same as in all real-life arithmetic problems*. Fifteen inches equals 1 foot 3 inches, not 1 foot 5 inches.

Two reasons come to mind as to why so many children missed this problem: 1) the automaticity standard, which required them to turn off their brains while computing, and 2) always computing in base ten, which has been the case in all of the textbooks for K-6 math since the demise of the New Math in the mid 1970s.[47] The problem with that is the many equivalences that involve time, money, measurement, and fractions that are encountered in daily living that are *not* based on ten. For example, …

o   60 seconds = one minute; 60 minutes = one hour; 24 hours = one day; seven days = one week; 52 weeks = one year; 12 months = one year; 365 days = one year.

---

[47] Computing in bases other than ten was a hallmark of the New Math. Since then, only computing in base ten is the norm in every textbook for K-6 math in America.

o Five pennies = one nickel; two nickels = one dime; five nickels = one quarter; 20 nickels = one dollar; four quarters = one dollar; five one dollar bills = one five dollar bill; two five dollar bills = one ten dollar bill; four five dollar bills = one 20 dollar bill; five 20 dollar bills = one hundred dollar bill.

o 12 inches = one foot; three feet = one yard; 1760 yards = one mile; 5280 feet = one mile.

o 16 ounces (in weight) = one avoirdupois pound; 2000 avoirdupois pounds = one ton.

o Three teaspoons = one tablespoon; 16 tablespoons = one cup; one cup = eight ounces (in capacity); two cups = one pint; two pints = one quart; four cups = one quart; 16 ounces = one pint; 32 ounces = one quart; four quarts = one gallon.

o All fractions except for tenths. For example, two halves = one whole; three thirds = one whole; four fourths = one whole; and so on.

If item 5 were administered today, the results would probably be as dismal as the ones I got more than 30 years ago because children are still restricted to just base ten in the Common Core arithmetic curriculum, which conditions them to ignore the *context* surrounding an arithmetic problem.

Closing Remarks

When I was parenting, I wanted to teach my children how a long-term goal could be achieved by reaching short-term goals that lead to the long term goal. To do so, I told them how I used to hike out of the Grand Canyon after having spent a night at the bottom of the canyon in a sleeping bag alongside the Colorado River, something I did several times in the late 1950s when I attended Arizona State College (now a university) in Flagstaff, Arizona as an undergraduate.

I told them that to hike out of the canyon, I took one step after another on the Bright Angel trail until I reached the rim, about a mile above me when I started out with the rising sun. I then told them that every time I hiked out of the canyon, I always looked back to where I had been at the start of the day and marveled at how far one step at a time had taken me, but added that had I wandered off the path I was on, I might have gotten lost or fallen off a cliff. So to achieve a long-term goal, I told them to accomplish it a step at a time in the right direction, just like I hiked out of the Grand Canyon.

Since my children had never been to the Grand Canyon when I **told** them that story, I knew that whatever picture had formed in their minds as they listened to me would not be the inspiring picture that had formed in my mind as I told them about my looking back whenever I hiked out of the Grand Canyon. Had I taken a picture of that moment, I could have **showed** them what I was talking about, but I had not taken one, so telling them about it was the best I could do in the moment. A year or so later, though, I had them experience what I was talking about by having them **do** it with me. I took them to the bottom of the Grand Canyon and hiked out with them a few days later. Then, as we mustered our willpower to take the last few steps to top out, I had them look back so they could see and experience for themselves how amazingly far a step at a time in the right direction could take them *and how good it **felt** to achieve a demanding goal.*

Teaching is communicating by **telling** (which is language-based), **showing** (which is image-based), and **coaching** (which is action-based). Learning is paying attention while **listening**, **seeing**, and **doing**; however, students differ in how well they learn by those behaviors. Some learn best by listening, others by seeing, and some by doing.

Educators refer to students with these learning styles as auditory, visual, and tactile-kinesthetic learners, respectively. That does not mean that auditory learners, for example, only learn by listening. It just means that they learn *best* that way and do not benefit as much from seeing and doing as visual and tactile-kinesthetic students might, who also learn from listening, but not as well as auditory learners.

A good lesson is a mixture of all three ways of communicating to students in order to capitalize on *every* student's best way of learning.[48] That is why hiking out of the Grand Canyon with my children was a way better lesson than my just telling them what I wanted them to learn. In having them hike out of the canyon, they could actually see and feel for themselves the wonder and feeling of accomplishment I had experienced when I hiked out of it long before they were born.

Kipp Shoecraft, age 11

Textbooks and computerized math programs "teach" by telling and showing. For example, a unit on fractions will define a fraction, show some pictures of fractions as equal parts of things, and conclude with some problems to solve, like identifying shapes that have been divided into fourths. Lessons like that are like the Grand Canyon lesson I told my children except for not being able to show them a picture of how far I had walked to exit the canyon. Said lessons cater to auditory and visual learners, but lack meaning for tactile-kinesthetic learners. They cannot "feel" what they are supposed to learn. That is why MOVE IT Math teaches *backwards*, so to speak.

When a topic in math can be modeled concretely, a MOVE IT Math lesson for that topic proceeds from the enactive (doing) stage to the iconic (seeing) stage to the symbolic (written) stage — the instructional sequence advocated by Jerome Bruner (1960), arguably, the guru of constructivism, the learning theory explained in Chapter 2. Thus the lesson begins with children **doing** things with concrete materials (like adding and subtracting with the MOVE IT Math base two and base three blocks) in response to instructions that they read or **hear** from their teacher and **seeing** the result.

That is why MOVE IT Math uses TouchMath to introduce the numerals 1, 2, 3, 4, 5, 6, 7, 8, 9 to young children, including pre-schoolers. The touchpoints on the numerals make numbers *real*. With TouchMath, children can **touch** and **see** why, for example, 2 + 2 = 4. They can touch and see the touchpoints on the numerals as they **hear** their teacher say *"two plus two equals four."*

---

[48] Both the National Council of Teachers of Mathematics and the Mathematical Association of America favor multi-sensory instruction.

$$2 + 2 = 4$$

That is why Tina in Chapter 3 began teaching fractions to her first graders by giving each of them a complete set of fraction cakes from halves to sixteenths. Having them solve problems with *real* fractions enabled every one of them to learn about fractions based on how they learned best. They all **heard** in varying degrees what she told them about adding and subtracting fractions, but they could also verify **hands-on** and **see** for themselves the truthfulness in what she told them.

References

Baroody, Arthur J. & Herbert P. Ginsburg. "The Effects of Instruction on Children's Understanding of the 'Equals' Sign," *Elementary School Journal*, University of Chicago, 1983.

Bruner, Jerome. *The Process of Education, A Landmark in Educational Theory*, Harvard University Press, Cambridge, Massachusetts and London, England, 1960.

CCSSM. See Common Core State Standards for Mathematics.

Common Core State Standards for Mathematics, *Common Core State Standards Initiative: Preparing America's Students for College & Career*, National Governors Association and the Council of Chief State School Officers, 2010.

Curriculum and Evaluation Standards for School Mathematics, National Council of Teachers of Mathematics,1989.

Gardner, Howard. *The Unschooled Mind: How Children Think & How schools Should Teach*, Basic Books, 1991.

Hock, Dee. Founder and former CEO of VISA, quotefancy.com/dee-hock-quotes.

MOVE IT Math, moveitmath.com. MOVE IT is an acronym for Math Opportunities, Valuable Experiences, Innovative Teaching.

NCTM Standards. See *Curriculum and Evaluation Standards for School Mathematics.*

Promising Practices, United States Department of Education (USDOE), 1994, 1995.

'Rithmetic in Residence, a summer math camp. For an accounting of the camp, go to moveitmath.com and click on the train at the top of the home page.

Shoecraft, Paul. "Equals" Means "Is the Same As," *The Arithmetic Teacher*, National Council of Teachers of Mathematics, April 1, 1989.

Shoecraft, Paul. *Math Games & Activities*, Volumes 1 & 2 (more than 600 pages of backline masters for making hundreds of math games out of cardstock), Dale Seymour Publications, 1984.

Chapter 9: Whole Number Arithmetic for All Children

Any approach that continually revisits topics year after year without closure is to be avoided. — National Mathematics Advisory Panel, 2008

Kindergarteners' Achievements in Arithmetic

Addition and Subtraction are treated in immediate connection; also Multiplication and Division. Thus their correlations are more clearly shown. — Robinson's First Lessons in Mental and Written Arithmetic, **1871**

Kindergarteners learn that "equals" means "balanced" or "is the same as" with balancing activities and math balances. They learn that addition, subtraction, multiplication, and division are abstractions of common everyday combining and separating events. To understand base ten numeration, they learn how to add and subtract with the MOVE IT Math arithmetic blocks in base two, base three, and base ten. They also learn how to count out the addition facts, subtraction facts, and the multiplication and division facts for 2, 5, and 10 on the TouchMath numerals.

$$0 \quad 1 \quad 2 \quad 3 \quad 4 \quad 5 \quad 6 \quad 7 \quad 8 \quad 9$$

Counting Out All 200 Addition and Subtraction Facts

To count out all 100 addition facts, kindergarteners learn how to <u>count forward</u> on all of the touchpoints on both numbers and how to <u>count on</u>. By either method, the answer is the last number in the count. For example, for 5 + 9, they can count 1-2-3-4-5; 6, 7-8, 9-10, 11-12, 13-<u>14</u>, or they can count on 5 from 9 by counting on the touchpoints of the 5 (10, 11, 12, 13, <u>14</u>) or count on 9 from 5 by counting on the touchpoints of the 9 (6, 7-8, 9-10, 11-12, 13-<u>14</u>). Thus by either method, 5 + 9 = <u>14</u>.

To count out all 100 subtraction facts, kindergarteners learn how to <u>count back</u> from the minuend on the touchpoints of the subtrahend or how to <u>count on</u> from the subtrahend to the minuend and keep track of the counts. For example, for 15 – 7, they can <u>count back</u> 7 from 15 on the touchpoints of the 7 (14, 13-12, 11-10, 9-<u>8</u>), in which case the answer is the last number in the count, or they can <u>count on</u> (without touchpoints) from 7 to 15 (8-9-10-11-12-13-14-15) and keep track of the counts on their fingers or with tally marks (||||| |||), in which case the answer is the number of counts. Thus by either method, 15 – 7 = <u>8</u>.

Counting Out the Multiplication and Division Facts for 2, 5, and 10

Kindergarteners also learn the skip counting songs and dances for 2, 5, and 10 and how to skip count by those numbers for ten counts each: 2-4-6-8-10-12-14-16-18-20, 5-10-15-20-25-30-35-40-45-50, and 10-20-30-40-50-60-70-80-90-100. That done, they learn how to count out the multiplication and division facts for 2, 5, and 10 by skip counting by those numbers. For example, …

○ <u>To count out a multiplication fact</u>, they skip count by one of the numbers on the touchpoints of the other number. For example, to count out 2 x 5, they <u>skip count</u> by 2 on the touchpoints of the five (2-4-6-8-<u>10</u>) or by 5 on the touchpoints of the two (5-<u>10</u>). The product is <u>10</u>, the last number in either count. Thus 2 x 5 = <u>10</u>.

○ To count out a division fact, they skip count by the divisor to the dividend and keep track of the number of counts. For example, to count out 35 ÷ 5, they skip count by 5 to 35 (5-10-15-20-25-30-35) while keeping track of the number of counts on their fingers or with tally marks (|||| ||). The quotient is 7, the number of counts. Thus 35 ÷ 5 = 7.

First Graders' Achievements in Arithmetic

Children can use numbers having nine figures as readily and intelligently as those having three figures, if Notation and Numeration be understood, [that is, if place value in a base ten system of numeration is understood]. — Robinson's First Lessons in Mental and Written Arithmetic, **1871**

First graders review what they learned in kindergarten and learn how to add and subtract in base four or five with colored counters. That done, they demonstrate *mastery* of addition and subtraction of whole numbers. To do so, they solve a certification problem for each operation that is so demanding that even high school math teachers and university math professors would have to accept as evidence that they had mastered the algorithms they used.

Certification Problems for Adding and Subtracting Whole Numbers

As shown on the next page, the certification problem for addition is adding ten ten-digit numbers, and the one for subtraction is subtracting a ten-digit number from a ten-digit number with zeros in the minuend. To solve the certification problem for addition, children use the Addition Facts Algorithm by L.B. Hutchings (1976), which is demonstrated in the appendix on pages 305-307. To solve the one for subtraction, they use the standard decomposition algorithm.

By the end of the year, *every first grader will have solved the certification problems for addition and subtraction of whole numbers*. That happens because of how the school year begins in a first grade MOVE IT Math class. At the start of the year, first graders are shown the certification problems for addition and subtraction that they will learn how to solve and the award certificates they will receive to document that they solved them.[49]

Notably, *the award certificates will already be filled out with the students' names and teacher's signature and posted on a wall in the classroom.* All that is missing on the certificates is the date when each certification problem is solved, which depends on the child and their teacher's assessment of when they are ready to solve it, *which is when they are able to think their way through it.*

Upon solving a certification problem, a child checks their answer against the number on the award certificate with their name on it for the problem they solved. If the numbers match, the child claims the certificate and is welcomed by their teacher and classmates into the Monster Math Club of America, of which I am the President by default. If the numbers do not match, the child will be shown why by their teacher and tutored by the teacher or a classmate until they can solve it on their own.

---

[49] All four certification problems and their award certificates for addition, subtraction, multiplication, and division of whole numbers are included full-size in the appendix with permission to duplicate.

The point of the drama in initiating children into the Monster Math Club of America is to signal to them, their teachers, and their parents that except for occasional review, *they are done with addition and subtraction of whole numbers*. They know how to add and subtract and have proved it! No more wasting away mentally in grades 2 and 3 just adding and subtracting two- and three-digit numbers as dictated in the Common Core arithmetic curriculum for K-3.

Counting Out the Multiplication and Division Facts for 3 and 4

First graders review the songs and dances for for 2, 5, and 10 and how to skip count by those numbers that they learned in kindergarten, after which they learn the skip counting songs and dances for 3 and 4 and how to skip count by those numbers for ten counts each: 3-6-9- ... 30 and 4-8-12- ... 40. That done, they learn how to count out the multiplication and division facts for 3 and 4 by skip counting by those numbers. For example, ...

○ To count out a multiplication fact like 3 x 4, they skip count by 3 on the touchpoints of the four (3-6-9-12) or by 4 on the touchpoints of the three (4-8-12). The product is 12, the last number in either count. Thus 3 x 4 = 12.

○ To count out a division fact like 32 ÷ 4, they answer the question *"How many 4s in 32?"* by skip counting by 4 to 32 (4-8-12-16-20-24-28-32) while keeping track of the number of counts on their fingers or with tally marks (||||| |||). The quotient is 8, the number of counts or 4s in 32. Thus 32 ÷ 4 = 8.

Second Graders' Achievements in Arithmetic

Second graders review what they learned in kindergarten and the first grade and how to add and subtract in base ten by adding and subtracting in base two, base three, and base ten with the MOVE IT Math arithmetic blocks and how to add and subtract in a base other than those bases with colored counters. That done, …

Counting Out the Multiplication and Division Facts for 6, 7, 8, and 9

Second graders review the songs and dances for for 2, 3, 4, 5, and 10 and how to skip count by those numbers, after which they learn the skip counting songs and dances for 6, 7, 8, and 9 and how to skip count by those numbers for ten counts each: 6-12-18- … 60,   7-14-21- … 70, 8-16-24- … 80, and 9-18-27- … 90. That done, they learn how to count out the multiplication and division facts for 6, 7, 8, and 9 by skip counting by those numbers. For example, …

○ To count out a multiplication fact like 6 x 7, they skip count by 6 on the touchpoints of the seven (6-12-18-24-30-36-42) or by 7 on the touchpoints of the six (7-14-21-28-35-42). The product is 42, the last number in either count. Thus 6 x 7 = 42.

○ To count out a division fact like 63 ÷ 9, they answer the question *"How many 9s in 63?"* by skip counting by 9 to 63 (9-18-27-36-45-54-63) while keeping track of the number of counts on their fingers or with tally marks (||||| |). The quotient is 7, the number of counts or 9s in 63. Thus 63 ÷ 9 = 7.

Certification Problems for Multiplying and Dividing Whole Numbers

Once second graders can count out all 100 multiplication facts and all 90 division facts by skip counting by the numbers 2 through 9 and have learned how to multiply and divide whole numbers with the Multiplication Facts Algorithm by L.B. Hutchings (Connors, 1982) and Chunk It division, as shown in the appendix, they demonstrate proficiency with multiplication and division of whole numbers, just like they demonstrated mastery of addition and subtraction of whole numbers in the first grade. They solve the certification problems for multiplication and division. For multiplication, they multiply the 10-digit number 2,586,710,394 by 83. For division, they divide the ten-digit number 7,842,843,096 by 6.[50]

When their answers to the certification problems match the numbers on their award certificates for those problems, they claim their certificates and are welcomed into the Monster Math Club of America by their teacher and classmates.

---

[50] Multiplying and dividing with big numbers used to be painfully tedious, but with the advent of inexpensive calculators, that is no longer the case. To acknowledge that while still teaching children how to multiply and divide with paper and pencil, the standard for being proficient with those operations with paper and pencil has been sensibly downgraded to being able to multiply with one- and two-digit multipliers and divide with one-digit divisors.

MONSTER MULTIPLICATION PROBLEM

NAME _____

DATE _____

2 5 8 6 7 1 0 3 9 4
x   8 3

MONSTER DIVISION PROBLEM

NAME _____

DATE _____

6 | 7 8 4 2 8 4 3 0 9 6

MONSTER MULTIPLICATION AWARD

This certifies that

_____

can multiply by any
1- or 2-digit number

Witness/Date

214,696,962,702

MULTIPLICATION
ADEPT

Paul Shoecraft
President, Monster Math Club of America™

MONSTER DIVISION AWARD

This certifies that

_____

can divide by any
single-digit number

Witness/Date

1,307,140,516

Paul Shoecraft
President, Monster Math Club of America™

By the End of Grade 2

By the end of grade 2, children in the MOVE IT Math arithmetic curriculum will have learned how to count out or mentally figure out all 390 number facts and how to add, subtract, multiply, and divide whole numbers and will have demonstrated that by solving the certification problem for each operation. *They will have achieved closure for whole number arithmetic* and can prove it with the award certificates they get that acknowledge their achievement. They *know* they know whole number arithmetic. *They own it.* It is now *their* math, not just their teacher's math.

Moreover, since kindergarten, they will have been adding, subtracting, multiplying, and dividing fractions with the fraction cakes shown in Chapter 8 and are ready to climb past the fraction and decimal rungs on the Math Ladder that was described in Chapter 3. By the end of grade 3, they will have climbed past the 17th rung on the Math Ladder, the rung for division of decimals, with but a few rungs to go to climb past all of the rungs for arithmetic.

In stark contrast, by the end of grade 2, children in the Common Core arithmetic curriculum will still be memorizing the addition and subtraction facts and mechanically adding and subtracting two-digit numbers, and all they will have done with fractions by then is identify pictures of common fractions such as half a banana. In grade 3, they will repeat grade 2 for most of the year by adding and subtracting three-digit numbers instead of two-digit numbers. Afterwards, they will begin the task of memorizing the times table. <u>By the end of grade 3, they will be hanging on to the seventh rung of the Math Ladder, the rung for the multiplication and division facts, as they struggle to memorize them</u>.

Noteworthy in MOVE IT Math is spreading out the teaching of the number facts for multiplication and division over three years, beginning in kindergarten:

○ **Kindergartners** learn the skip counting songs and dances for the 2s, 5s, and 10s and how to skip count by those numbers to count out the multiplication and division facts for 2, 5, and 10.

○ **First graders** learn the skip counting songs and dances for the 3s and 4s and how to skip count by those numbers to count out the multiplication and division facts for 3 and 4.

○ **Second graders** learn the skip counting songs and dances for the 6s, 7s, 8s, and 9s and how to skip count by those numbers to count out the multiplication and division facts for 6, 7, 8, and 9.

In contrast, in the Common Core arithmetic curriculum for K-3, all 190 multiplication and division facts are dumped on children in their entirety in grade 3, the same as they were dumped on me and my classmates in grade 4 in the 1940s. Why anyone ever thought doing that was a good idea is beyond my understanding, but I admit to never questioning it until I learned how the number facts for those operations could be counted out by skip counting by the numbers 2 through 9.

The Arithmetic Curriculum 150 Years Ago

MOVE IT Math is not radically innovative in completing the teaching of whole number arithmetic by the end of grade 2. As shown on the next page in *Robinson's First Lessons in Mental and Written Arithmetic* that was published in **1871**, MOVE IT Math's goal to get arithmetic done in K-2 is reflected in the arithmetic curriculum in America before the Progressive Education movement in 1880 that mandated that arithmetic should be memorized, the same as today. The problems on pages 78-79 in *Robinson's First Lessons* indicate early mastery of addition and subtraction of whole numbers, as do those on pages 95 and 141 for early mastery of multiplication and division of whole numbers.

As you can see in Robinson's book, early closure for whole number arithmetic for all four operations (+, −, x, ÷) was the goal of the math curriculum prior to the Progressive Education movement. When the operations were introduced back then, *they were taught*, the same as in MOVE IT Math.

Robinson's First Lessons in Mental and Written Arithmetic, © **1871**

When I bought Robinson's book for a few dollars in an antique shop, I just glanced at it, so I was surprised when I finally took the time to leaf through it and saw that MOVE IT Math's arithmetic curriculum is a "New & Improved" version of Robinson's arithmetic curriculum.

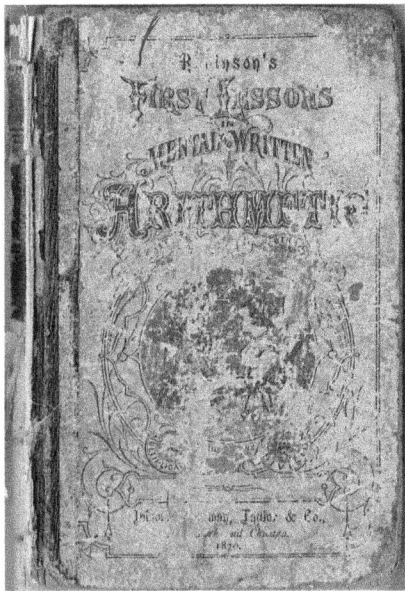

Throughout the book the pupil is taught to use his reason and common-sense, and, as a rule, is not required to memorize until he perceives the truth of what is to be committed to memory. The aim has been not to load down the pupil with Arithmetic <u>as a burden from without</u>, but to cause it to spring up <u>within him</u> by a natural and healthful process; that, growing and unfolding with his intellect, it may be an organized, vital and <u>indestructible part of himself</u>. — Robinson's First Lessons in Mental and Written Arithmetic, 1871

Second Book in Arithmetic, © **1882**

As shown on the next page in Harper & Brothers' *Second Book in Arithmetic* published in **1882**, teaching base ten numeration was a topic in its own right before the arithmetic curriculum was changed to the current one during the Progressive Education era. Note that it was taught with pictures of base ten blocks like MOVE IT Math's base ten blocks that were addressed in Chapter 8: Ten one-blocks make a ten-block, ten ten-blocks make a hundred-block (not 100 one-blocks), and ten hundred-blocks make a thousand-block (not 1,000 one-blocks). That is, *"Ten the same make one of the next bigger thing."*

SECOND BOOK IN ARITHMETIC

8   *SECOND BOOK IN ARITHMETIC.*

7. To express an integer not greater than nine, only one figure is used.

8. To express an integer greater than nine, two or more figures are used.

Ten ones   are   one ten.

9. *Ten ones taken together are a ten.*

Ten is written 10

Two tens are written	20	Six tens are written	60
Three tens " "	30	Seven tens are "	70
Four tens " "	40	Seven tens " "	80
Five tens " "	50	Seven tens " "	90

10. When an integer is expressed by two figures, the left-hand figure expresses the *tens*, and the right-hand figure the *ones*.

1 ten and 6 ones are written 16

3 tens and 5 ones, written	35	7 tens and 0 ones, written	70
5 tens " 6 "	58	9 tens " 1 one, "	91

Ten tens   are   one hundred.

11. *Ten tens taken together are a hundred.*

One hundred is written 100

Two hundred is written	200	Six hundred is written	600
Three hundred " "	300	Seven hundred " "	700
Four hundred " "	400	Eight hundred " "	800
Five hundred " "	500	Nine hundred " "	900

10   *SECOND BOOK IN ARITHMETIC.*

Ten hundreds   are   one thousand.

13. *Ten hundreds taken together are a thousand.*

One thousand is written 1,000

Two thousand,	2,000	Six thousand,	6,000
Three thousand,	3,000	Seven thousand,	7,000
Four thousand,	4,000	Eight thousand,	8,000
Five thousand,	5,000	Nine thousand,	9,000

14. *Ten thousands taken together are a ten-thousand. Ten ten-thousands taken together are a hundred-thousand.*

One ten-thousand is written	10,000
Two ten-thousands are twenty thousand, written	20,000
Eight ten-thousands are eighty thousand, "	80,000
One hundred-thousand is written	100,000
Five hundred-thousands are written	500,000

15. When an integer is expressed by more than three figures, the figure at the left of hundreds expresses *thousands*, the figure at the left of thousands expresses *ten-thousands*, and the figure at the left of ten-thousands expresses *hundred-thousands*.

16. Every three figures in an integer, counting from the right, are a *period*. Periods of figures are separated by commas. The first or right-hand period consists of   8 74,2 3 5

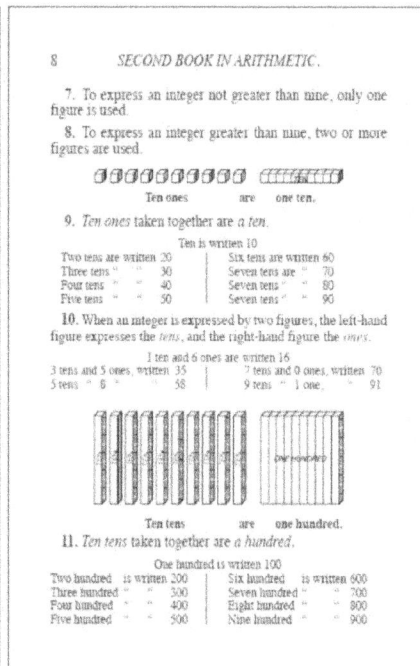

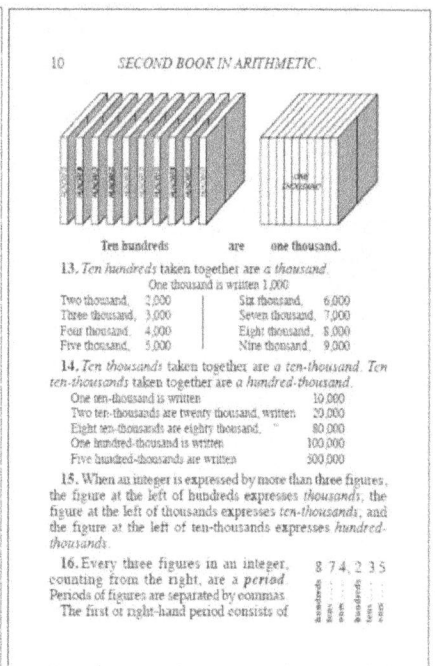

Closing Remarks

One of the most prominent merits of [MOVE IT Math] is its unparalleled capability of instilling in early primary students a math-related self-confidence and the notion that math is fun and easy. When a first grader correctly solves upper grade math problems such as subtracting 2,540,159 from 8,002,011; multiplying 7,586,423 by 3; and dividing 7,842,843,096 by 6 and ends with a shrug, a smile, and words of pride, 'It's simple,' one cannot resist appreciating the effectiveness of MOVE IT Math. — Kwame Opuni, 1999

Knowing whole number arithmetic by the end of the second grade allows for accelerating the teaching of science by two or more years, *which means that many more years spent on science in K-12!*

References

Connors, John Byron, A Comparison of Hutchings' Low Stress Algorithm and Three Other Methods in Multidigit Multiplication for Calculation Speed and Accuracy (1982). Master's Theses. 1702. https://scholarworks.wmich.edu/masters_theses/1702.

Hutchings, L.B. "Low-stress algorithms." *Measurement in School Mathematics, 1976 Yearbook.* National Council of Teachers of Mathematics, 1976.

National Mathematics Advisory Panel FINAL REPORT, U.S. Department of Education, 2008.

Opuni, Kwame A. Project GRAD: Graduation Really Achieves Dreams. Program Evaluation Report (1998-99). Houston Independent School District, Houston, Texas, Dec. 20, 1999. ERIC Number ED443779.

Robinsons First Lessons in Mental and Written Arithmetic, edited by Samuel D. Barr, A.M. New York: Ivison, Blakeman, Taylor, & Company, 138 & 140 Grand Street. Chicago: 133 & 135 State Street. © **1871**.

Second Book in Arithmetic, Harper & Brothers, © **1882**.

Chapter 10: MOVE IT Math in the News, 1989 to 1993

Seeing is Believing. Thirty years ago, I could have invited you to visit elementary schools that were using Multimodality Math/MOVE IT Math so you could have judged for yourself the effectiveness of the program. You would have been able to talk to the children and their teachers about their experiences with it. You could have seen for yourself how advanced the children were in math compared to their peers in Common Core math classes. Fortunately, reporters for the local media back then did visit Multimodality Math/MOVE IT Math classes, and the following accounts of what they witnessed will enable you to accompany them vicariously on their visits.

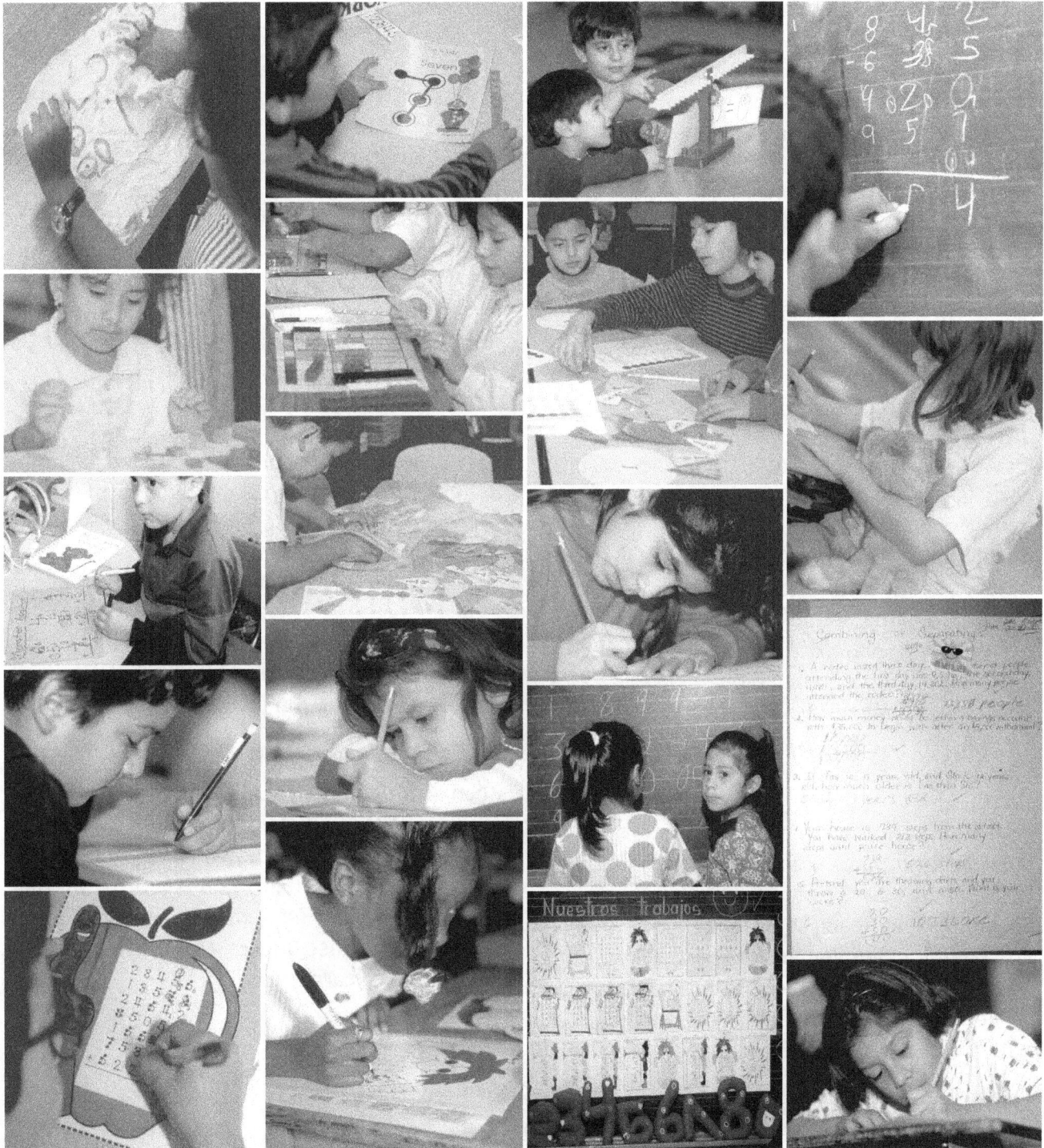

Kindermath Program Teaches Algebra, Geometry

Paul Fortnew, Managing Editor, *The Wave*, Port Lavaca, Texas, May 21, 1989

PORT O'CONNOR — Kindergarten pupils doing algebra and fractions, then teaching older students what they have learned? The Kindermath Program being used at the Port O'Connor School by Judy Anderson is accomplishing just that. Using a program [Multi-Modality Math aka MOVE IT Math™] developed by Dr. Paul Shoecraft of the University of Houston–Victoria, kindergarten and first grade students at Port O'Connor are learning — enthusiastically — fractions, algebra, basic geometry and place values.

Anderson was introduced to the program last summer while taking post-graduate courses at UHV from Dr. Shoecraft, and she decided to implement the program in her kindergarten class. The program is currently being used by both Anderson and First Grade Teacher Linda Fielder, and curricula are being developed for use in the second and third grades.

Dr. Shoecraft's program uses the natural curiosity of the students to introduce them to math concepts and then uses specially developed games to teach the basics. The kindergarten and first grade pupils are teamed for the games as "math buddies." *"The games enable teachers to enrich the basic curriculum,"* Anderson said. *"The kids gained confidence and increased knowledge. They were challenged counting money and developing numbers,"* she said.

Anderson said the program uses tools such as balance beams to teach algebra and cakes [pie-sized circles cut into halves to sixteenths] to teach fractions. Place values are taught using games called Bank It and Clear It that employ blocks with bases of [2, 3, and] 10. Plane geometry is also taught with a series of games using figures.

Dr. Shoecraft came to Port O'Connor last week to observe the implementation of the program. He was accompanied by several guest teachers, Dr. Toni Turk, assistant superintendent for curriculum in the Calhoun County Independent School District, and School Board Member Naomi Albrecht.

Dr. Turk said he was very impressed with the program, especially the work with the older students. *"It works,"* Dr. Turk said.

The older student aspect of the program involves the use of what are called "peer tutors" from the fifth and sixth grades. The peer tutors help the younger students with the math and, in return, the math skills of the older students are improved. Working with the younger students improves the math of the peer tutors by, among other things, helping them to "brush up" on skills learned in earlier grades and possibly forgotten.

Dr. Turk said the Port O'Connor School is the only school in the county using the program in a total immersion concept, but parts of the program are being used in elementary schools across the district, and the results are being shared with other schools.

The teachers and students visiting the Port O'Connor School with Dr. Shoecraft are all enrolled in education programs at UHV.

Since this is the first year the program has been implemented in a total immersion environment in an elementary school, the end results are still unknown. But, the early returns indicate Kindermath is a program that enables younger students to master math concepts and also increases their desire and eagerness to learn.

New Math Formula Motivating Calhoun Kids

Linda Hetsel, Staff Reporter, *The Victoria Advocate*, Victoria, Texas, May 21, 1989

PORT O'CONNOR — It began as a simple challenge, but that challenge has come full circle for a Port O'Connor teacher who is teaching kindergarten students second- and third-grade level math. Students in her class — some not yet five years of age — eagerly count whole and fractional units, work algebraic equations to compute area, multiply and divide, graph coordinates and beg for "monster problems" to challenge them more.

To these students, math is exciting as they play "Touch Math," "Bank It and Clear It," and "Tug of War," games played with balance beams and other "hands-on" teaching aids introducing them to basic math concepts.

This by Judy Anderson, a teacher who readily admits that in the fourth grade she made a "D" in math and was convinced that she just wasn't smart enough to understand math. *"In our society, if you can do math, you are perceived by yourself and by others as a 'smart' person because math is a high prestige subject,"* Mrs. Anderson said. Like many elementary school teachers, she once was more comfortable with her language art skills and relied heavily on worksheet and rote memory assignments to teach math.

To increase her math competency, Mrs. Anderson attended a graduate-level University of Houston–Victoria course taught by Dr. Paul Shoecraft in Port Lavaca last August. At first she was skeptical, Mrs. Anderson said. *"Much of what we were being taught seemed inappropriate for kindergarten students, but I continued the class to raise my own math competency."*

She did agree with Dr. Shoecraft that teaching aids that provide students with hands-on experiences were better for her young students than pencil and paper.

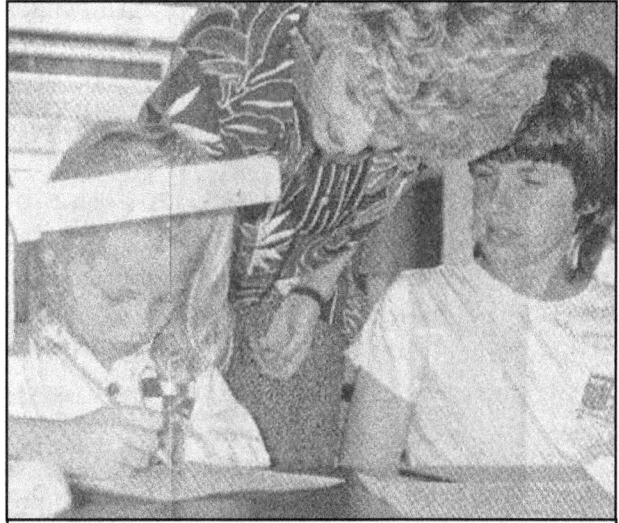

Judy Anderson, Port O'Connor kindergarten teacher, works with Jessica Young and Justin Tigrett on solving a problem using the "Math Modality" teaching techniques she learned in Dr. Shoecraft's class.

At the conclusion of the course, Dr. Shoecraft challenged the teachers to try a few of the course's math games with their students just to test their response, Mrs. Anderson said. *"But he also told us something more important and that is that students need to experience success."*

Mrs. Anderson first tried the balance beam teaching aid with her students. Students hang markers on each end of a balance beam to work simple addition problems. When the beam balances, they know they have the correct answer. By the end of the lesson, the students had determined on their own that if $a = b$ and $b = c$, then $a = c$.

Mrs. Anderson then introduced "Bank It and Clear It," a game that employs wooden blocks to teach number bases. For this game, students may pretend that they are in "Two Land," "Five Land," "Ten Land" or "Twelve Land" as they learn base 2, 5, 10 or 12 numbering systems.

Another "Bank It and Clear It" technique soon has the young students adding columns of three- and four-digit numbers faster than most adults. "TouchMath," a program that assigns touchpoints to each number, allows children to add and subtract even if they know their numbers only by rote, she said.

From there, students progress to "Tug of War" to learn positive and negative integers as they move a marker on a linear gameboard with holes drilled for numbers from -10 to +10. At the toss of a die, the students move their marker in turn, forward and backwards across the "zero" line at the middle.

Fraction cakes [cake-sized circles cut into halves, thirds, fourths … sixteenths] soon have the children understanding that $3/8 + 2/8 = 5/8$, Mrs. Anderson said, adding that the students move to using pencil and paper as soon as they understand the underlying math concept.

"Geoboards" — pegboards marked into square units — teach students how to determine the area of a shape bounded by an elastic band as they count whole and fractional units. From this the students soon learn to use algebraic formulas to determine area, she said.

"The most exciting thing is that we're giving these students 'math eyes.' That is, they're learning to look at and to solve problems mathematically. They're becoming independent thinkers, not memorizers."

In November, Mrs. Anderson introduced her techniques to Port O'Connor first grade teacher Linda Fielder and they began combining their classes on Tuesdays and Thursdays. Peer tutors from the classes of Port O'Connor fifth-grade teacher Elizabeth Bell and sixth-grade teacher Sue Schmaltz soon joined them.

Peer tutors were students who needed a boost to their self-esteem and the opportunity to review material learned in an earlier grade but also possibly forgotten. Teamed with the younger students as "math buddies," the peer tutors quickly learn new math concepts as they teach, then return to their own classes to share their knowledge.

Initially we needed the extra hands," Mrs. Anderson said, *"but now our students are teaching themselves and others, leaving us free to roam from student to student to check out what they are doing and to give further instruction as needed."*

Kindergarten and first grade students under Mrs. Anderson and Mrs. Fielder are now learning essential elements that are normally taught in the third grade. *"We're between the last of the second grade and the first of third grade because we're doing multiplication and division that we know as 'skip counting,'"* Mrs. Anderson said.

"For Multi-Modality Math to be successful, it must be taught at all grade levels, and teachers must reduce their reliance on textbooks which research is showing are inadequate," Mrs. Anderson said.

Random testing of the students has confirmed results, and Port O'Connor principal Marilyn Bratcher expects that the TEAMS scores of students who have taken classes employing Math Modality techniques will show a dramatic increase.

"It's a little frightening because these kids are so advanced, but it's also exciting because it's a challenge," Mrs. Bratcher said, adding that she and all of her teachers were planning to take the Multi-Modality Math course this summer. Mrs. Bratcher's enthusiasm is also shared by CCISD Assistant School Superintendent, T.R. Turk, who summed up his support with the comment: *"Because it works."*

Port O'Connor Elementary School Demonstrates Multi-Modality Math Program

Charlyn Finn, Staff Reporter, *The Wave*, Port Lavaca, Texas, September 27, 1989

Some Port O'Connor Elementary School students demonstrated Tuesday night the Multi-Modality Math program piloted by the school last year before members of the Calhoun County Independent School District board of trustees. The same program on Oct. 25 will be inspected by the National Science Foundation and the program may be set up as a national training program to educate other teachers.

The students demonstrated the *balance beam*: The learner solves [for] an unknown number in an algebraic equation using a math balance beam. The first and second graders, assisted by sixth graders, demonstrated how very young students are capable of learning algebra. Plus, the students obviously loved learning the math that usually is withheld from students until high school.

The students occupied the floor space of the school board chambers and demonstrated geoboards, a system whereby the learner finds the area of polygons in square units using methods called "box it" and "rectangle it" on the geoboards. Using real money, the youngsters showed members of the audience, who joined them on the floor, how to bank it. The youngsters added pennies, nickels and quarters in base five.

Apparently, the Port O'Connor students think Monster Math is fun math. The first and second graders actually added seven 8-digit numbers using the Hutching's Algorithm and checked their answers in the millions with calculators. Finally, the students Robbed The Bank. They drew cards and made numbers on the card with Base-10 blocks. A second card was drawn and that number was subtracted from the first, again with Base-10 blocks.

Paul Shoecraft, Ph.D., of the University of Houston – Victoria, introduced the program to Port O'Connor at the invitation of Principal Marilyn Bratcher. Shoecraft told the CCISD board members that in 1986, United States eighth graders compared last in math to students around the world. *"This is truly a nation at risk,"* he stated. *"Children today will live in a worldwide marketplace and must be able to compete effectively. Now U.S. students can't.*

Shoecraft believes there is too much repetition in [math] and that U.S. students are not being given the opportunity to learn more math. He said 70 percent of the Grade 2 textbooks are devoted to what was already covered in Grade 1.

Port O'Connor students were tested last year with a Woodcock Johnson Norm Test to discover the impact of the Multi-Modality Math program. In first grade the average grade was 3.3 and the lowest score 2.1. In kindergarten the average grade was 3.0 and the lowest grade 2.2.

Sixth grade students were tested through the California Achievement Test and the average sixth grade student was on 8.4 grade level in math knowledge. The highest level was 11.4.

After the demonstration, the school board awarded certificates to the students, their teachers, and principal Bratcher.

Teachers were Linda Fielder, Paula McCauley, Susan Schmaltz and Judy Anderson.

First graders were Michael Carey, Kim Watley, Zac Giessel, Eric Melsom, Tim Gonzales, Cole Munsch, Deanna Williams and Kevin Lewis.

Second graders were Cherice Apostalo, Kim Klamm, Kacie Skalak, Karie Skalak, Kristen Weathersby, Jessica Young, Rene Trevino, Charles Harper, Lauren Tigrett and Devon Vasquez.

Sixth graders were Evan Dierlam, Carl Collins, April Bourg, Bryan Logue, Jake Lucey, April Brown, Carly Trousdale, Brandy Ogg, James Roark, Carisa Shaw, Jamie Harper, Curtis Gosnell, Gary Klamm, Darin Vasquez and Amanda Bullock.

Federal Funding for Math Program: UHV Receives Grant of $67,000

The Victoria Advocate, Victoria, Texas, December 17, 1989

The University of Houston-Victoria has received a $67,000 federal grant to continue a program designed to upgrade math-teaching practices in area school districts.

Plans call for more than 100 teachers in grades kindergarten through six to participate in the program in 1990. Participating teachers will represent school districts in Cuero, Gonzales, Hallettsville, Nordheim, Waelder, Yoakum, Vysehrad and Columbus.

Teachers from the Rice Consolidated school district and from Hallettsville's Sacred Heart school will also participate.

The federal grant for the program will be matched with more than $33,000 from UHV and more than $33,000 in tuition fees and contributions from the participating schools.

The program is called Move It! — an acronym for "Math Opportunities, Valuable Experiences, Innovative Teaching."

Move It! classes will be held at Hallettsville High School next spring and in Cuero, Hallettsville, Gonzales, Columbus and at Rice Consolidated next summer.

Why train practicing teachers to teach math?

"The paper-and-pencil mathematics that adults learned in the paper-and-pencil era is not appropriate for children who live in a computer based society," said UHV education professor Paul Shoecraft, who has been operating the program for more than two years.

"With the program, we are looking to reform the way mathematics is being taught by aligning our teaching of math with research on how children learn, rather than try to mold the students to the curriculum," he said.

In past years, the program has been taught in schools in Victoria, Calhoun and Wharton counties.

A model classroom has been established in Port O'Connor where kindergarten students are doing second- and third-grade level math problems.

The federal grant, from the Eisenhower Math and Science Program, is the fifth grant received for the UHV program, which was originally known as Multi-Modality Mathematics.

Summer Volunteers

Daily Inquirer, Gonzales, Texas, July 12, 1990

Their students would never believe it, but 32 Pre-K through sixth-grade teachers willingly became students themselves for a three-week course in a new method of teaching math.

Dr. Paul Shoecraft's program, called MOVE IT, has gained national recognition as one of the most innovative teaching methods and GISD has the largest teacher commitment of any of the four districts involved in the project.

GISD has also been asked to serve as a model district for teachers from throughout Texas and the US to come and observe the program's effectiveness.

Classes are being held in the North Avenue cafeteria.

Math Program Inservice Project Funded

UH – Victoria Newsletter, January, February, March 1990

A very successful new way of teaching mathematics to kindergarten through sixth grade students, developed at the University of Houston – Victoria, has received federal funds of $67,380 for a $134,210 in-service project involving 10 area school districts and more than 100 elementary teachers.

UH–Victoria Education Professor Paul Shoecraft, who is directing the program and research, said the Eisenhower Math and Science Program funds will support the program in the Hallettsville area during the spring and second summer sessions of 1990.

This spring, Shoecraft is working with 25 Victoria area elementary teachers at Hallettsville High School during a semester-long session. They will prepare the curriculum for the 100 teachers to be trained in the summer program.

For the summer session, the participating school districts are Columbus, which will sponsor up to 21 teachers; Cuero, 15; Gonzales, 15; Hallettsville, 15; Nordheim, 6; Rice Consolidated, 15; Sacred Heart, 6; Vysehrad, 2; Waelder, 4; and Yoakum, 3 [for a total of 102 teachers].

The summer program's classes will meet in five locations: Cuero, Hallettsville, Gonzales, Columbus and Rice Consolidated.

The federal funding will be matched with $33,458 from UH-Victoria and $33,372 from the participating school districts in tuition fees and in-kind contributions.

Math Program Attracts National Notice

Interest in the innovative mathematics teaching techniques developed by Dr. Paul Shoecraft (center) is now being shown by educators on a national level.

The headquarters of the American Federation of Teachers in Washington, D.C. sent Marilyn Routh and Bob Nielsen [a PhD mathematician and former mathematics professor] to observe the Port O'Connor model program and to visit with teachers and administrators involved with the program in the school and its district.

The two took excellent reports of the program back to the AFT, which is interested in supporting in-service programs to improve mathematics instruction in the public schools.

Anyone Can Learn Math: New Programs Show How

Robert Nielsen. PhD mathematician and Assistant to the President, American Federation of Teachers, *American Educator*, AFT, Spring 1990

The assertion that "all kids can learn math" is probably the most radical idea associated with the education reform movement. It is likely that the majority of mathematics teachers don't believe it. Moms and Dads don't believe it. Kids don't believe it.

That all kids can learn math implies, of course, that all minorities, all women, all adults — indeed, *anyone* can learn mathematics. Furthermore, if anyone can learn mathematics, then ability grouping and tracking in school mathematics — arrangements meant to teach more to some, less to others — are obsolete and counterproductive.

Shoecraft's MOVE IT Mathematics

Most of us are so inured to the idea that mathematics is necessarily a tedious, demanding chore that it is difficult to imagine a classroom in which all kids learn mathematics — or, even more remote, a classroom in which all kids enjoy their mathematics lessons.

The idea that learning mathematics ought to be enjoyable is one of several principles in which University of Houston–Victoria Professor Paul Shoecraft grounds his MOVE IT Math program. At the Port O'Connor Elementary School in Texas, not only are all kids learning math, all of them are nuts about it.

Nicole, a first grader at POC and in her second year of the program, sat on the floor writing the answer to a problem she had just worked out. She gripped her pencil, fist like, her head well ahead of her motor skills.

"What are you doing, Nicole?" I asked. *"Adding fractions,"* she said, glancing at me briefly but squarely in the eye. And, indeed,

Using the Balance Beam to Understand Algebra: With the balance beam, even first graders can begin dealing with variables and algebra. Algebra is, after all, simply a matter of making the two sides of an equation balance. This student at Port O'Connor Elementary School is solving $3 + X = 11$.

there in front of her was a set of problems involving the addition of fractions.

"Why, you're too little to add fractions," I teased. She looked at me hard and said, *"Am not. I'll show you."* And with that, she started to work on $2/3 + 1/6 = \underline{\qquad}$.

Reaching in the box in front of her, she hunted up two pie-shaped wedges, one labeled 2/3 and the other 1/6. She carefully fitted them together, then shuffled the pieces in the box around until she found two labeled 1/3, again carefully fitting them together.

She puzzled a moment, then quickly traded in the two 1/3s for four 1/6s. She placed them together, then counted quickly, pointing to each separate wedge as she did, *"One, two, three, four, five. Five-sixths! See?"* She looked at me and grinned, picked up her pencil and scrawled a legible, albeit crude, 5/6 on her paper.

The amazing thing about Professor Shoecraft's program is that once one sees the students doing their math, and liking it, the idea doesn't seem remarkable at all. Indeed, one is struck by its ordinariness.

The essence of the program is to introduce each mathematical concept in at least three different ways through the use of manipulatives and to replace memorization [of all 390 number facts] with three skills: count up, count down, and skip-count. In many schools, manipulatives are used as supplementary materials and grafted onto the old curriculum; in Shoecraft's schools, the manipulatives drive and make possible a new, more advanced curriculum.

In recent tests, POC kindergartners achieved at an average grade level of 3.0 on the [norm-referenced] Woodcock-Johnson test, with the lowest performer reaching the 2.1 grade level. "Special education" first graders achieved at grade level 2.1 on the same standardized test. These are unexpected results. POC is in a poor, rural community where the students have not been known for academic excellence or good high school graduation rates.

Still, Shoecraft complains that these tests don't reveal how much more the students have really learned. The use of tiles, coins, and balance beams, among other manipulatives, has given these children a strong sense of number, magnitude, place value, equality, area, and variable.

More importantly, when these students do their additions, subtractions, multiplications, and divisions, they are not blindly performing operations, they are understanding what they are doing. Nicole knows that "eighteen divided by two equals nine" means that eighteen "things" can be put into "nine piles of two." She also knows that six times four equals four times six equals twenty-four because $4+4+4+4+4+4 = 24$ and $6+6+6+6 = 24$, or as Nicole put it, *"four, eight, twelve, sixteen, twenty, twenty-four!"*

"Incidentally," Shoecraft volunteered, *"I let them count on their fingers."* As he said that, I felt my hands automatically go rigid at my side. *"Oh?"* I grunted suspiciously. Sensing my discomfort, he asked, *"Is that a problem for you?"* I said, *"Well, it's not the way I learned math."*

Patiently, Shoecraft explained, *"You have to ask yourself, what are children trying to do when they use their fingers?"* Tentatively I offered, *"Trying to figure the answer out?"* Smiling at me, he asked, *"Do you really have a problem with that?"* Looking around that happy classroom, I found I didn't anymore.

As at the University of Chicago [and its School Mathematics Project, resulting in the textbook series, *Everyday Math*], algebra and geometry are integral to POC's elementary school curriculum. Here, students determine the area of irregularly shaped geometric figures using "geoboards" and elastic bands. Algebra is introduced first through the balance beam as a means of establishing the equality of different equations.

Shoecraft's program works — and it works for kids from a variety of socioeconomic backgrounds. Both Port O'Connor and neighboring Wharton Independent School District have completely eliminated ability grouping and tracking through Shoecraft's program. The Wharton student population is a mix of Hispanics (30 percent), blacks (30 percent), and whites. The crucial ingredient in the success of MOVE IT Mathematics is the total Commitment to the philosophy that, given a multiplicity of learning approaches, every kid can learn math. If one way doesn't work, then another one is tried.

Toward a New Learning

A major goal of Paul Shoecraft is to demonstrate to teachers that memorization is not the most effective methodology.

In many ways, the school reform experience in mathematics is a metaphor of the whole school reform-restructuring agenda — and it is instructive to look at it that way.

For example, the principle that all kids can learn, that curricular content ought to be developed and presented from the base of students' previous knowledge and experience, and that classrooms structured as laboratories and workshops work better are clearly not unique to the study of mathematics.

In spite of the multitude of such programs, very few ever become integral parts of the overall school curriculum. The reasons for this failure are many, however several should be flagged. Teacher burnout is high in special programs; there is seldom paid time for program development; external program funding runs out; and there is rarely any program followup or networking among teachers in isolated programs.

Second, there is a good deal of anecdotal evidence that learning-disabled students perform exceptionally well in the new learning environments. One suspects that research will validate this result and that the reason for students' success lies in the multiplicity of conceptual presentations.

Third, there is anecdotal evidence that self-confidence and self-esteem accrue to students who achieve in mathematics and that this accounts for higher student achievement across the board. This result may well be rooted in the belief that mathematics is "hard" and that if students learn this most difficult subject, they begin to believe they can learn anything.

Finally, it must be observed that the reconstruction and restructuring of the school mathematics curriculum is among the most important problems in education reform.

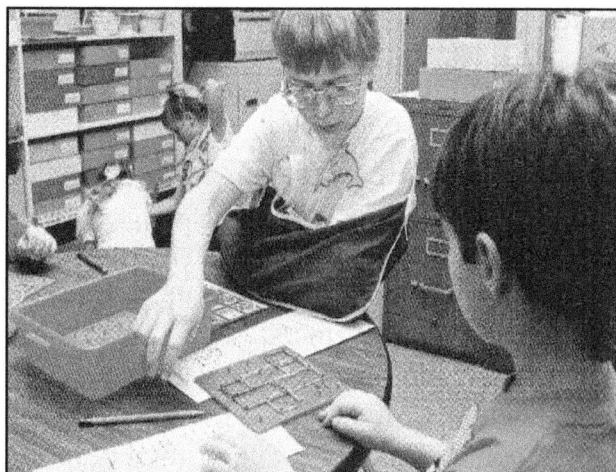

Finding Area and Perimeter with the Geoboard: With the geoboard, students can begin working with area and perimeter in the first grade. To begin, students wrap rubber bands around nail studs in fanciful ways, creating different polygons. To calculate the polygons area, students count the number of whole and half-squares contained within the rubber bands. To calculate the perimeter of a polygon with only vertical and horizontal sides, students count the number of horizontal and vertical nail-to-nail segments that make up the figure. To calculate the perimeter of a polygon with slanted sides, students must first learn about square roots and the Pythagorean Theorem, both of which they learn easily using the geoboard.

Without mathematics reform, there can be no science reform — without science reform, no real education reform.

But in this, we are fortunate, because, far from being the intractable problem it often seems, we know a great deal about "the trouble in mathematics" and what can be done to fix it. *All kids can learn math, and it's time they begin to do so.*

Monster Math for Little Kids" Workshop [in New York City]

UFT Math, Vol. III, No. 1, New York, New York, September 1990

Kindergartners adding fractions? Calculating the area of a triangle? Oh, yes, and loving it! Learn how all students can enjoy math, doing more earlier than you might have imagined possible, from kindergarten up. This approach to elementary math involves cooperative small-group learning, manipulatives and multiple approaches to computation and problem solving.

The Math Teachers Committee will sponsor an all-day workshop at UFT [United Federation of Teachers] headquarters on Saturday 13 October. With the support of the AFT [American Federation of Teachers]. Paul Shoecraft will come from Texas to explain Monster Math.

Little kids in kindergarten and first grade, in a very poor school district in south Texas, have learned how to add ten-digit numbers, ten rows deep, and to subtract as well. By second grade, they are ready for fractions. Their methods are very exciting, both for the methodology itself and for the strong mathematical base on which it is built.

The full day's activities will include breakfast, lunch at a restaurant, and three two-hour sessions. All participants will leave with a class set of FRACTION CAKES [cake-sized circles cut into halves, thirds, fourths to sixteenths] to be ready for immediate implementation.

A separate mailing will go to all on the Math Teachers Committee's mailing list. Those not on the mailing list should send in the tear-off found on p. 5 and can register for the workshop using the stub to be found in The New York Teacher. New Teacher Workshop credit will be granted to those who attend.

Teachers in a MOVE IT Math workshop at Ser Niños, a charter school in Houston, c. 2010

Making Math Fun: How a Texas Professor Is Turning Math Instruction on its Head

On Campus, American Federation of Teachers/AFL-CIO, Vol. 10, No. 2, Oct. 1990

You never really know how tight your shoes are until you wear a bigger size. The new shoes feel so good that you wonder why you didn't make the change sooner.

That's what Move It! math, a professional development program in math instruction that places understanding before procedure, is like. It feels right for the kids who are now really understanding, learning and remembering the ideas and processes of mathematics and for the teachers who, having been frustrated with a bad fit for too long, are finding math a joy to teach.

University of Houston – Victoria Associate Professor Paul Shoecraft ("Dr. Paul" to his students) has searched for 20 years for the right fit. Through research, experimentation and pure success, he has put together Move It! math, a reformed math instruction program for K-6 teachers and students that allows first graders and even kindergartners to understand addition and subtraction in less that a year. Shoecraft's program, in essence, compresses into a few months what normally takes years of traditional math instruction.

In regular math programs, says AFT associate member Shoecraft, we focus on getting kids to memorize. "Until those kids have memorized the basic facts, they can't do" addition and subtraction and basic arithmetic. "Monster" addition, subtraction, multiplication and division (which involves calculations in the millions) and mastery of fractions, decimals and percents can all be wrapped up by the fourth grade in the Move It! math program, says Shoecraft. "We can do it if we start back in kindergarten and do it gracefully, in a fun, entertaining way. And the kids love it.

"By grade four, we've completed what is now called K through 8 math or at least K through 7," says Shoecraft. *"We've bought two and a half years from the standard*

Gonzales teacher Lynn Gescheidle has a group of *pre-kindergarten* students hop their way through a touch-count exercise. Teachers reinforce the application of the learning technique in a variety of settings.

curriculum. Middle school math will have to become something entirely different."

Using a system that gives kids power over numbers, namely touch math, or as Shoecraft calls it, the "tap and tally" system, children are able to count and understand what they are doing, as opposed to merely memorizing by rote. Touch Math provides kids with an imaginary system of dots superimposed over the Arabic numerals.

"Our kids only have to learn four things with tap and tally," says Shoecraft, while sitting at a local Hallettsville, Texas, barbecue house where *On Campus* finally caught up with him for an interview. After learning where the "touchpoints" are on numbers one through nine, which takes about [two weeks] for kindergartners, the children also have to know how to count forward to 18, how to count backward from 18, and how to skip count [e.g., 2-4-6 … 20, 3-6-9 … 30, 10-20-30 … 100].

With that information understood and learned, even kindergarten kids are *"70 to 80 percent of the way home free on the 390 basic math facts."* This, says Shoecraft, *"removes a major barrier to kids' success in math."*

When Memory Fails

The key to success is giving kids enough information to let them figure out math problems by themselves. Memorization — the mainstay of current elementary math instruction—can only go so far and favors those children who have good memories. Shoecraft's approach favors everyone.

For example, young children have difficulty grasping the concept of what a number is. Their developmentally immature minds can only grasp numbers up to 7. Therefore, explains Shoecraft, if you put 10 objects together and ask children to count them, they will count 10. If you then spread the objects apart and ask them to count them again, they'll count 10 again. But if you ask them which is the larger group, they will tell you that the spread apart group is larger.

Piaget was the one who discovered this inability to grasp, or "conserve," large numbers, a fact that is all but ignored in the traditional elementary math curriculum. It is the natural limitation that forces children to rely on memorization instead of deductive skills to learn math. Shoecraft gets around the barrier with a game called "Fair Lands," which puts children in a world of numbers that includes "Two Land" and "Three Land" (i.e., the base 2 and base 3 systems), instead of in the abstract base-10 system.

Through Fair Lands, children learn to add and subtract by borrowing and carrying. At the end of their first year, all the children are able to add ten 10-digit numbers. They also are able to understand mathematical concepts that their traditionally educated brothers and sisters have learned by rote.

The games, objects, examples and hands-on activities make numbers real to children. The manipulatives that Move It! math uses are developmentally appropriate, says Shoecraft. And there are plenty of them: arithmetic blocks that teach the concept of "fair trades" in different number bases and math balances that look like colorful weight scales with a number line and "tags" that can be hung on either side to illustrate equality.

Figuring Out What Works

The declining mathematics achievement of America's preadolescents was what triggered Shoecraft's concern about the quality of math teaching. Why, he wondered, were kids having so much trouble? What he found was that kids weren't the only ones floundering. Their teachers were, too.

"The average elementary school teacher has had less that one course in math or science," explains Robert Nielsen, special assistant to the AFT president and a former professor of mathematics, *"and that course was probably a methods course. These teachers will tell you that the reason they became elementary school teachers is because they were weak in math or science."*

"Shoecraft sees that they learn enough mathematics—both the Move It! math and the math theory behind it—to discover math themselves, in much the same way that the kids discover math. Then they develop the enthusiasm to teach their students and their colleagues."

Nielsen adds that the teachers-teaching-teachers aspect of the program is vital to its success. *"You can't get all the elementary teachers to go back to school, so teaching teachers to teach each other has a nice multiplier effect."*

Shoecraft's program is one of several cropping up around the country that could revolutionize the teaching of math, says

Glenda Lappan, National Science Foundation (NSF) director for teacher preparation programs. The programs that NSF is funding range from those that could reform instructional practice to bring coherence to both math and science teaching, to those where the computer is central. These latter programs use computer technology to bring real classes of children and their teachers [together] in the university classroom.

Lappan, who also sits on the National Council of Teachers of Mathematics' board, explains that reform efforts are focused on changing mathematics curricula and instruction standards. *"K-12 change is pushing the university to think about its role in teacher preparation,"* says Lappan. *"I'm pleased to see the university responding to the challenge. The university community is taking this seriously."*

Indeed, the evidence about the current state of math achievement in America shows our country trailing behind most of our industrialized counterparts. Reform is "critical," says University of Houston – Victoria president Glen Goerke, who lauds and supports Shoecraft's changes. *"I think we've ignored the fact that what we're doing just does not work, and we can't continue. Math is a base for so many other things. If we're going to have American students achieve at a [higher] level in math and science, we've got to change some things."*

What Teachers Say about Move It Math

This summer, *On Campus* visited Shoecraft's three-week "multi-modality, sensory-based" math program as it was being presented to teachers from 10 rural school districts in Texas. The program also provided real live kids on which the teachers could practice their new techniques. The teachers' enthusiasm was catching.

"I saw teachers change their opinion of math in the first day," says Sue Gottwald, director of elementary education in the Gonzales, Texas,

A six-year-old works with the balance beam to learn about elements of algebra.

independent school district. *"They no longer think math has to be memorized."*

Principal Don Rainey of the North Avenue Elementary School in Gonzales says he is sold on the concept and has seen it improve the performance of third-grade students who previously scored poorly on achievement tests. *"If my teachers are willing to commit the time to [learn how to use the program], so am I."*

Pre-K teacher Lynn Gescheidle of Gonzales, Texas, is trying out the methods. *"The touch-counting method gives kids a basis for problem solving,"* says Gescheidle. *"They don't just memorize the answer and the numbers. They have a method for figuring things out."*

Debbie Tumis, a trainer from Hallettsville, Texas, says she took "Dr. Paul's" approaches in math class and practiced what she learned on her own children: Her 12-year-old, she found, learned how to add and subtract in any base in 45 minutes.

"This is hands-on," says Tumis. All of the methods and tools and games that illustrate concepts *"get you away from the math sheets. Math is now 'Do It!'"*

Area Leaders to View [Video of] Shoecraft's Math Concepts

The Victoria Advocate, Victoria, Texas, October 28, 1990

The University of Houston – Victoria will meet with area business and industry leaders to preview a video of an innovative mathematics education program that could revolutionize how the subject is taught.

Instead of 2 + 2, this concept would allow pupils in first and second grade to do "monster math" problems with confidence and accuracy, and it would leave them anxious to tackle more, according to UHV mathematician Paul Shoecraft.

Shoecraft has been working on the development of the unique mathematics education program for years. The success of the program already has been demonstrated in several area schools, according to UHV.

Shoecraft's concept of teaching math is an attempt to eliminate repetition. *"Math books,"* he said, *"have never been very good. By grade two, they're 70 percent redundant."*

He said today's typical teacher of first grade pupils spends nine months teaching addition and subtraction of whole numbers. Nine months is also blocked for teaching the same thing to second-graders. Third-graders and fourth-graders devote four months to adding and subtracting. The fifth-grader has two months of repetitive review, and sixth-grade students have at least one month of basic review.

"That's 29 months of repetition," Shoecraft said. *"It equals nearly 3¼ years just doing addition and subtraction of whole numbers."*

Shoecraft said his addition and subtraction concept can be mastered by students in four months, and they don't have to go back year after year for repetitive review. Teaching multiplication and division can also be speeded up.

months, and they don't have to go back year after year for repetitive review. Teaching multiplication and division can also be speeded up.

We can wrap it all up — addition, subtraction, multiplication, and division — by the end of the second grade. That's years saved," he said. *"It buys time to teach other subjects."*

With his program, Shoecraft said *"children proceed from a base of confidence. They know how to get the right answer. We've taken the drudgery out of math. Children who have learned this mathematics concept come away from the program feeling like they're smart — and they are. And smart kids stay in school."*

UHV President Glenn Goerke said the approach to teaching mathematics could take a quantum leap forward if Shoecraft's teaching concept is pursued. *"Achievement in math, supposedly the toughest subject, could directly impact achievement in other subjects,"* Goerke said.

"I think the thing that has impressed me the most is to watch the enthusiasm and pride of achievement by first-graders who have mastered the concept. That same enthusiasm can be transferred to their behavior, to other subjects."

Geri Nielsen monitoring adding and subtracting in Ten Land with arithmetic blocks

New Math of '60s Now "MOVE IT" in the '90s

Janet L. Whitehead, Staff Reporter, *The Daily Tribune*, Bay City, Texas, November 9, 1990

There was a day when ciphering was the most dreaded part of book-learning to the majority of children in elementary classrooms. But like everything else, that's changing.

One college professor is teaching kindergarten through sixth grade Bay City teachers that children don't have to just look at an inundation of numerals on a page that have to be added together to get the correct sum and then go on to the next set of numbers.

Dr. Paul Shoecraft, education teacher at the University of Houston's Victoria campus, for 19 years has been working on a new way to look at math — a program called Move It.

Each Thursday night at the Bay City High School library this semester he has been "moving it" into the curriculum of the city's elementary school math classes.

Mary Brown, math coordinator for the Bay City Independent School District, pointed out that the United States has been far behind many other countries in their math and sciences education.

"I wanted Shoecraft here for an awareness that math and science are important," She said. She explained how the program uses "the concrete" for young children so they can understand by seeing what the older students have been learning on paper in the "abstract" sense.

"The kids are probably a lot more interested, We'll just have to wait and see about results. It's just another crutch to get them to learn to add. Problem solving is hard because children have to think and they have to be taught to think," Brown said.

The district coordinator said recent changes in math-teaching techniques have been the biggest, and she predicts that by the time textbooks for 1995 and 1996 are chosen there will be even greater changes.

Part of Shoecraft's program is called "touch math" or "tap and tally." The students have points on a number that they tap out with their fingers and count in such a way that it makes it easy for first- and second-graders to add several large numbers called "monster math."

Shoecraft uses pictures, games and stories, some of the things the younger students enjoy, in teaching the math.

The Fair Lands game [with arithmetic blocks] makes a hands-on experience with [the structure of a numeration system with place value based on a particular number]. For example, [in Three Land or base three,] three of one block equals one of the next bigger block, and three of those blocks make [one of the next bigger block,] and so on.

With fraction cakes, children can see that three-fourths is bigger than two-thirds and they can put the pieces on a whole circle, the "1," until the same space is filled.

"Ninety percent of kindergarten through sixth-grade teachers' strength is in language arts," Shoecraft said. *"What I wanted to do is capitalize on that."* He went on to say that previously, when teachers got past story time in their rooms, smiles dimmed on teacher's and children's [faces] as the paper-and-pencil task of addition came around.

"We don't improve on the same old same old," said Shoecraft. *"We blow it out of the water and push the standard curriculum right out the front door."*

After first graders learn the same kind of thinking that goes into the tedious math of higher grades, he explained, they are ready for something more advanced when they reach those grades. He stressed the program be pursued into the higher grades so children aren't repeating anything.

Since the program has been taught, which is only for about two years tops in any classroom, according to Shoecraft, some students are able to do math at least two grade levels higher. *"They're just doing everything above grade level."*

Fay Gilzow, first-grade math teacher at Pierce Elementary, said she is just starting her students with the new program. *"They enjoy it. They think it's great that they can add so many numbers and get the right answer."* She said she thinks it will be a big help, especially when they find out they can do the math and are not afraid of math.

An elementary teacher at Linnie Roberts said one of her students, whose performance was very poor in math, has come up to excellent scores with the new math concept and his confidence and new thinking has spilled over into other of his classes. *"All of a sudden this child has really come to life,"* she said.

Other teachers said it was a "no-fail" type of learning and now children can hardly wait to get to math class. And while youth of today learn math the fun way, teachers who learned through the old system are being taught "add" some pizazz to math and "subtract" the pencil and paper approach.

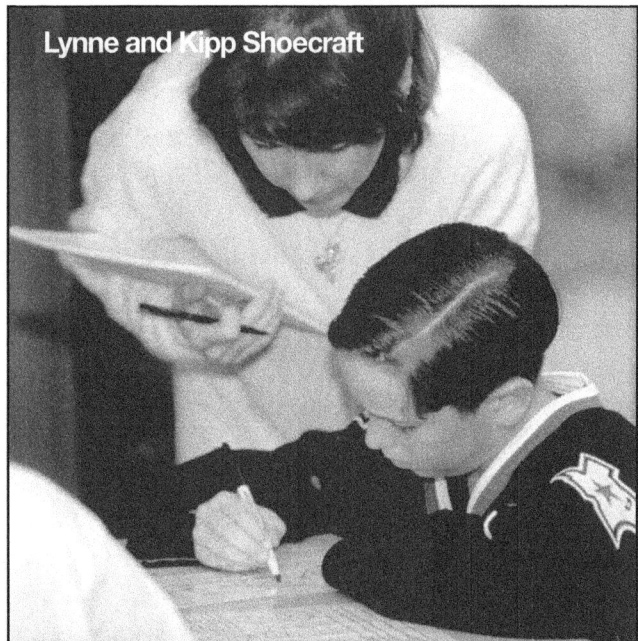

Lynne and Kipp Shoecraft

Lynne Shoecraft watching Fair Trades Up game in Five Land with counters

Geri Nielsen visiting with children adding and subtracting in Five Land with counters

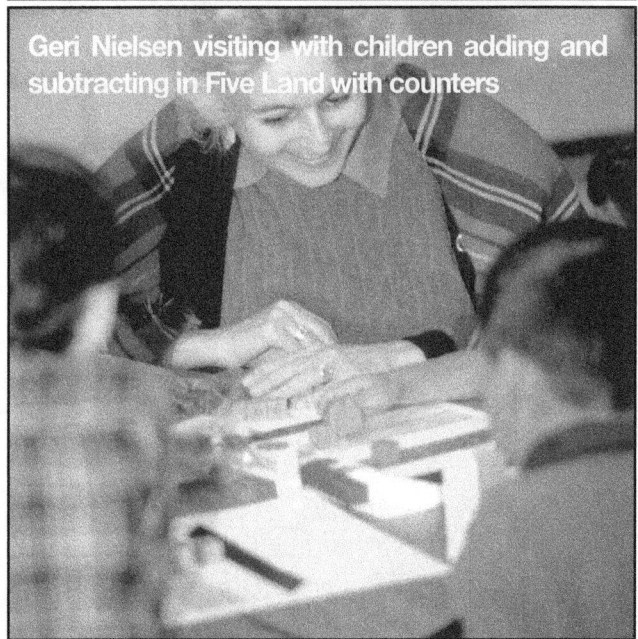

New UHV Math Video Screened

Don Brown, Staff Reporter, *The Victoria Advocate*, November 21, 1990

A group of Victoria area educators and business people previewed a new University of Houston – Victoria video Monday introducing "Move It," an innovative mathematics education program aimed at elementary teachers. A first-grade class from O'Connor Elementary School demonstrated their Move It math skills for the group, answering problem after problem and asking for more.

UHV professor Paul Shoecraft, who developed the "Move It" math program, said *"I'd like to see half the students in this country able to do monster math by the year 2000,"* he said. *"It's possible."*

He said 15 to 20 schools in the area already are trying the program. Calhoun County ISD has made terrific strides with Move It. Other districts, such as Comal, Hallettsville, Gonzales, Waelder, Victoria, Beeville, El Campo, Bay City and Palacios, have initiated Move It into some of their classrooms.

University President Glenn A. Goerke said UHV received a $75,000 grant from Union Carbide, which helped get initial introduction to several classrooms started this year. He said he would like to see a $1 million endowment to help extend the outreach of the Move It program. *"Move It doesn't make what we now do better. It replaces it."*

Goerke said UHV is on the cutting edge of revolutionizing how math is taught at the elementary level and cited recent headlines in which the National Endowment for the Humanities raps training, testing, and textbooks, or others, which say the U.S. is running short of scientific minds.

"Nationally, the number of college freshmen majoring in the physical sciences has dropped from 3.3 percent in 1966 to 2.2 percent last year."

"Math and statistics have fared even worse. Only 0.6 percent of freshmen plan to go into those majors this year," he said. *"If you look at our programs in engineering and math and science right now, you will find there are very few non-nationals moving through our programs,"* he said. *"Something has got to change and not just at the university level."*

Move It is an acronym for "math opportunities, valuable experiences, innovative teaching." The program focuses on the skills and thought processes children must acquire to succeed in mathematics. The program meets children on their terms, and results in children demanding more, bigger and harder problems.

Move It projects mastery of whole number arithmetic by grades two or three [and facility with] fractions, decimals, and percent by grades three or four.

Pam Dolezal, a teacher at Hallettsville Elementary, said her first grade students are doing algebra. *"They solve equations like $3 + X = 5$ and $10 = Y + 6$ and do monster addition and subtraction problems, adding any ten 10-digit numbers and subtracting any two 10-digit numbers and do them correctly."*

According to Shoecraft, this achievement alone means a savings of more than two years of instructional time. *"The conventional textbook-based curriculum guarantees boredom and punishes learning,"* he said.

"We're asking teachers to retrain," said Shoecraft. *"The longer they stay in the standard curriculum, the greater their students' risk of becoming math damaged."*

Move Over for New MOVE IT Math Program

Kaleidoscope, *Herald–Zeitung*, New Braunfels, Texas, December 2, 1990

MOVE IT (Math Opportunities, Valuable Experiences, Innovative Teaching) Math is a professional development program for teachers of students in kindergarten through grade six. The program results in rapid student understanding of equality and base and place value; early ability to solve and graph algebraic equations; natural facility with fractions; mastery of whole number addition and subtraction by grade one; and command of whole number multiplication and division by grade two.

In addition, MOVE IT Math centers on children's strengths and focuses on the kinds of skills and thought processes they must acquire to succeed in mathematics. These include having a literal feel for equality; being skilled at counting forward and backward and skip counting; understanding how a numeration system works; and knowing the function of fractions.

The brainchild of Dr. Paul Shoecraft from the University of Houston–Victoria, the program was introduced to CISD [Comal Independent School District] teachers in an intensive training program during the summer. Shoecraft was back in the district in November to conduct a follow-up MOVE IT Math workshop for the elementary school educators.

In a recent issue of American Educator, Shoecraft's MOVE IT Math program is described in the article, "Anyone Can Learn Math, New Programs Show How." The article says, *"Shoecraft's program works — and it works for kids from a variety of socio-economic backgrounds. In Dr. Shoecraft's schools, the manipulatives make possible a new, more advanced curriculum."*

As in other school districts that have implemented MOVE IT Math, Comal ISD elementary students as young as first graders will become proficient in skills such as adding and subtracting 10-digit numbers and will work with fractions.

"More importantly, with the MOVE IT Math program, when students perform addition, subtraction, multiplication, and division, they are understanding what they are doing," said Carol Hall, assistant superintendent for curriculum at Comal ISD. *"Also, algebra and geometry are an integral part of the elementary curriculum. Students determine the areas of irregularly shaped geometric figures using geoboards and elastic bands."*

Algebra is introduced first through the balance beam as a means of establishing the equality of different equations.

"Through acceleration of the elementary math curriculum by eliminating unnecessary repetition and by teaching the students for true understanding of mathematics concepts, Shoecraft's program allows the majority of students to master the mathematics skills necessary for completion of Algebra 1 by grade 8," Hall said. *"According to Shoecraft, this must become the standard expectation if we are to truly prepare our students for future success in mathematics and for the students to be truly competitive in future international job markets."*

Comal ISD plans to continue the MOVE IT Math training in the summer of 1991 when advanced training will be conducted for those teachers trained last August, and more teachers will receive the initial training. Also, Shoecraft will be working with the teachers to realign the mathematics curriculum to assure that all students meet success and acquire skills that have resulted in rapid student mathematics achievement.

Mathematics Made Easy with MOVE IT!

Barbara Gorzycki, Staff Reporter, *UHCLIDIAN*, University of Houston – Clear Lake, Feb. 4, 1991

Travel back for a moment. You're in third grade and the room is deathly quiet. Heads are down and eyes focus blankly on anything — anything, that is, but the teacher. You know to avoid her eyes because if you look up, she just might call on you. Oh, no. A churning gut. Fear. Sweat. Dizziness.

It's the dreaded mathematics lessons. Remember?

If you're like millions of Americans who have an aversion to math, you've probably experienced at least some of these emotions. But numbers don't have to be such scary business. In fact, Dr. Paul Shoecraft, professor of math education at UH-Victoria, is currently proving that learning mathematics in today's classrooms can be downright fun.

For the past 20 years, Shoecraft worked to put together a professional development program for elementary school teachers and students that emphasizes understanding math concepts instead of memorizing basic math facts. His MOVE IT (Math Opportunities, Valuable Experiences—Innovative Teaching) Math program may well be revolutionizing the teaching of math.

In the standard memory-based curriculum now taught statewide, students don't learn basic number facts until grades seven or eight, but in the MOVE IT Math program, students master these number facts by grades two or three through the use of manipulatives [tangible objects], games, and the kinds of activities that move students from abstract ideas of numbers to real-world applications of math.

"We remove the hurdle of the basic facts," said Shoecraft. *"There are 390 of those to learn. We use a Tap and Tally approach, [the number facts component of TouchMath]. With Tap & Tally, the children learn to figure out all 390 basic facts. By grade one students can master the 200 addition and subtraction facts. By grade two it's easy for all students to master [the 190] multiplication and division facts."*

According to Shoecraft and his teachers, the students not only learn math, but they love the whole process. Robert Haas, superintendent of Hallettsville I.S.D., saw firsthand the advantages of MOVE IT Math when the district implemented the program.

"I had observed a true breakthrough as far as students' interest and ability to do math at the lower grade levels," said Haas. The success rates included students from all socioeconomic levels.

Essentially the MOVE IT Math program uses manipulatives and the three skills of count up, count down and skip count. Once these are mastered, students gain a rapid understanding of equality and base and place value, an early ability to solve and graph algebraic equations, a natural facility with fractions, the mastery of whole number addition and subtraction by grade one and the mastery of whole number multiplication and division by grade two. By grades three and four students master fractions, decimals and percentages.

Approximately 20 schools in this area have implemented MOVE IT Math successfully. But Shoecraft operates with no illusions about the mechanics that propel school districts. Currently seven textbooks are on the TEA [Texas Education Agency] state adoption list until 1995. These texts represent *"probably 90 percent of what most children experience in material in grades one through six,"* said Shoecraft. Most of the material is redundant.

Interested in all aspects of student learning outcomes, Shoecraft combed through these

180

seven texts, counting the number of pages devoted to [just] addition and subtraction of whole numbers for grades one through six.

The numbers are revealing. Scott, Foresman now devotes 534 pages to just addition and subtraction; Silver Burdett and Ginn, 574; Addison-Wesley, 584; Harcourt Brace and Jovanovich, 515.

Within the MOVE IT Math program students master whole number addition, subtraction, multiplication and division by third grade. Years saved with no redundancy of curriculum.

According to Shoecraft, the key to a successful relationship between a student and his or her math is to identify that which doesn't work and try that which will work. Since students today will be competing for opportunities in a global economy tomorrow, school districts must address the declining performance of U.S. students in the fields of math and science.

One recent survey of 13-year-olds from six countries, including the United Kingdom and Canada, placed U.S. students last in math and below average in science. In another recent survey of 13 countries, U.S. high school seniors placed twelfth in algebra, eleventh in geometry and chemistry, last in calculus and ninth in physics.

In his recent video, Shoecraft challenges his viewers. *"You pick the most impoverished school district in this country, and pick from that some grade one children who heretofore have had no hope of succeeding in the school system, and you give us 10 weeks, three hours a day with those children, and we'll show you children that will knock the socks off any other first graders in this nation."*

Richards, Sharp to Eye Math Class
Victoria Advocate, March 19, 1991, page 3A

Governor Ann Richards and State Comptroller John Sharp will visit the University of Houston – Victoria and O'Connor Elementary School Friday to see and hear about a new concept for teaching math.

They will be briefed on "Move It Math," a teaching concept developed by UHV professor Paul J. Shoecraft, and will then see the concept demonstrated by kindergarten and first grade students in the classroom.

Richards and Sharp, who will be traveling separately, will be accompanied by …

o Mannie Justiz of Austin, dean of the University of Texas' School of Education;

o Bonnie Floyd, chairman of the Creative Learning Center in Dallas;

o Cynthia Canevaro of EDS Corp. in Dallas;

o Joe Ramsey of Southwestern Bell Telephone in Dallas;

o Mike Lunsford of Dallas, representing Mary Kay Cosmetics' Adopt-a-School program; and

o Ralph Pahel, executive director of the Boys and Girls Club of Greater Dallas.

Skip counting by three in Spanish in a MOVE IT Math class

Move It Math Whizzes Cheer Governor Richards on City School Tour

Don Brown, Staff Reporter, *The Victoria Advocate*, Mach. 23, 1991

FIRST ARTICLE IN CHAPTER 1

Governor Ann Richards found more than she bargained for while visiting classrooms at O'Connor Elementary School Friday.

Richards was in Victoria at the invitation of university of Houston-Victoria President Glenn Goerke to see firsthand the Multimodality Math/Move It Math program developed by Paul Shoecraft, a professor at UHV.

She admitted she knew very little about Shoecraft's concept before arriving in Victoria, but before leaving, she said she was "extremely impressed."

Richards received a briefing and watched a short video at the UHV Petroleum Training Institute, then went to the O'Connor campus to see first-graders breeze through "monster" addition, subtraction, multiplication and division problems.

Move It, an acronym for math opportunities, valuable experiences and innovative teaching, focuses on the skills and thought processes children must acquire to succeed in mathematics. *"Move It Math is a quantum leap ahead of traditional math instruction concepts now taught in Texas classrooms,"* Richards said.

"I was impressed by the fact these children have not only learned a great deal about numbers, but also can work independently in the classroom," she said. *"I think it's remarkable. Obviously, these children are excelling far beyond the standardized requirements of Texas public schools."*

Prior to entering politics, Richards taught at the intermediate and high school levels. *"I have been in a lot of classrooms, but never in one where independent work was going on as successfully as what I have seen taking place here,"* she said.

She said she plans to urge newly appointed Education Commissioner Lionel "Skip" Meno and members of the State Board of Education to *"do as I did and come to Victoria to see a program that really works. Any exposure to this program would certainly convince people at the Texas Education Agency that it's worth emulating statewide. The question will be how much of it we can do and how fast we can put this program on track."*

Richards said she always asks children how they are doing in school and if they like math. *"Younger children usually say they like math, but the older they get, the more they say they're not good in math or don't understand it."* She said the education system *"has to assume blame when something happens to change a child's opinion of math from 'I can do it' to 'I can't do it.'"*

Richards said that while state leaders have been focusing a lot of attention on how Texas will finance public education, she is equally concerned about the "quality" of the product. *"I think the quality discussions about what children are actually learning and what we're turning out of the public school system will be the next big debate in Austin,"* she said.

Richards praised Shoecraft and the support given his math concept by the UHV faculty. *"The University of Houston–Victoria is doing a remarkable job, and their campus here is truly important in the higher education system of this state."*

She also commended teachers at schools like O'Connor *"who are willing to go out on a limb to allow their students to excel. A lot of times the big thing that holds back education is the fear of trying something new."*

MOVE IT Math Gets Governor Richards' Attention and Endorsement

UH–Victoria Newsletter, September 1991

Impressed by her March visit to Victoria to see Dr. Paul Shoecraft's MOVE IT Math project in action, Governor Ann Richards later endorsed funding of the program as one of her top priorities. In addition, State Comptroller John Sharp, in a video produced by his office for television, touted MOVE IT Math as one of two innovative programs in the state that represent his idea of positive change.

During the Governor's visit here, she was briefed on Shoecraft's program at the university, then driven to O'Connor Elementary School to see the teaching concepts put into practice in the classroom. As quoted in the Victoria Advocate, Richards said, *"MOVE IT Math is a quantum leap ahead"* in math instruction for kindergarten and elementary school children.

Paul and Lynne Shoecraft greet Gov. Ann Richards during her March visit to Victoria.

Over the summer, Shoecraft was responsible for MOVE IT Math training conducted in New Braunfels, Dallas, Rosenberg, San Antonio, and Beeville. In addition, MOVE IT Math workshops directed by Shoecraft were held in Kerrville, Schulenburg, Gladewater, Goliad, Waller, Lohn, Devine, and Santa Fe.

Assisted by area teachers trained in MOVE IT Math, Shoecraft developed over the summer a K-6 curriculum guide for incorporating the program into the math curriculum.

Port O'Connor Math Program at Cutting Edge of a Revolution in Education

George Macias, Texas Co-op Power, *Swisher Electric Edition*, March 1992

SECOND ARTICLE IN CHAPTER 1

A small rural school on the Texas Gulf Coast is helping revolutionize the way educators think about mathematics and education in general.

The Port O'Connor School is a public school with classes from kindergarten through sixth grade. The school has a modest budget and limited resources, and students are typically from working class families, mostly the sons and daughters of shrimpers and farmers.

For two years in a row, third graders at Port O'Connor have scored 100 percentile on the state's standardized third grade math test. These young mathematicians not only add, subtract, multiply, and divide bigger numbers than much older students in other schools, they also enjoy the learning process. Astonishingly, they are grasping and excited about algebra concepts as early as kindergarten.

"Nothing sets these kids apart except for this new way of instruction. Any child from any background can profit from this way of being taught," says Mrs. Marilyn Bratcher, the principal of Port O'Connor School. She says the Port O'Connor experience is proving that what schools and teachers usually expect of students is far, far below what they can accomplish.

"Math books in the state of Texas do the same thing over and over. They're boring," says Mrs. Bratcher. *"We don't even use textbooks except as a resource because they're behind. In our country, intelligence is equated with math. Good math skills open doors for so many good occupations,"* she says.

The significance of the Port O'Connor "experiment" became clear to Principal Bratcher one day while sitting on the floor with kindergarten students in Judy Anderson's class. *"I was assessing Judy for the career ladder. When I realized what was happening, I was so excited I crawled up to her and said, 'Judy, you are teaching algebra to kindergarten kids — and it's only October!'"*

Mrs. Anderson began rethinking her approach to teaching after attending a three-week class presented by Dr. Paul Shoecraft at the University of Houston at Victoria. The program was funded with federal assistance made possible by Eisenhower Math/Science Grants and the Education for Economic Security Act.

Dr. Shoecraft promotes a revolutionary system of teaching math that is shaking the foundation of the educational hierarchy and drawing the interest of prestigious think tanks such as the National Science Foundation. His program, or parts of his program, have been tried or adopted by several school districts, including Calhoun, Comal, Hallettsville, Gonzales, Waelder, Victoria, Beeville, El Campo, Bay City, and Palacios ISDs.

Dr. Shoecraft, who has been working on his educational revolution since the early 1970s, says several South Texas Schools are part of this new cutting edge reform movement, but the Port O'Connor School was the first school to fully commit an entire school to this innovative teaching style. Port O'Connor's Mrs. Anderson was one of the first teachers to take elements of Dr. Shoecraft's math program and adapt them to a kindergarten classroom and make them work on a sustained basis.

Mrs. Anderson has also developed a program called "Math Buddies," which brings older children from higher grades together with younger kids. Principal Bratcher explains that Math Buddies serves several needs, such as reducing the amount of physical work for the teacher and reinforcing basic mathematical concepts for older students.

The math program at Port O'Connor includes many different modes of learning, using objects (called "manipulatives"), such as arithmetic blocks. The key to this learning process is that it attracts the student's attention by asking the child to do more than just listen to a teacher talk. Instead, the students are actively involved in the process.

Mrs. Anderson says the program grew out of total frustration: *"We had to do something to help these kids want to stay in school. So few graduated from high school. We had to make learning challenging and enjoyable."*

Her excitement for the program is apparent, both in the classroom and just talking about it. *"I like the fact that my students like math, but more important than that is that they love to study. They love to learn."*

In Texas, the general guidelines for kindergarten are that students should be able to compare things, to recognize numbers one through 20, and to be familiar with circles, squares, and triangles. [In contrast,] by the end of kindergarten, most of the Port O'Connor kids are familiar with addition and subtraction of whole numbers, fractions, coordinate graphing, positive and negative integers, solving for one unknown in an algebraic equation, area and perimeter of polygons, symmetry, and place values of numbers into the thousands.

Dr. Robert Nielsen, a [former math professor], member of the Mathematical Sciences Education Board in Washington, DC, and an assistant to the president of the American Federation of Teachers, has visited the Port O'Connor School twice and is a strong advocate of both the Port O'Connor School and of Dr. Shoecraft's work.

Dr. Nielsen says, *"It's the direction that school mathematics ought to go everywhere. Port O'Connor has demonstrated that all kids can learn math. I hadn't realized how dramatically things can change until I visited* Port O'Connor School. *I'm convinced that this stuff is so radically different from the way you and I learned math that the only way to convince people is for them to see it happen. If you haven't seen it, you can't believe it."*

"There's something wrong when math is a subject that almost all kids learn to hate. The Port O'Connor School is dramatic proof that it doesn't have to be that way," says Nielsen, and he encourages parents and teachers to ask questions about their own math programs in their local areas. He says, *"If you walk into a classroom and the kids are having fun, then the school has got the right stuff; if not, then it's just teaching math the old-fashioned way."*

Dr. William Kirby, head of the Texas Education Agency for many years, says he thinks the program is *"super because it sets high expectations for kids, and it utilizes teachers with adequate training who believe in themselves."*

Dr. Kirby says students, through the use of concrete objects, are actually *"experiencing mathematical concepts first hand. They are actively engaged with their hands and their eyes, not just their ears. Too often kids sit passively, listening to a lecture."*

He says, *"If we're going to get where we need to be in this country, we need to change the model that's existed for the last 200 years of the teacher as merely a dispenser of facts."*

Dr. Kirby is also impressed with Port O'Connor's "Math Buddies" program because of the team effort it inspires. He says, *"Sometimes older kids are helping younger kids, and sometimes the reverse is true."*

He says the Port O'Connor School has convinced him that teachers should be guides, mentors, and assistants to students. *"Teachers should be directors of learning, like symphony conductors. Not everyone in a symphony gets to sit in the first chair, but everybody participates and has an important role."*

For education reform to work, he says, it must come from local schools, unlike traditional "top down" reform. *"Give people at local districts the freedom to be creative."*

Principal Bratcher is convinced that the math program at the Port O'Connor School is the tip of the iceberg, and teachers in all grades at the school are applying the successful methods used in Mrs. Anderson's class, such as crossing grade levels and using manipulatives, in teaching other subjects.

"This has really changed the way we do things in our whole school, and because Port O'Connor has hardly any student turnover, the long-term success of the school will be easy to monitor," she says.

So far, with kindergarten students regularly testing out at second and third grade levels, and some as high as junior high levels, the success is fairly obvious.

The University of Houston at Victoria honored the young mathematicians at the Port O'Connor School in 1989 by offering UHV scholarships to all of the students on the condition they finish junior college. Dr. Shoecraft says the university wanted to acknowledge the important role these students played in making this type of innovative program work.

One Port O'Connor student, Devon Vasquez, wrote President Bush while attending second grade at the Port O'Connor School: *"Teachers from all over Texas have come to watch us do our work because they want their kids to do the fun things we do. We would be happy if you came to Port O'Connor, Texas, to our math class so we can show off our brains to you."*

New Math Program at Ciavarra Elementary School
Ruth M. Whitaker, *The Devine News*, Devine, Texas, December 17, 1992

A new math program, "MOVE IT Math," started at Devine's John J. Ciavarra Elementary School last year, is improving students' test scores dramatically.

At the regular monthly board meeting at 7:30 p.m., Monday, December 14, 1992, Principal Linda Stanton and teachers Melissa Lyles and Sandra Jolley and Counselor Nadine Sczech, presented a video program on Move It Math.

Mrs. Sczech told board members that math test scores, at 56.2% mastery in 1990-91 came up to 73% in the 1991-92 school year and stand at 82% this year.

The program uses songs, games and a "Tap and Tally" system that helps students learn [all] 390 number facts in grades K, 1, 2 and 3.

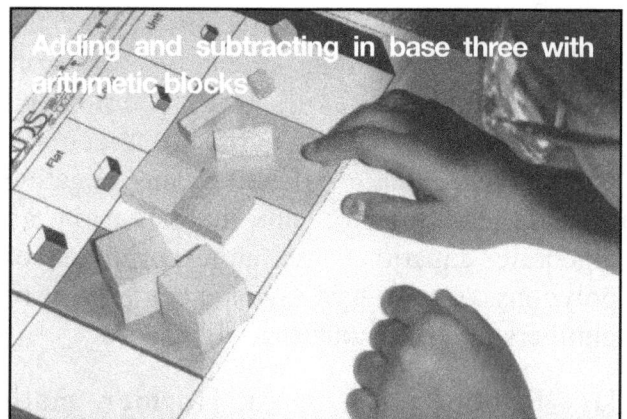
Adding and subtracting in base three with arithmetic blocks

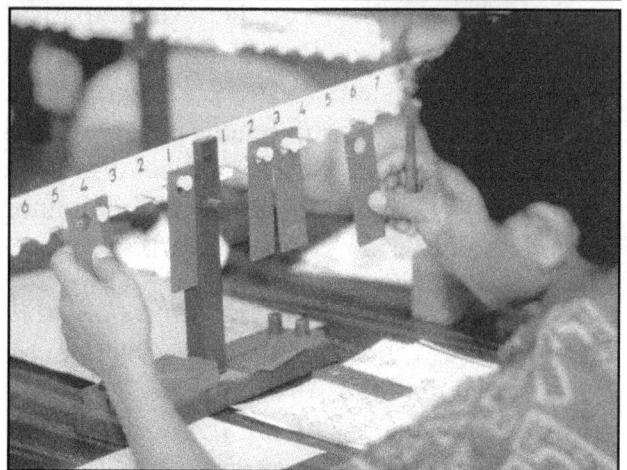

It All Adds Up

Starita Smith, *Austin American-Statesman,* Austin, Texas, February 25, 1993

The numbers are as large as Desiree Quintero, 7, is tiny. The first-grader at Waelder Elementary School easily subtracts 169,396 from 176,543, borrowing three times as she works on each column of numbers.

Desiree isn't the prodigy of her class. The problem she did came straight off the assignment sheet that every other first-grader in her class had to complete last week. Next door, kindergartners are practicing multiplication of fives; in another room, fifth graders are adding and subtracting fractions.

These students are learning arithmetic through an innovative curriculum developed by Paul Shoecraft, a professor at the University of Houston at Victoria. The curriculum, called MOVE IT Math, uses children's love of playing games, singing and dancing to teach them arithmetic.

The 130 students at the school make up roughly half of the Waelder Independent School District in a small town about 55 miles southeast of Austin, said Norman Woolsey, the school's principal. The students are 60 percent Hispanic, 30 percent black and 10 percent white. Most are from low-income families, he said.

It isn't hard to find schools with largely minority and poor enrollments in Austin and throughout the nation. And it is at those schools where a growing achievement gap in test scores between white students and minority students has been documented. But in Waelder, that scenario is changing.

"In the 18 months the school has used MOVE IT Math, scores on national and statewide standardized tests have risen significantly. In 1991, 65 percent of third-graders mastered the math section of the Texas Assessment of Academic Skills exam. In 1992, that figure jumped to 93 percent. The one student who did

Paul Shoecraft, developer of MOVE IT Math, and Lynne Shoecraft work with first-graders Darwin Jones, left, Juan Cabarello, and Brandon Hull at Waelder Elementary School. Test scores for students have improved using the new system.

not pass had transferred from another district," Woolsey said.

Pupils are performing at least a grade level higher than their peers on the Comprehensive Test of Basic Skills, a national exam. For example, second-graders' math skills were at the mid-fourth-grade level.

Gov. Ann Richards has praised MOVE IT Math, said Leticia Vasquez, a spokeswoman for the governor's office. *"She likes that program. She saw it in action in Victoria about two years ago and was very impressed, as was (state Comptroller) John Sharp,"* Vasquez said.

They don't learn the meaning behind the symbols. For example, the equal sign means "balanced" or "is the same as." In MOVE IT classes, first graders play with a balance beam on which they can hang plastic sticks. The beam, which resembles a miniature seesaw, is balanced only when equivalent weights are on both sides.

MOVE IT Math also teaches students to see addition, subtraction, multiplication and division as combining and separating actions and to use a set of counting skills [to figure out the number facts] instead of memorizing them.

As word about the program has spread, it has attracted admirers who praise it as the way to turn around the sluggish test scores, boredom and anxiety that math spells for many students, and detractors, who are skeptical and criticize it as sort of a glorified finger counting system.

"It's very impressive when you first see it because the kids are manipulating large numbers," said Timy Baranoff, who observed MOVE IT Math when the Austin Independent School District was considering the curriculum several years ago. *"It's very showy."*

Baranoff, who retired last month, is former AISD [Austin Independent School District] director of elementary instruction and curriculum. She said she did not like the curriculum because she was concerned that young students did not have a firm grasp of basics that they would need later on. *"Are the kids truly understanding what's happening when they are manipulating those huge numbers?"* she asked.

Ann Powell, AISD math coordinator for grades kindergarten through 12, shares Baranoff's feelings. She said she did not see the children being taught problem solving; instead, they were being taught computation.

"Computation is when you give the child a page of basic problems to work in addition or subtraction. Problem Solving is a real life situation: If you had eight marbles and John had 10, how many marbles are there? They need to use no gimmicks to work these problems to show they can come up with the answer because we don't (use gimmicks) in everyday life," she said.

Mary Lester, director of mathematics for the Dallas Independent School District where 30 elementary schools have volunteered to use

the program, disagrees. *"I contend there is a lot of problem solving in MOVE IT Math,"* she said. *"You have to first decide what is a problem. 'Mary had 75¢ and John gave her 25¢, how much is that?' is not a problem."*

Problems involve more skills," Lester said. Her example: Mary has eight coins, and the value of the coins is 78¢. What types of coins does she have? *"Then you have to do math first. You may use trial and error. You have to investigate and come up with one arrangement,"* she said.

Lester said teachers in Dallas are enthusiastic about the program. They have told her that they especially enjoy having more control over tailoring their classes to their students' needs.

Students in the program learn far more than the arithmetic textbooks for their grades cover, she and Shoecraft said. To compensate, Shoecraft has developed a library and a database of lesson plans and activities tied to the state-mandated essential elements of mathematics.

Waelder third-grade teacher Doris Richards said being freed from the textbook is *"the best thing that has ever happened in my career."* And because they can do math so well, the children have a view of themselves not often expressed. It was summed up by Lionel Elias, a 10-year-old fifth-grader: *"Everybody in our school is smart."*

Mexico School Principal Learns about MOVE IT Math

Scott Reese Willey, *The Victoria Advocate, Victoria, Texas,* March 27, 1993, page 3A

Teresita Sainz, principal of Antonio Caso, the largest private school in Guadalajara, Mexico, spent the afternoon Friday with University of Houston-Victoria professor Paul Shoecraft. The two toured a Victoria elementary school where students as young as 5 years old were having fun solving algebra problems through an exciting new program known as "Move It Math."

The program developed by Shoecraft over three decades recently received nationwide recognition for improving student interest and performance in mathematics. Move It Math also captured the attention of Gov. Ann Richards who visited a school two years ago to get a first-hand look at the program.

The innovative program teaches children complex math problems through the use of games instead of relying on memorization and bookwork. Sainz praised the program Friday as Richards praised it. *"Perhaps we can get Doctor Shoecraft to come to Mexico and teach us [Move It Math],"* she said, as she circulated among students at Dudley Elementary School.

It wouldn't be the first time Shoecraft visited the Guadalajara elementary school since he introduced the concept to some of the school's second-graders two years ago. Sainz was sent to Texas to find out more about the program. She said she hoped to find a way of rejuvenating middle-school students who have grown bored of math or who have given up on math altogether. The smiles, laughter and interest displayed by math students at Dudley convinced her she had found what she was looking for, she said.

Eric Castillo, an 11-year-old fourth grader at Dudley, was one of the key players in convincing her of that. He taught the principal addition and subtraction through a game in which dice are rolled and different-sized

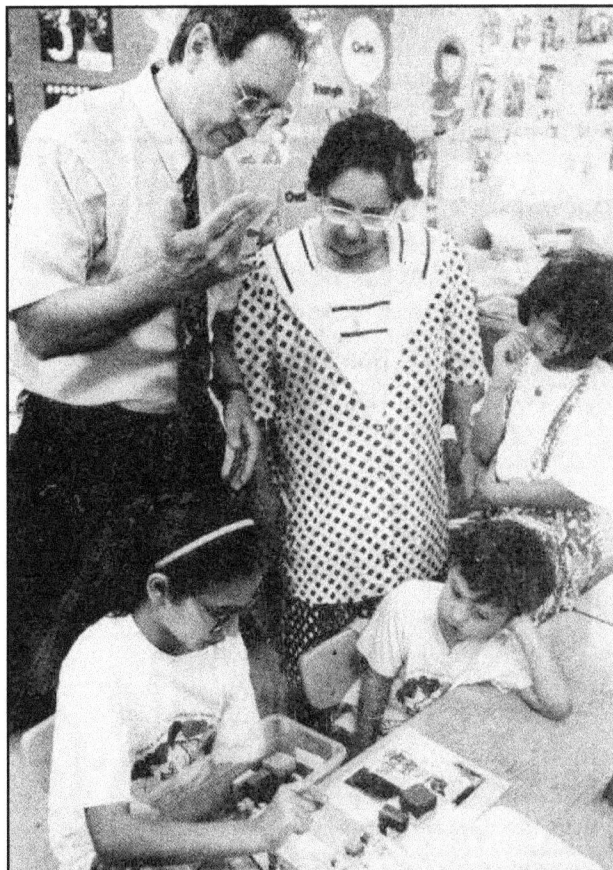

UHV professor Paul Shoecraft explains the Move It Math program to Teresita Sainz, the principal of a Guadalajara elementary school in Mexico.

blocks are [placed on or removed from] a gameboard. Each of the blocks are sized so that if they are put together, they form a block one size larger.

To [subtract], blocks of different sizes are placed on the gameboard. To win the game, all of the blocks must be removed. That is accomplished through a roll of a die. If the die lands on the number three, for instance, then three [unit] blocks must be taken off the board. If three [unit] blocks are not on the board, then the player may exchange the next larger block for smaller blocks [and so on until they have three or more unit blocks]. *"It's like exchanging a quarter for five nickels,"* said Shoecraft.

Move It Math Found Exciting

Lee L. Allbright, Letters to the Editor, *The Victoria Advocate*, April 11, 1993, page 3

On March 27, your staff writer, Scott Reese Willey, caught my attention and has been on my mind ever since. Something wonderful was happening in our public schools in Victoria.

We were taking the lead on something really exciting and perhaps turning the corner in math with "Move It Math," developed by Dr. Paul Shoecraft, a professor with the University of Houston–Victoria.

As a past school trustee, I know how vigorously we looked for new and better ways for educating the youth in our community. Last year I was at a meeting where Dr. Shoecraft presented a brief program about Move It Math. Its potential seems very encouraging to me.

Recently, members of the VISD [Victoria Independent School District] board were concerned about lagging grades and how we might improve or increase the learning among students who are having difficulty retaining what they are exposed to in the classroom. I think perhaps they found one of the prime movers to educating our young people.

Quite frankly, I believe our children are bored. To dish out more of what they are bored with isn't of any benefit. In 1991 the law was changed to increase the school year from 175 to 180 class days. A new bill is being sent to the full House Public Education Committee asking for a reversal of that action. Five teacher training and planning days were moved into class days without increasing payroll costs. More is simply not the answer.

I am sure the teachers on the front lines of educating our youth feel that it is beyond reason to be using methods from the past. We need to be concerned about rejuvenating students who have grown bored of math or who have given up on math altogether.

I believe this is the way for the future. It is very exciting to have its beginning here in Victoria. I feel confident that we will find that not only math knowledge will show improvement, but other subjects will show improvement as well. The program has been developed over several decades and received nationwide recognition for improving interest and performance in the field of mathematics.

It has caught the attention of Gov. Ann Richards, who is very interested in the direction our education is taking. I am pleased that the local board has initiated the program at Dudley Elementary School. It must be exciting for board members when they visit a school and see students as young as age five who are having fun solving algebra problems through an exciting new program. Of course, the dream won't be complete until you are able to visit any school in the district and find the same excitement.

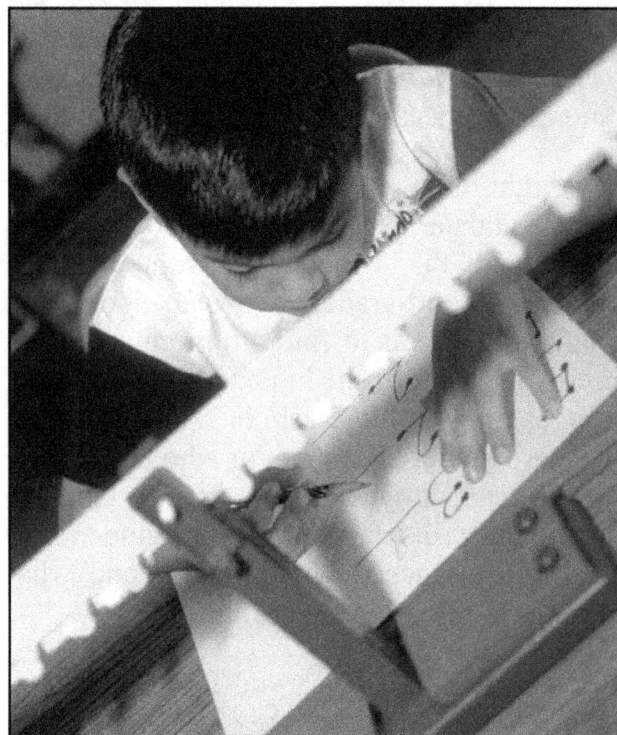

MORE Documented Results for Multimodality Math/MOVE IT Math

Bay City Independent School District, Bay City, Texas

Roberts Elementary began using MOVE IT Math in 1990. Cherry Elementary and Holmes Elementary have been using a K-8 basal mathematics series. Historically, Cherry Elementary has been the highest scoring elementary school in the district. The math scores for grade 2 were Cherry + 5.3, Holmes + 0.7, Roberts + 7.1, and District + 4.5. Those for grade 4 were Cherry – 2.9, Holmes + 0.5, Roberts + 3.3, and District + 0.2. — Bay City ISD Student Productivity Index, COGAT and ITBS (NAPT), Bay City, Texas, 1992

Bloomington Independent School District, Bloomington, Texas

Communication received from the Bloomington ISD Curriculum Coordinator:

I would like to take the time to thank you for all the valuable MOVE IT Math training our teachers have received and the genuine concern you have shown towards the students in our district. I have enclosed a graph comparing the results of the math TAAS [Texas Assessment of Academic Skills] scores from 1990-91 and 1991-92. As you can see the third grade scores have gone up dramatically [from 43.5% to 90.0%] and I attribute much of this to our new math approach. — B. Williams, Curriculum Coordinator, Blooming ISD, 1992

Palacios Independent School District, Palacios, Texas

In February a report was included in the board packet that explained a planned evaluation of the MOVE IT Math program for Kindergarten and First Grade. We have completed the evaluation and have prepared the following reports to explain the results. The reports include (1) the original evaluation design, (2) the procedures that were used to randomly select the students to be evaluated, (3) score distribution tables for both grades, and (4) percentile distribution tables for both grades. The results of the evaluation were generally positive and reinforce the California Achievement Test Data.

* *The mean national percentile for Kindergarten was the 92nd percentile.*

* *The median national percentile for Kindergarten was the 96th percentile.*

* *The mean grade equivalent for Kindergarten was First Grade, Second Month.*

* *The mean national percentile for First Grade was the 71st percentile.*

* *The median national percentile for First Grade was the 78th percentile.*

* *The mean grade equivalent for First Grade was Second Grade, Second Month.*

The administration recommends that the board authorize the expansion of the MOVE IT Math program from Kindergarten and First Grade to include Second Grade for the 1991-92 school year. — Mike Maxwell, Curriculum Coordinator, to William E. Reaves, Superintendent, Palacios ISD, Palacios, Texas, May 30, 1991

Chapter 11: Why the MOVE IT Math Methodology Is Effective

The curriculum of a subject should be determined by the most fundamental understanding that can be achieved of the underlying principles that give structure to that subject. — Jerome Bruner, 1960

In science, whenever an experiment yields a surprising result, like light being both a particle and a wave, scientists hold off on believing it. For the result to become a fact, other scientists must replicate the experiment and get the same result. With that in mind, Multimodality Math/ MOVE IT Math was an experiment that yielded a surprising result.

Children's test scores in math soared! In particular, scores on norm referenced tests, such as the Woodcock Johnson, California Achievement Test, and Stanford 9, were 2-3 years above grade level, and students who were at-risk (one year below grade level in math and/or reading) caught up with their peers. Moreover, *this occurred in virtually every class with every teacher in every elementary school in every city in every state in which the program was implemented.* Thus it is a *fact* that Multimodality Math/MOVE IT Math is effective with *all* children, and how it came to be explains why.

MOVE IT Math is not the result of trial and error until I got it right. Nor did it come to me in a dream. It is the result of asking the right question about teaching arithmetic and being able to answer it because of a book I had written. The question was *"What must children understand about arithmetic to excel in the subject?"* The book was *The Arithmetic Primer* (Addison-Wesley, 1979), now out of print.

The Arithmetic Primer was a self-help, self-instructional arithmetic program for middle school students and up, including adults, who needed to quickly fine-tune their computational skills in order to meet some competency requirement, such as being able to test out of basic math classes in high school and college. The book identified the main skills in arithmetic and dissected each one into its component subskills. For example, it dissected adding, subtracting, multiplying and dividing whole numbers into 37 component subskills, as shown on the next page from the book's table of contents. For fractions, decimals, and percent, the book identified an additional 62 component subskills. Thus by my count, arithmetic consists of 37 + 62 = 99 subskills.[51]

---

[51] Authors differ in the number of subskills that constitute arithmetic. Bryce Shaw, for instance, identified 123 such subskills for an individualized computational skills program that Houghton Mifflin published in 1972, just in time for me to use at 'Rithmetic in Residence, the summer math camp I created and directed in 1974 and 1975 while I was an Assistant Professor in the math department at Arizona State University. The program came in several large cardboard boxes with enough practice sheets for each of the 123 subskills for an entire class, and my success with it at the camp motivated me to write *The Arithmetic Primer* to individualize instruction in arithmetic for single users.

Subskills for Adding, Subtracting, Multiplying, and Dividing Whole Numbers

Adding Whole Numbers

Pretest

1. Addition Facts, No Trading
2. Addition Facts, Trading Ones for Tens
3. Three Single-digit Addends, Trading Ones for Tens
4. Addition, No Trading
5. Two Addends, Trading Ones for Tens
6. Two Addends, Trading Tens for Hundreds
7. Two Addends, Trading Ones for Tens and Tens for hundreds
8. Three Addends, Trading Ones for Tens and Tens for hundreds
9. Addition, Trading Ones for Tens, Tens for Hundreds, Hundreds for Thousands, and so on

Post Test

Subtracting Whole Numbers

Pretest

1. Subtraction Facts, No Trading
2. Subtraction Facts, Trading Tens for Ones
3. Subtraction, No Trading
4. Subtraction, Trading Tens for Ones
5. Subtraction, Trading Hundreds for Tens
6. Subtraction, Trading Tens for Ones and Hundreds for Tens
7. Subtraction, Trading Tens for Ones, Hundreds for Tens, Thousands for Hundreds, and so on
8. Subtraction, Trading Across Zeros

Post Test

Multiplying Whole Numbers

Pretest

1. Multiplication Facts
2. Single-digit Multiplier, No Trading
3. Single-digit Multiplier, Trading Ones for Tens
4. Single-digit Multiplier, Trading Tens for Hundreds
5. Single-digit Multiplier, Trading Ones for Tens and Tens for Hundreds
6. Single-digit Multiplier, Trading Ones for Tens, Tens for Hundreds, and Hundreds for Thousands
7. Two-digit Multiplier
8. Three-digit Multiplier
9. Multiplier with Terminal Zeros
10. Multiplier with Non-terminal Zeros

Post Test

Dividing Whole Numbers

Pretest

1. Division Facts
2. Single-digit Divisor, Two-digit Dividend
3. Single-digit Divisor, Three-digit Dividend
4. Single-digit Divisor, No Zeros in Quotient
5. Single-digit Divisor, Zeros in Quotient
6. Two-digit Divisor, No Zeros in Quotient
7. Two-digit Divisor, Zeros in Quotient
8. Divisor with Three or More Digits
9. Fractional Remainder
10. Decimal Remainder

Post Test

Origin of the Four Golden Keys that Unlock the Doors to Success with Whole Number Arithmetic

After writing *The Arithmetic Primer*, I enjoyed the freedom of being able to think of other things. Years passed before I thought about the book again. What prompted me to do so was learning about TouchMath. Until then, I had never questioned the need to memorize the number facts. However, in realizing that the number facts could be taught quickly and effectively to *all* children by teaching them how to count them out, *which amounts to teaching them with understanding*, I revisited *The Arithmetic Primer*.

In viewing the book's table of contents, I could see why so many adults reach for a calculator for even the simplest of calculations. They lack confidence in their ability to compute. They know they have forgotten some of the arithmetic that they knew when they were in school. *Ninety-nine subskills and 390 number facts are a lot for a person to remember after no longer practicing them or taking a math class after graduating from high school or college.*

Knowing that forgetting is unavoidable because of the ongoing process in the human brain to "prune" unused synapses (the connections between neurons that are formed and activated in storing information), I knew that adults who did not use arithmetic routinely would likely forget some of the 99 subskills that constitute the subject. So to teach arithmetic so it would not wither away in the minds of adults, I wanted another way of presenting the subject, and the only way aside from memorizing it was to understand it. Accordingly, I imagined a second edition of *The Arithmetic Primer*, except in terms of what about arithmetic must be **understood** versus **known** (remembered/memorized) to do well in the subject.

With the book in hand, I could answer my question about teaching arithmetic with understanding. All I had to do was answer it for each of the 99 subskills that I had identified that make up arithmetic. With that in mind, I began by answering it for the following nine items on the pretest in *The Arithmetic Primer* for adding whole numbers.

Pretests for Adding, Subtracting, Multiplying, and Dividing Whole Numbers

Pretest for Adding Whole Numbers

1. $3 + 4 = ?$ 2. $5 + 8 = ?$ 3. $7 + 9 + 6 = ?$

4. 37	5. 46	6. 276	7. 473	8. 487	9. 31098
+ 12	+ 37	+ 352	+ 289	29	28736
				+ 315	+ 10596

As you can see, the items on the pretest for adding whole numbers correspond to the subskills for adding whole numbers. The first item ($3 + 4 = ?$) corresponds to the first subskill (Addition Facts, No Trading), the second item ($5 + 8 = ?$) corresponds to the second subskill (Addition Facts, Trading Ones for Tens), the third item ($7 + 9 + 6 = ?$) corresponds to the third subskill (Three Single-digit Addends, Trading Ones for Tens), and so on. To begin, I started by answering the question for $3 + 4 = ?$, the first item on the pretest for adding whole numbers.

Assuming that the numbers themselves are meaningful and represent amounts in the minds of children, as established by numerous counting and ordering activities, ...

○ The equal sign (=) must be **understood** as meaning "balanced" or "is the same as," not "tell me the answer," as on a calculator.

○ The addition sign (+) must be **understood** as signifying a combining action.

○ The addition fact 3 + 4 = 7 must be **known**. Children must remember it or know how to count it out or figure it out mentally.

I then answered the question for 5 + 8 = ?, the second item on the pretest for adding whole numbers, and found one more thing that must be understood and one more thing that must be known.

○ How to add (combine numbers) in a numeration system with place value that is *based* (built) on the number ten must be **understood**, that is, *base ten numeration must be understood* (which means the answer 13 to 5 + 8 must be understood as 10 + 3), and the addition fact 5 + 8 = 13 must be **known**.

Next, I scanned the remaining items on the pretest for adding whole numbers and found that nothing else had to be understood. All that changed from item to item were the columns in which trading ("carrying" or regrouping) would occur[52] and the addition facts that had to be known. Once children learn that to add in base ten, *"ten the same must be traded for one of the next bigger thing,"* the column in which trading occurs becomes irrelevant. *Thus all that must be understood to solve 2347860159 + 8906153427 is no different than what must be understood to solve 5 + 8!*

Continuing, I answered what must be understood and what must be known to solve the items on the pretests in *The Arithmetic Primer* for subtracting, multiplying, and dividing whole numbers.

Pretest for Subtracting Whole Numbers

1. 7 – 3 = ?   2. 14 – 8 = ?

3. 746	4. 73	5. 537	6. 752	7. 52371	8. 70500
– 425	– 28	– 154	– 386	– 14796	– 32954

Pretest for Multiplying Whole Numbers

1. 8 × 7 = ?

2. 34	3. 284	4. 152	5. 247	6. 13248	7. 3615	8. 7083	9. 42859	10. 7814
× 2	× 3	× 4	× 3	× 5	× 24	× 235	× 35000	× 2006

---

[52] For example, in solving the pretest item for subskill 4 for adding whole numbers (46 + 37), the trading is in the ones column, whereas in solving the pretest item for subskill 5 for adding whole numbers (276 + 352), the trading is in the tens column.

Pretest for Dividing Whole Numbers

$$1. \; 42 \div 6 = \, ?$$

2. $7\overline{)86}$ 3. $6\overline{)452}$ 4. $4\overline{)5126}$ 5. $5\overline{)25308}$ 6. $39\overline{)846}$ 7. $48\overline{)9723}$ 8. $801\overline{)65902}$

9. Determine $510 \div 24$ and write the remainder as a fraction. Simplify if you know how.

10. Determine $317 \div 7$ to three decimal places.

As you can see in scanning the items on the pretests for these operations, all that changes from operation to operation is the meaning of the operation and the number facts that must be known for it. For instance, all that changes from adding whole numbers to subtracting whole numbers is that subtraction must be understood and the subtraction facts must be known. Thus to understand whole number arithmetic, all that must be understood and all that must be known are the following, which are called Golden Keys in MOVE IT Math because they unlock the doors to success in arithmetic:

○ Golden Key #1: The equal sign (=) must be **understood** as meaning "balanced" or "is the same as" in math, not "tell me the answer," as on a calculator.

○ Golden Key #2: The signs for the four operations must be **understood** as signifying combining or separating actions. The addition sign (+) signifies combining any amounts. The subtraction sign (−) signifies separating an amount into any two amounts. The multiplication sign (x) signifies combining *equal* amounts. And the division sign (÷) signifies separating an amount into *equal* amounts.

○ Golden Key #3: All 390 number facts must be **known** (remembered/memorized) or **accessible** by counting them out or figuring them out mentally.

○ Golden Key #4: To add, subtract, multiply, and divide whole numbers with understanding, base ten numeration must be **understood**, which amounts to knowing that *"ten the same must be traded for one of the next bigger thing and vice versa."*

Thus all 37 subskills that constitute whole number arithmetic in *The Arithmetic Primer* can be taught by teaching children four things: 1) how the equal sign means balanced or "is the same as," 2) how to read (interpret) arithmetic problems as combining and separating events, 3) how to count out or figure out a number fact if forgotten or unsure of it, and 4) how to make and record trades in a numeration system with place value that is based on the number 10.

Origin of the Golden Key that Unlocks the Doors to Success with Fractions, Decimals, and Percent

Having answered what must be understood versus known for whole number arithmetic, I determined the same for fractions. As shown on the next page, *The Arithmetic Primer* dissects adding, subtracting, multiplying, and dividing fractions into 18 component subskills, each of which is tested for in the following pretests for adding, subtracting, multiplying, and dividing fractions.

Subskills for Adding, Subtracting, Multiplying, and Dividing Fractions

Adding Fractions

Pretest

1. Adding Fractions with Equal Denominators
2. Adding Fractions with Unequal Denominators
3. Adding Mixed Numbers, No Trading
4. Adding Mixed Numbers with Trading

Post Test

Subtracting Fractions

Pretest

1. Subtracting Fractions with Equal Denominators
2. Subtracting Fractions with Unequal Denominators
3. Subtracting Mixed Numbers, No Trading
4. Subtracting Mixed Numbers with Trading

Post Test

Multiplying Fractions

Pretest

1. Multiplying Fractions
2. Multiplying Fractions and Whole Numbers
3. Multiplying Fractions and Mixed Numbers
4. Multiplying Whole Numbers and Mixed Numbers
5. Multiplying Mixed Numbers

Post Test

Dividing Fractions

Pretest

1. Dividing Fractions
2. Dividing Fractions and Whole Numbers
3. Dividing Fractions and Mixed Numbers
4. Dividing Whole Numbers and Mixed Numbers
5. Dividing Mixed Numbers

Post Test

Pretest for Adding Fractions and Mixed Numbers

1. $\frac{1}{8} + \frac{3}{8} = ?$ 2. $\frac{2}{3} + \frac{4}{5} = ?$ 3. $3\frac{1}{3} + 2\frac{1}{6} = ?$ 4. $2\frac{1}{2} + 3\frac{4}{5} = ?$

Pretest for Subtracting Fractions and Mixed Numbers

1. $\frac{5}{6} - \frac{1}{6} = ?$ 2. $\frac{3}{8} - \frac{1}{5} = ?$ 3. $3\frac{2}{3} - 1\frac{1}{4} = ?$ 4. $5\frac{1}{2} - 2\frac{7}{8} = ?$

Pretest for Multiplying Fractions and Mixed Numbers

1. $\frac{2}{3} \times \frac{3}{4} = ?$ 2. $2 \times \frac{5}{6} = ?$ 3. $\frac{2}{3} \times 3\frac{2}{3} = ?$ 4. $3\frac{5}{8} \times 4 = ?$ 5. $1\frac{1}{2} \times 2\frac{1}{5} = ?$

Pretest for Dividing Fractions and Mixed Numbers

1. $\frac{3}{4} \div \frac{1}{2} = ?$ 2. $\frac{4}{5} \div 2 = ?$ 3. $4\frac{2}{3} \div \frac{7}{8} = ?$ 4. $3\frac{1}{4} \div 5 = ?$ 5. $3\frac{1}{8} \div 4\frac{2}{5} = ?$

To add, subtract, multiply, and divide fractions, one computes with their numerators and denominators, which are whole numbers, so nothing new needs to be understood or known to do that. However, <u>fraction notation</u>, <u>equivalent fractions</u>, and <u>the rules for computing with fractions</u> need to be **understood**. To that end, the following must be **known** in order to find common denominators and reduce fractions to their lowest terms: Any non-zero number divided by itself equals the number 1, and the number 1 times any number equals that number. (Symbolically, for $N \neq 0$, $N \div N = N/N = 1$, and for all N, $1 \times N = N \times 1 = N$.)

For instance, to reduce a fraction, you were probably taught to divide the numerator and denominator with a number that would divide them evenly. For example, to reduce 8/12, you were probably taught to divide the 8 and 12 by 4, resulting in ⅔. In other words, you were taught to "cancel" the number 1 written 4/4, as shown below:

$$\frac{8}{12} = \frac{4 \times 2}{4 \times 3} = \frac{4}{4} \times \frac{2}{3} = 1 \times \frac{2}{3} = \frac{2}{3}$$

Understanding Fractions and Fraction Notation

For children to understand fractions and fraction notation, they must see and handle fractional parts of things. Beginning in kindergarten, every child should have or have access to a set of fraction cakes from halves to sixteenths made out of cardstock or a commercially available set of fraction cakes made out of plastic from halves to twelfths. Fraction cakes that only show from halves to sixths or eighths are too limited in the number of problems that can be solved with them.

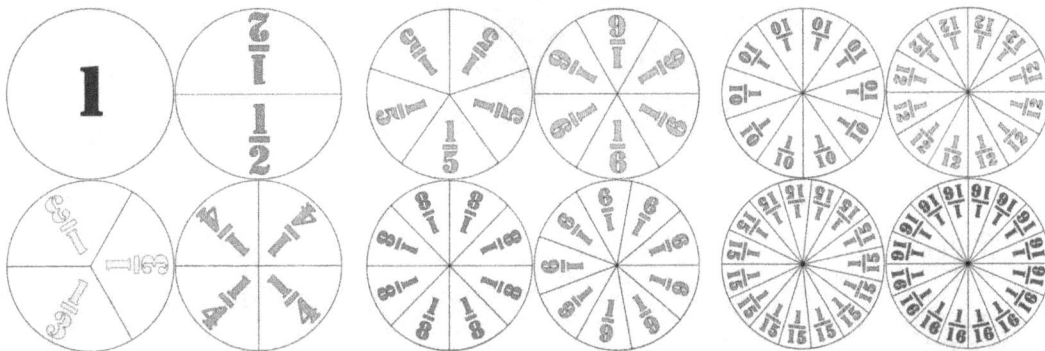

With the fraction cakes, children can literally *see* that fractions are equal parts of things, that the denominator (bottom number) of a fraction denotes the number of parts, and that the numerator (top number) of a fraction denotes the number of **parts** under consideration. For example, 7/12 signifies the shaded portion of the circle below that has been partitioned into 12 equal parts.

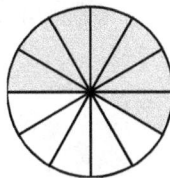

Understanding Equivalent Fractions

To understand equivalent fractions with the fraction cakes, children match fractional parts of them in accordance with the following rule: If the fraction cakes are differentiated by color, only those that are the same color may be combined to make a match; otherwise, only those with the same denominator may be combined to make a match. The following are some of the matches that can that can be made with the fraction cakes that are equivalent to ½.

Not okay would be a match like the following:

Understanding the Rules for Computing with Fractions

To understand the rules for computing with fractions, children solve fraction problems with the fraction cakes. They acquire answers that they can literally *see* are correct. They then examine the problems they solved to determine what they could have done with just the numbers in the problems to get the same answers. In so doing, children can discover and verify all four non-intuitive rules for computing with fractions, as shown with the following task cards. For children still learning to read, their teacher can direct them through the activities on the task cards.

Adding and Subtracting Fractions with Like Denominators

Solve ⅛ + ⅜ with the fraction cakes. On the paper provided, draw a picture to show how you solved it. Then, show how to get the answer you got with the fraction cakes with just the numbers in the problem. Repeat for ⅚ – ⅙. Make a rule on how to add and subtract fractions with the same denominators.

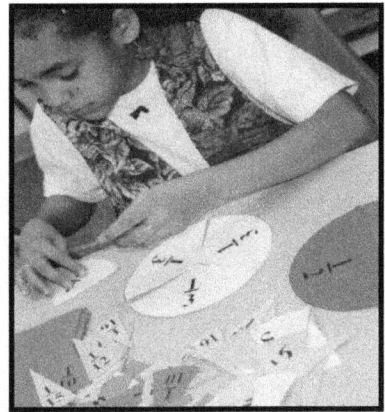

Adding and Subtracting Fractions with Like Denominators

Name **Listening**

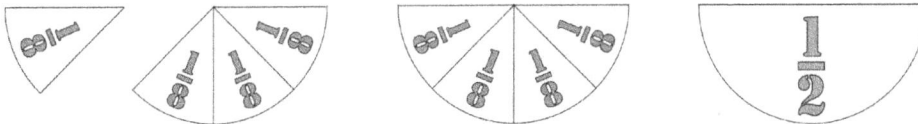

$$\frac{1}{8} + \frac{3}{8} = \frac{4}{8} = \frac{1}{2}$$

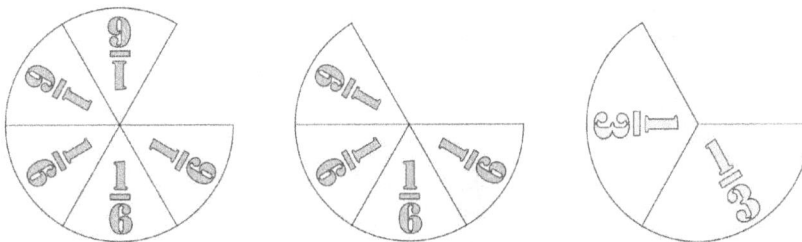

$$\frac{5}{6} - \frac{1}{6} = \frac{4}{6} = \frac{2}{3}$$

Rule: **Add or subtract the numerators but not the denominators. Put the answers over the denominators. So 1/8 + 3/8 = (1 + 3)/8 = 4/8 = 1/2, and 5/6 – 1/6 = (5 – 1)/6 = 4/6 = 2/3.**

Adding and Subtracting Fractions with Unlike Denominators

Solve ⅓ + ¼ with the fraction cakes. On the paper provided, draw a picture to show how you solved it. Then, show how to get the answer you got with the fraction cakes with just the numbers in the problem. Repeat for 2/5 – 3/10. Make a rule on how to add and subtract fractions with different denominators.

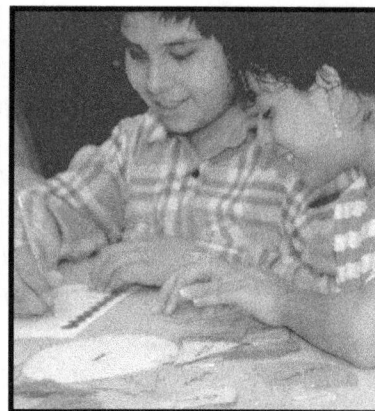

Adding and Subtracting Fractions with Unlike Denominators

Name **Thinking**

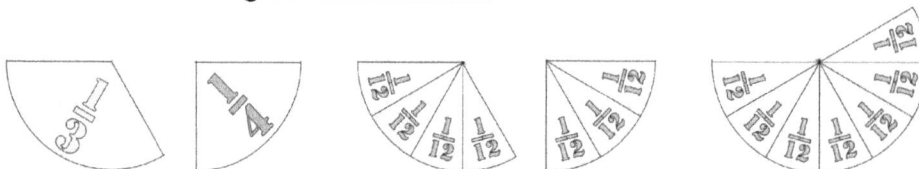

$$\frac{1}{3} + \frac{1}{4} = \frac{4}{12} + \frac{3}{12} = \frac{7}{12}$$

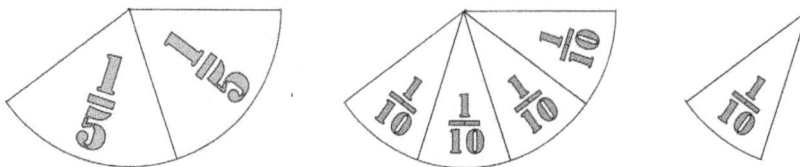

$$\frac{2}{5} - \frac{3}{10} = \frac{4}{10} - \frac{3}{10} = \frac{1}{10}$$

Rule: Make the denominators the same. Then add or subtract the numerators but not the denominators. Put the answers over the like denominators. So 1/3 + 1/4 = 4/12 + 3/12 = (4 + 3)/12 = 7/12, and 2/5 – 3/10 = 4/10 – 3/10 = (4 – 3)/10 = 1/10.

Multiplying Fractions

Solve ⅔ x ¾ (read *"How much is two-thirds of 3/4?"*) and ½ x ⅗ (read *"How much is half of 3/5?"*) with the fraction cakes. On the paper provided, draw a picture to show how you solved them. Then, show how to get the answers you got with the fraction cakes with just the numbers in the problems. Make a rule for multiplying fractions.

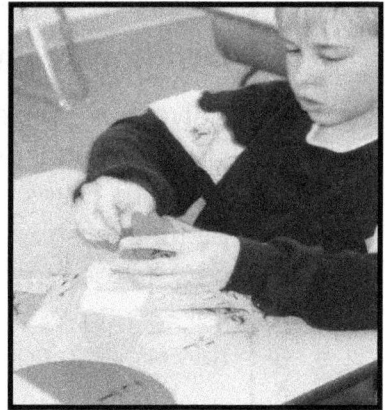

Multiplying Fractions

Name _Questioning_

Two-thirds of 3/4 is 2/4, which is 1/2, as you can see. So 2/3 x 3/4 = 1/2.

Half of 3/5 is half of 6/10, which is 3/10, as you can see. So 1/2 x 3/5 = 3/10.

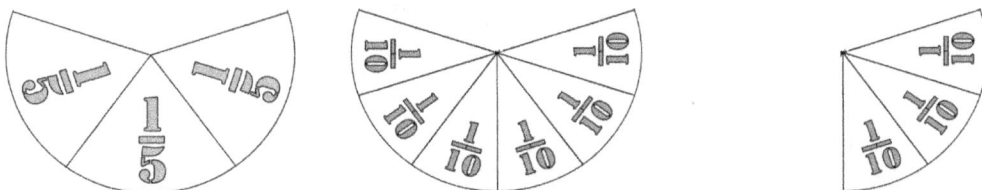

Rule: Multiply the top numbers and the bottom numbers. Put the answer for the top numbers over the answer for the bottom numbers. So 2/3 x 3/4 = (2 x 3)/(3 x 4) = 6/12 = 1/2, and 1/2 x 3/5 = (1 x 3)/(2 x 5) = 3/10.

Dividing Fractions

Solve 1 ÷ ¹⁄₁₀ (read *"How many tenths in 1?"*) and ¾ ÷ ½ (read *"How many halves in ¾?"*) with the fraction cakes. On the paper provided, draw a picture to show how you solved them. Then, show how to get the answers you got with the fraction cakes with just the numbers in the problems. Make a rule for dividing fractions.

Dividing Fractions

Name **Learning**

There are ten tenths in 1, as you can see. So 1 ÷ 1/10 = 10.

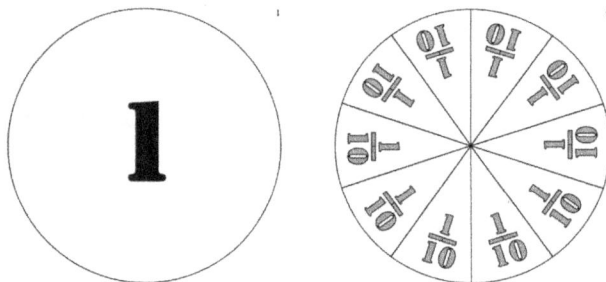

There are one and a half halves in ¾, as you can see. So ¾ ÷ ½ = 1 ½.

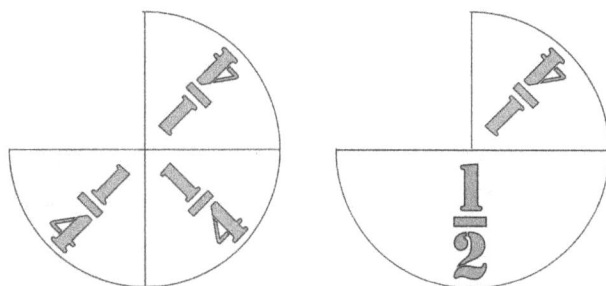

Rule: *"Ours is not to reason why; just invert [the divisor] and multiply."* Author unknown. So 1 ÷ 1/10 = 1 x 10/1 = 10, and 3/4 ÷ 1/2 = 3/4 x 2/1 = 6/4 = 3/2 = 1-1/2. Why the divisor is turned upside down and becomes a multiplier is explained in the appendix.

Thus the Golden Key for understanding fraction notation, equivalent fractions, and the rules for computing with fractions is adding, subtracting, multiplying, and dividing *real* fractions — fraction cakes from halves to twelfths or sixteenths.[53]

As for understanding decimals, nothing new must be understood because they are a logical consequence of base ten numeration. Adding with base ten blocks is guided by the rule that ten blocks the same must be traded for one of the next bigger block, and subtracting with the blocks is guided by the opposite of that rule, that any base ten block may be traded for ten of the next smaller blocks. Thus a base ten unit block may be traded for ten of the next smaller blocks, which would be blocks one-tenth the size of the unit block, and a one-tenth block may be traded for ten of the next smaller blocks, which would be blocks one-hundredth the size of the unit block, and so on.

The decimal point is a marker. Numbers to the left of it are whole numbers, and numbers to the right of it are fractions whose denominators are powers of ten (10, 100, 1000, and so on). With that in mind, the following equivalences between decimals and fractions become apparent: $0.1 = 1/10$, $0.01 = 1/100$, $0.001 = 1/1000$, and so on, which imply, for instance, that $0.3 = 3/10$, $0.24 = 24/100$, and $0.057 = 57/1000$.[54] As for computing with decimals, they are added, subtracted, multiplied, and divided as if they were whole numbers, but where to put the decimal point in the answers must be known, which is easily taught.

As for understanding percent, nothing new must be understood because a percent is a fraction. The word "percent" literally means "per one hundred," like "per the number of *cents* in a dollar." Thus 50 percent (50%), for example, means 50 per one hundred, which equals 50/100, which equals 5/10, which equals 0.5 and 1/2.

Percents common-size fractions as parts per 100. Thus how to turn a fraction into hundredths must be known. To do so, divide the numerator by the denominator out to three decimal places and round off to two decimal places. Then read the answer as so many "hundredths" and substitute the word "percent" for the word "hundredths."

For example, to turn 5/12 into a percent, divide 5 by 12 out to three decimal places ($5 \div 12 = 0.416$) and round off to two decimal places (0.42). Then read 0.42 as 42 "hundredths" and substitute the word "percent" for the word "hundredths." Thus $5/12 = 0.42 = 42$ "hundredths" = 42 "percent" = 42%. In other words, with a calculator, enter 5, tap the division key, enter 12, tap the equals sign, round off to two decimal places, and so on as just shown.

---

[53] Colored rods that are 1 cm, 2 cm, 3 cm … 10 cm long can also be used to represent fractions by making the ten rod the unit rod, in which case the five rod, for example, would represent 5/10 or half of the ten rod; however, without the ten rod available for comparison, the fraction represented by the five rod cannot be determined. In contrast, regardless of its size, half of a circle is clearly half of a circle without the need to have the whole circle present for comparison.

[54] The zeros in front of the decimal points in the examples are optional. They serve no mathematical purpose. They are used in writing about decimals to draw a reader's attention to the decimal points.

Closing Remarks

To summarize, to understand arithmetic, all that needs to be understood are the four Golden Keys for whole number arithmetic and the Golden Key for fractions: adding, subtracting, multiplying, and dividing fractions with *real* fractions, like the fraction cakes from halves to twelfths or sixteenths. If you visit the MOVE IT Math website @ moveitmath.com, you will see that it presents arithmetic in terms of the five Golden Keys just delineated: Equals (Golden Key #1), ASMD Actions (Golden Key #2), Number Facts (Golden Key #3), Fair Trades (Golden Key #4), and Real Fractions (Golden Key #5).

Note: MOVE IT Math guides students to discover the rules for adding, subtracting, multiplying, and dividing fractions but does not require them to memorize them. It acknowledges that the rules will probably be forgotten if not used often because they are counterintuitive. Accordingly, it teaches students how to "remember" the rules if/when they forget them. It has students pretend they have forgotten the rules and has them draw pictures of the fraction cakes that will remind them of the rules.

For instance, if not sure of the rule for how to add fractions with like denominators, it teaches them to draw two halves, thirds, or fourths of a circle and draw them combined, as shown below for two halves. As you can see, ½ + ½ = two halves = 2/2, not 2/4, so the rule is to add just the numerators when the denominators are the same.

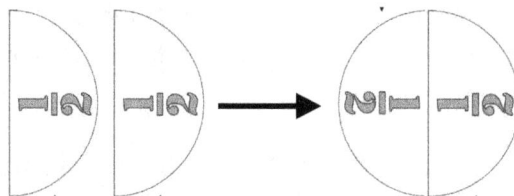

Teaching children to understand arithmetic is achieved by asking questions to get them to think about what they know until they connect it with something that their teacher wants them to know. For teachers to do that, they, themselves, must understand arithmetic. MOVE IT Math provides teachers with that understanding. It shows them how to actually *teach* arithmetic — *how to instill it in their student's minds* — instead of just talking about it.

References

Bruner, Jerome. *The Process of Education, A Landmark in Educational Theory*, Harvard University Press, Cambridge, Massachusetts and London, England, 1960.

Shaw, Bryce. *Individualized Computational Skills Program*, Houghton Mifflin, 1972.

Shoecraft, Paul. *The Arithmetic Primer*, Addison-Wesley Publishing Company, Inc., 1979.

CHAPTER 12: Project GRAD

Project GRAD [Graduation Really Achieves Dreams] was founded by Jim and Kathryn Ketelsen on a vision of equity — that ALL youth deserve access to an education that leads to a rewarding future. GRAD's belief is that all individuals can improve their lives when they gain access to guidance, resources, and ongoing support to develop and achieve goals for post-secondary attainment or workforce training. — GRADcafé, 2022

In 1988, Davis High School was the lowest performing high school in Houston, Texas. Only ten percent of its graduating class enrolled in college, but Jim Ketelsen was about to change that. He attributed the low enrollment to the vast majority of Davis students being from low-income families. Accordingly, while a Regent of the University of Houston and the CEO of Tenneco, a Houston-based conglomerate, he used his influence with the university and the business community to initiate and direct a program to provide monetary incentives from local businesses for Davis students to graduate and enroll in college. His goal was to increase the number of enrollees to at least half the size of their class when they began high school in the ninth grade.

Within three years of the start-up of Mr. Ketelsen's program, *the number of Davis graduates enrolling in college had quadrupled,* but that was not enough to meet his goal. The number of Davis graduates enrolling in college was still less than half the size of their class when they began high school. Nonetheless, the program was clearly working, so you might think that he would have stuck with it for a few more years until he reached his goal, but he realized that his program was limited in the number of students it could incentivize.

Financial incentives to graduate and attend college was a boon for students who began high school with the reading, writing, and math skills they needed to pass the college preparatory courses (like Algebra II) that most colleges require for entrance. Conversely, it was a bust for those who did not enter Davis with those skills because their counselors would advise them not to take college preparatory courses. Thus the number of Davis graduates who would be motivated by financial help to graduate and attend college was limited to the number of ninth graders who were proficient with the three Rs when they began high school: Reading, 'Riting, and 'Rithmetic. So for at least half of a graduating class to enroll in college, at least half of the class had to be proficient with the three Rs when they began high school, which was not happening at Davis at the time.

The Project

Mr. Ketelsen retired from Tenneco in 1992 and could have given up on Davis. Instead, he created Project GRAD, the same cash for college program as before except augmented with a K-8 program to improve the teaching of reading, writing, and math in the "feeder" schools for Davis High School — the seven elementary schools[55] and one middle school whose students would eventually "feed into" (attend) Davis. To make that happen, he used his friendships and connections with people like George H.W. and Barbara Bush to raise *millions* in contributions with which to fund the project. National supporters of Project GRAD included Continental Airlines, Ford Foundation, KnowledgeWorks Foundation, Lucent Technologies, U.S. Department of Education, and Verizon Foundation.

---

[55] **Jefferson** (preK-grade 5), **Lamar** (preK-grade 6), **Lee** (preK-grade 5), **Looscan** (preK-grade 5), **Martinez** (preK-grade 5), **Ryan** (preK-grade 6), **Sherman** (preK-grade 6)

Project GRAD consisted of five components: 1) scholarships to enroll in college, 2) social services/parental involvement, 3) classroom management, 4) literacy, and 5) mathematics. Each component is described below.

Scholarships

The promise of a $1,000 scholarship for each of four years to enroll in college for Davis graduates who met certain criteria was what energized Project GRAD from kindergarten on. To apprise the community served by Davis high school of the scholarships, Project GRAD initiated Walk for Success. Every year, hundreds of volunteers knocked on literally every door in the community to enlist community support for Project GRAD, explain the scholarships, and announce to parents that with their encouragement for their children to stay in school,[56] their children were on track to graduate from Davis and receive a Project GRAD scholarship to enroll in college and receive one-on-one help before and after enrolling to acquire financial assistance in order to complete college.

Compared to the cost of higher education even 30 years ago, $1,000 for each of four years to go to college may not seem like much, but, as pointed out earlier, within only three years of financial incentives to enroll in college after graduating, the number of Davis graduates enrolling in college had quadrupled. *Where there's a way, there's a will* — a will that could not be denied in spite of the rigorous eligibility requirements for a Project GRAD scholarship.

To receive a Project GRAD scholarship, a Davis student had to take the SAT (Scholastic Aptitude Test) as a sophomore or junior and meet the academic requirements of the Texas High School Recommended or Distinguished Graduation Plan, which required them to take a minimum in math of one year of geometry and two years of algebra (Algebra I and II) and graduate with a 2.5 overall grade point average (GPA). Additionally, they had to attend two Project GRAD six-week summer institutes on how to apply for college scholarships and enroll and succeed in college.

Social Services/Parental Involvement

This component was entrusted to Communities in Schools:

> *At Communities In Schools® (CIS®), we believe that every student, regardless of race, gender, ability, zip code, or socioeconomic background has what they need to realize their full potential in school and beyond. We walk by their side, in their communities, to challenge the systems and barriers that stand between them and their success in life.* — communitiesinschools.org/about-us.

CIS linked teachers, counselors, and parents with community resources, both public and private, that provided assistance to students with personal problems that affected their classroom performance. To establish that linkage, Project GRAD employed a full-time CIS social worker to reside in each Project GRAD school.

---

[56] The parental role is vital in ensuring the success of educational reform. When Davis graduates who earned a Project GRAD scholarship were asked to rank the top ten factors that motivated them to graduate from high school and pursue a college education, 71% of them identified parental encouragement as the strongest factor (Opuni, 1995).

Classroom Management

The classroom management component of Project GRAD was the responsibility of H. Jerome Freiberg, a professor in the Education Department at the University of Houston. Dr. Freiberg authored Consistency Management and Cooperative Discipline (CMCD), a nationally recognized classroom management program that is explained below. His job was to implement CMCD in the Davis feeder schools and Davis itself.

> *__Consistency Management__ focuses on classroom organization and planning by the teacher and other school staff. The teacher is trained to organize all classroom activities — from making seating arrangements, to passing out papers, sharpening pencils, taking attendance, and providing equal opportunity to participate in class — to create an orderly and supportive environment in which all students can participate and learn.*
>
> *__Cooperative Discipline__ ... [allows] students to share in the classroom management role of teachers and paraprofessionals. All students are given an opportunity to serve as leaders. As they progress through school, students assume responsibility for classroom management functions that range from passing out papers to assisting substitute teachers. Jobs are posted in the classroom, and students submit applications based on interest. Each position is rotated every four to six weeks. Students also are allowed to assume responsibility for resolving disputes, solving problems, and making decisions. In this way, students gain the experience necessary to become self-disciplined and to act as responsible citizens of the school community.*
> — American Federation of Teachers in recommending the program to its members, 1999

From prekindergarten through grade 12, Project GRAD students evolved under the same CMCD classroom management principles. The result was a school environment that developed mutual respect and consideration between students, teachers, staff, and administrators.

Literacy

The literacy component of Project GRAD was delivered with Success For All (SFA), a nationally recognized reading and writing program. SFA promoted a comprehensive restructuring of school resources to provide concentrated instructional time for reading. Its promise was that every child would be reading at the third grade level or higher by the end of the third grade, *and it was relentless in making that happen*. Every eight weeks, it assessed children's reading levels and provided one-on-one student-paced tutoring to assist those in need of it to move to the next level.

Mathematics

In 1993, Mr. Ketelsen met with me to discuss fulfilling the math component of Project GRAD with MOVE IT Math. As a former Regent of the University of Houston, he knew of the proven success of the program. At the time, I did not know about his initiative to use financial incentives to motivate high school students to take college preparatory courses at Davis High School and enroll in college. Nonetheless, as I listened to the comprehensiveness of his vision for the project, I enthusiastically agreed to make MOVE IT Math a part of it.

In learning about Project GRAD, I realized that my vision for disseminating MOVE It Math had been too narrow. My model had been to replace the existing elementary school math curriculum with the MOVE IT Math curriculum, which was like replacing a burnt out lightbulb with a new one. The new one lit up the classroom, but just during math time, *whereas Project GRAD's vision was to light up the classroom <u>and keep it lit</u>!* What started out as financial incentives for graduates of Davis High School to enroll in college became that plus a K-8 program to improve the teaching of math and reading in K-8.

For the next six years, I was responsible for the implementation of MOVE IT Math in the seven elementary schools and Marshall Middle School (grades 6-8) whose students would someday attend Davis High School. The arrangement was 1) that I would retain my position with the University of Houston – Victoria (UHV) but be assigned to work full-time with Project GRAD, 2) that UHV would offer both courses of MOVE IT Math at the Tenneco building in Houston a year apart for three hours of graduate credit each, and 3) that Project GRAD would cover the cost of enrolling teachers in the courses and equipping their classrooms with the hands-on materials they would need to implement what they learned in the courses. The arrangement was a win/win/win — a win for UHV, Project GRAD, and the Project GRAD teachers.

The arrangement was a win for UHV because it banked a sizable amount in tuition and fees for two three-hour graduate courses with about 200 elementary school teachers and middle school math teachers in each course. It was a win for Project GRAD because it got MOVE IT Math for its math component and my management of the component free because I was still employed by UHV. And it was a win for the Project GRAD teachers because they received six hours of university graduate credit that they could apply toward an advanced degree or use to satisfy the state requirement to update their teaching credentials with additional coursework every so often.

Project GRAD / MOVE IT Math Results in Houston, Texas

Project GRAD 2002: Math Success Closing the Gap, a Project GRAD PowerPoint presentation. The following results from the presentation show that after Project GRAD was initiated in the Davis, Yates, and Wheatley high school feeder schools in Houston beginning with Davis in 1994, the number of students passing the TAAS (Texas Assessment of Academic Skills), the state achievement test for math, increased substantially.

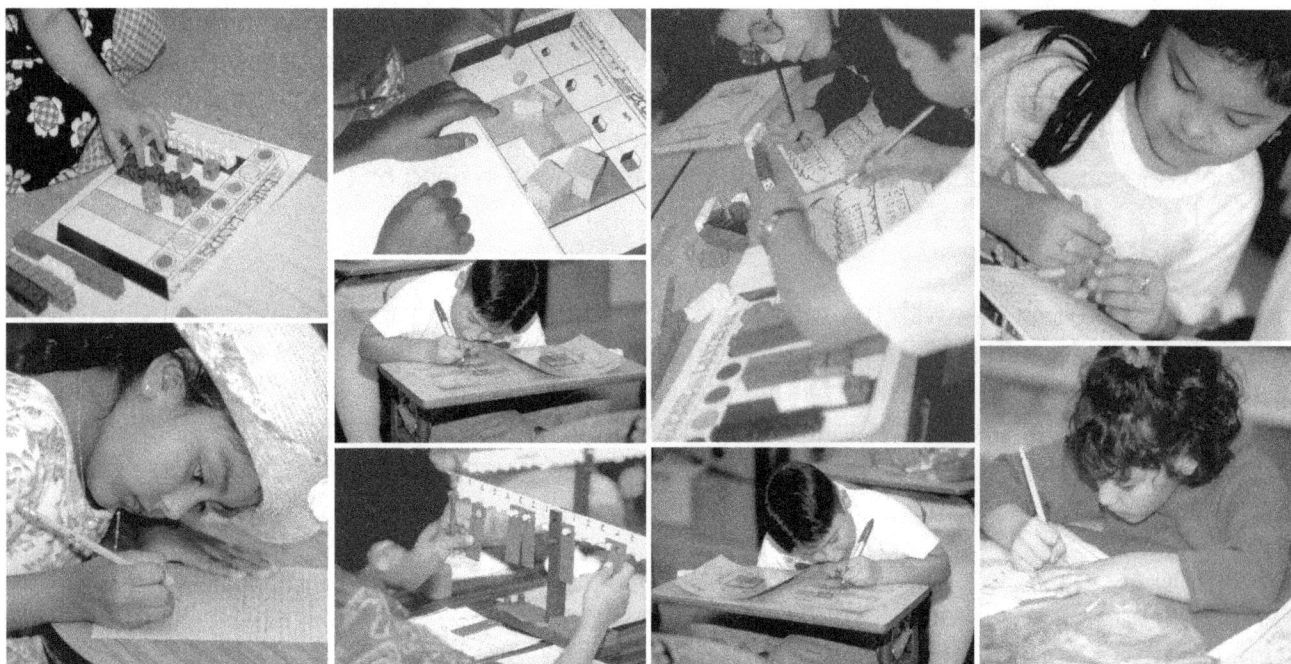

Texas - Houston ISD - Davis High School Feeder System
Fifth Grade TAAS Math Results: 1999-2002
All Students Not in Special Education

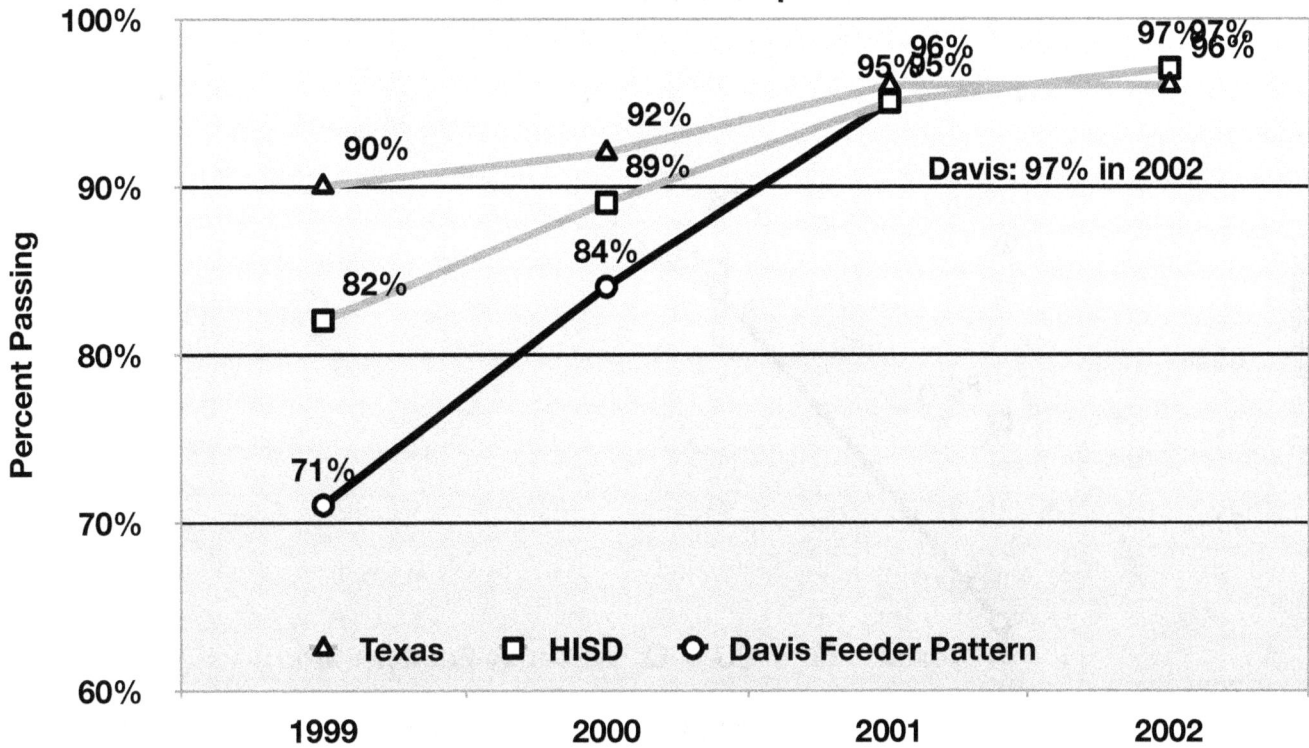

Davis: 97% in 2002

△ Texas　□ HISD　○ Davis Feeder Pattern

90% 82% 71% 92% 89% 84% 95% 95% 96% 97% 97% 96%

Texas - Houston ISD - Yates High School Feeder System
Fifth Grade TAAS Math Results: 1999-2002
All Students Not in Special Education

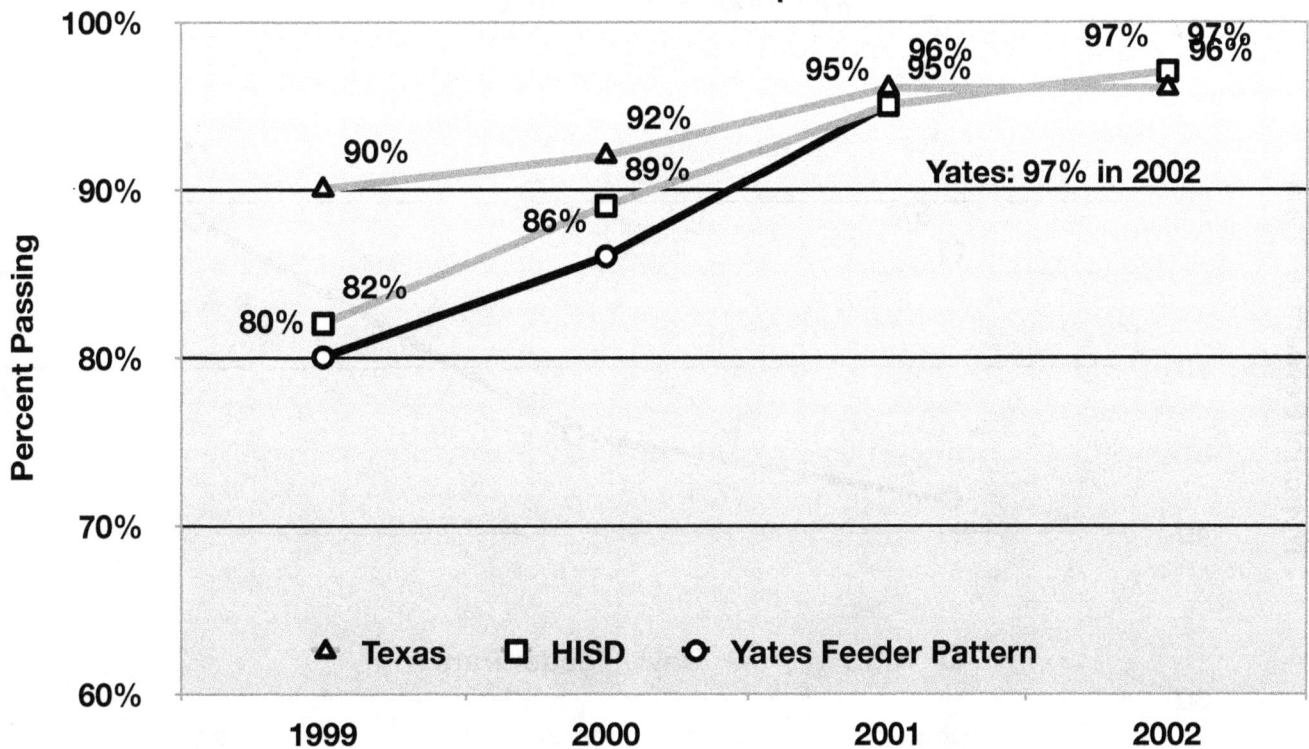

Yates: 97% in 2002

△ Texas　□ HISD　○ Yates Feeder Pattern

90% 82% 80% 92% 89% 86% 95% 95% 96% 97% 97% 96%

211

Texas - Houston ISD - Wheatley High School Feeder System
Fifth Grade TAAS Math Results: 1999-2002
All Students Not in Special Education

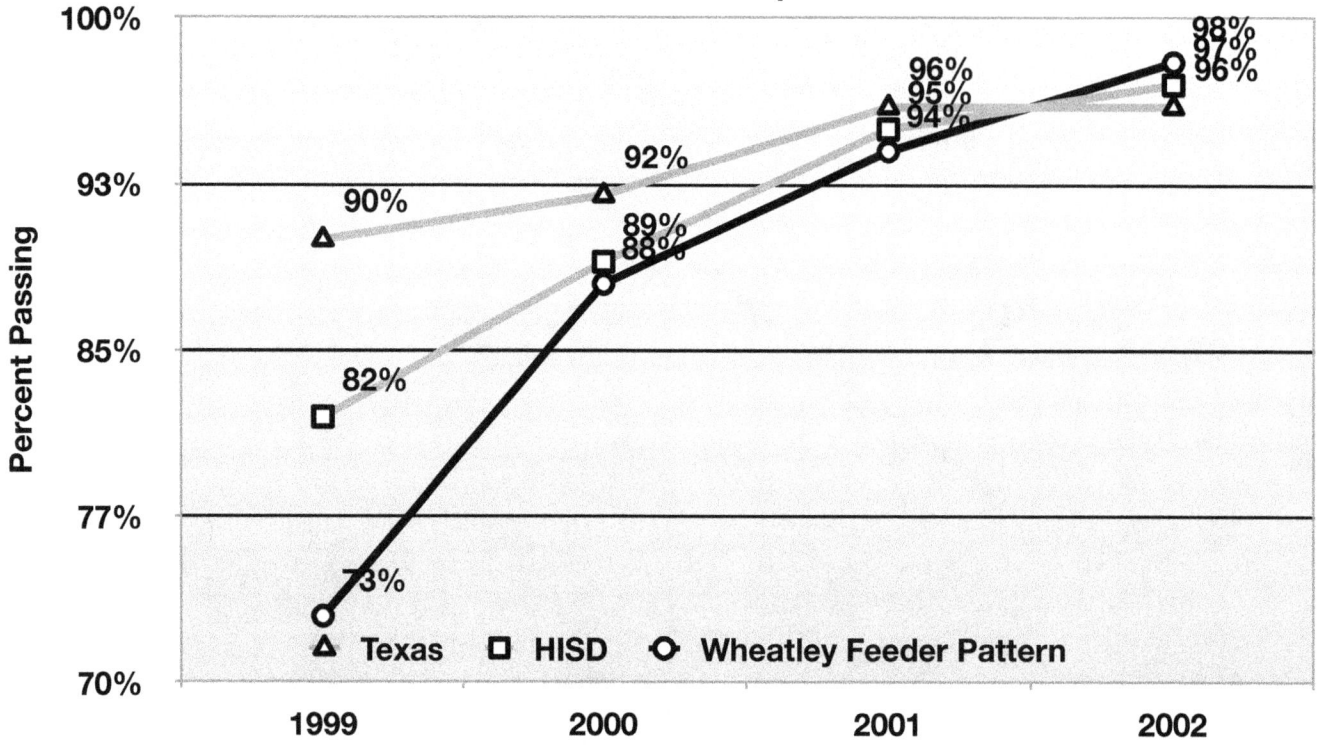

Percent Passing

100%

93%

85%

77%

70%

98%
98%
97%
96%

96%
95%
94%

92%

90%
89%
88%

82%

73%

△ Texas   ☐ HISD   ◇ Wheatley Feeder Pattern

1999   2000   2001   2002

Houston ISD vs. Davis High School Feeder System
Fifth Grade STANFORD 9 Math Results: 1999-2001
All Students Not in Special Education

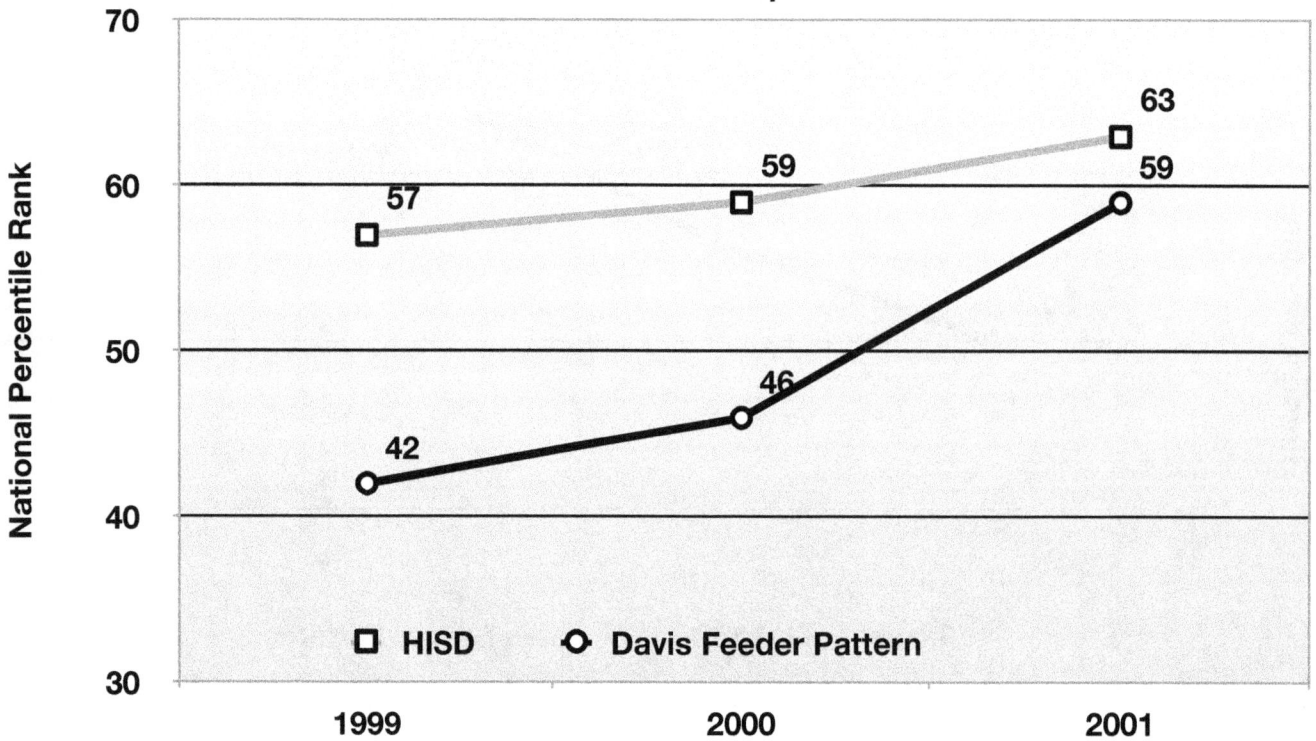

National Percentile Rank

70

60

50

40

30

63
59

59

57

46

42

☐ HISD   ◇ Davis Feeder Pattern

1999   2000   2001

Houston ISD vs. Yates High School Feeder System
Fifth Grade STANFORD 9 Math Results: 1999-2001
All Students Not in Special Education

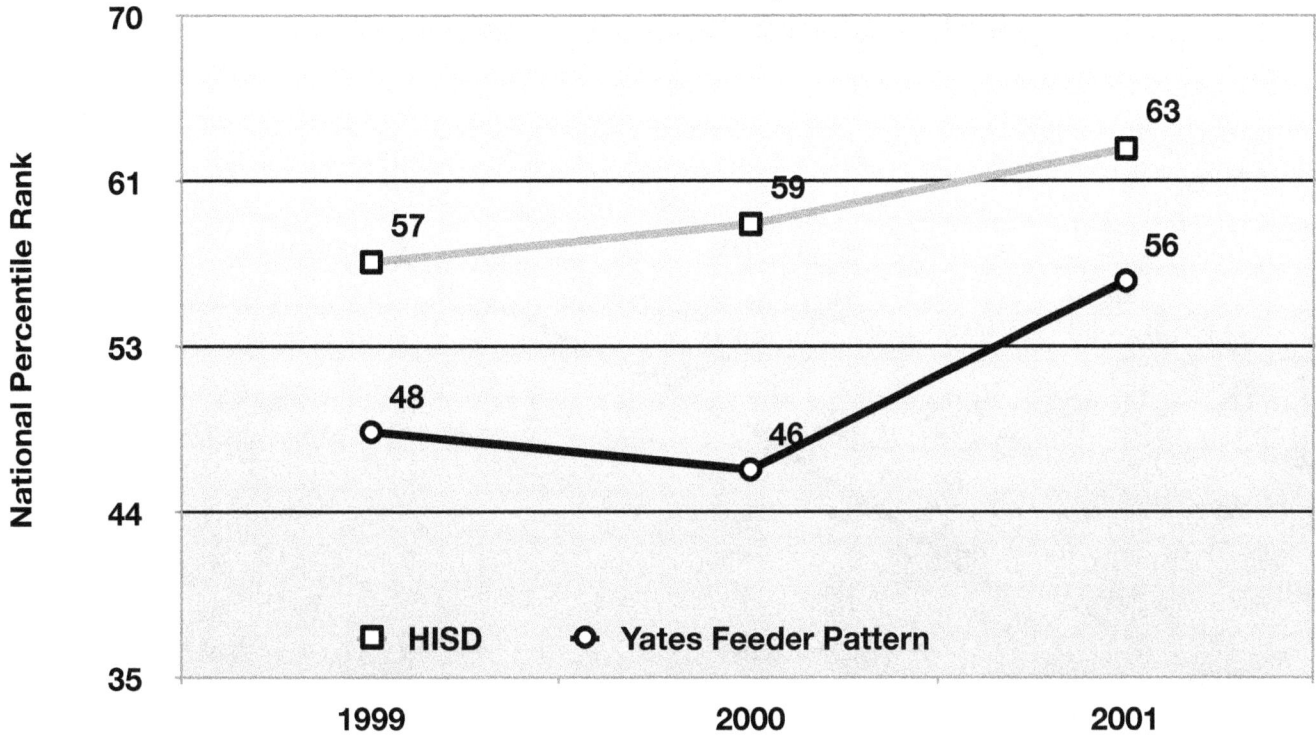

Houston ISD vs. Wheatley High School Feeder System
Fifth Grade STANFORD 9 Math Results: 1999-2001
All Students Not in Special Education

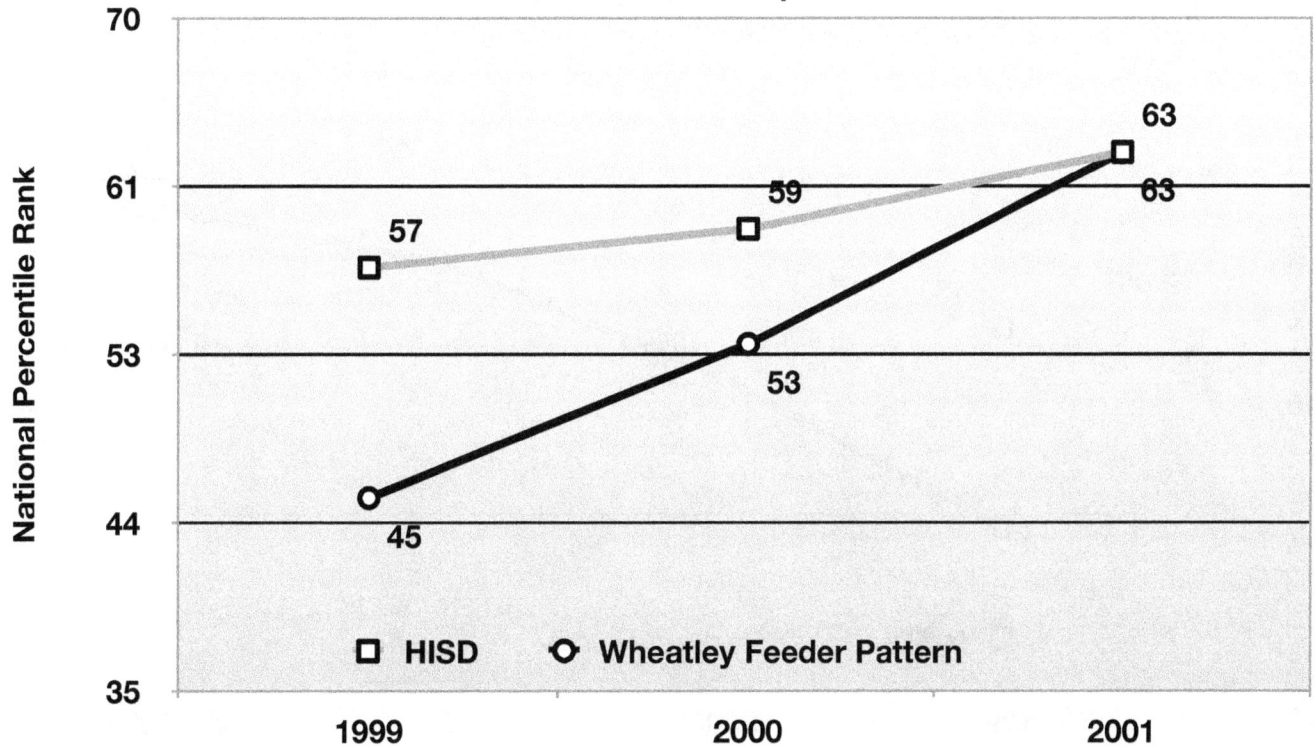

Texas vs. HISD vs. Davis, Yates, and Wheatley High School Feeder Systems
TAAS Math Results: 1999-2002
Grades 3-5, All Students

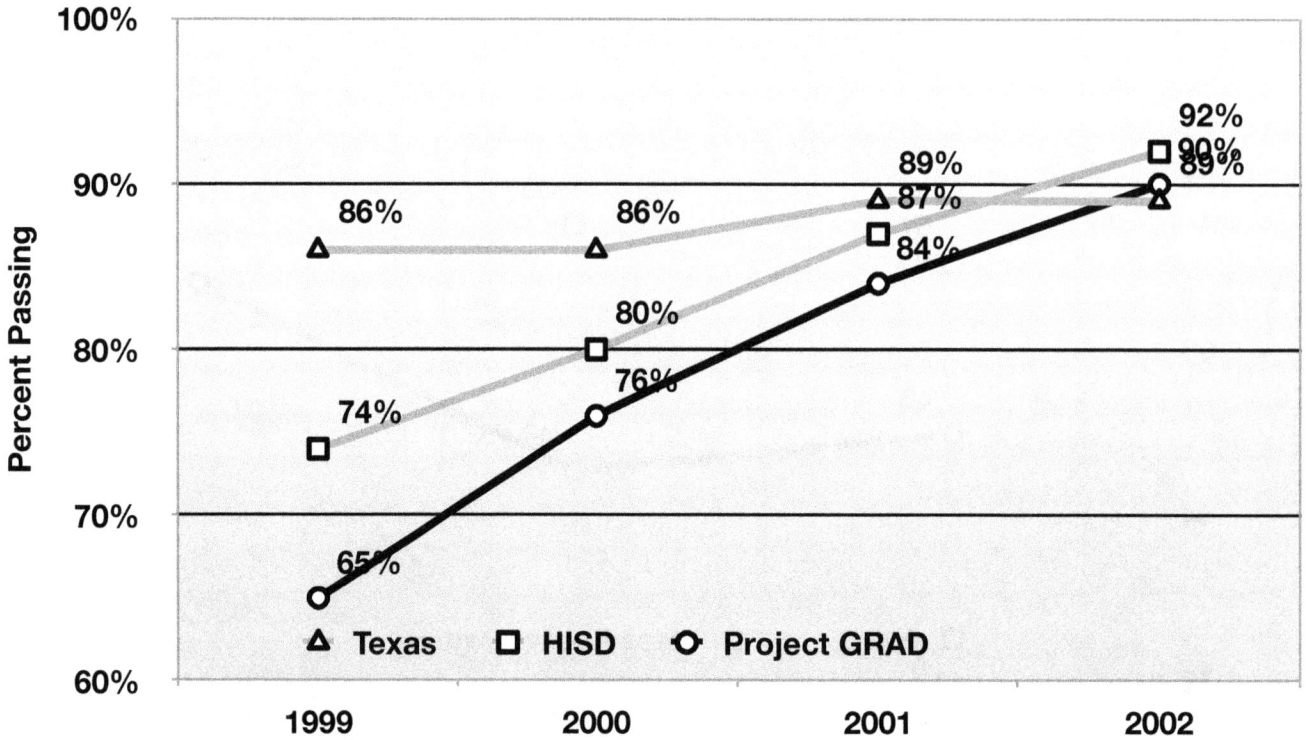

Houston ISD vs. Davis, Yates, and Wheatley High School Feeder Systems
STANFORD 9 Math Results: Percent at or Above 50th Percentile Rank Nationally
Grades 1-5, All Students Not in Special Education

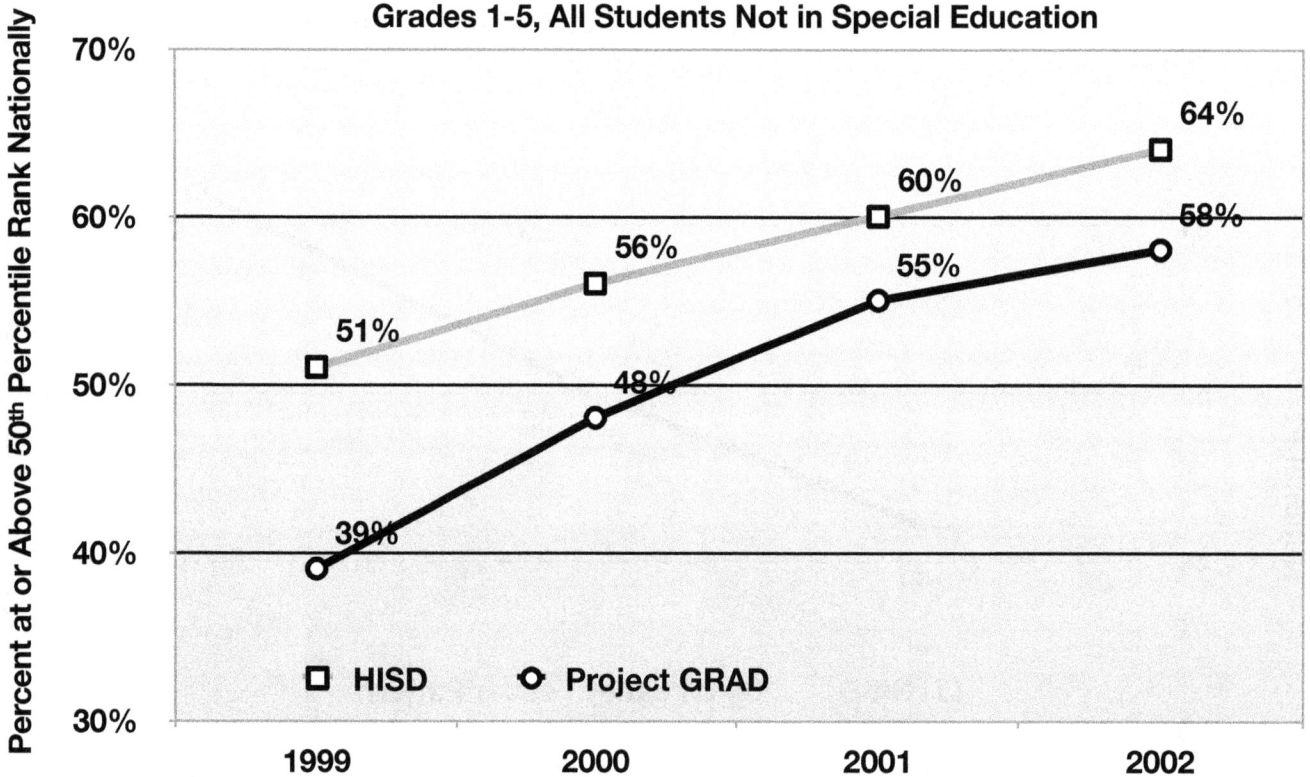

Davis High School Feeder School System
All Elementary Schools
TAAS Math Results for All Students Not in Special Education
1994-2002

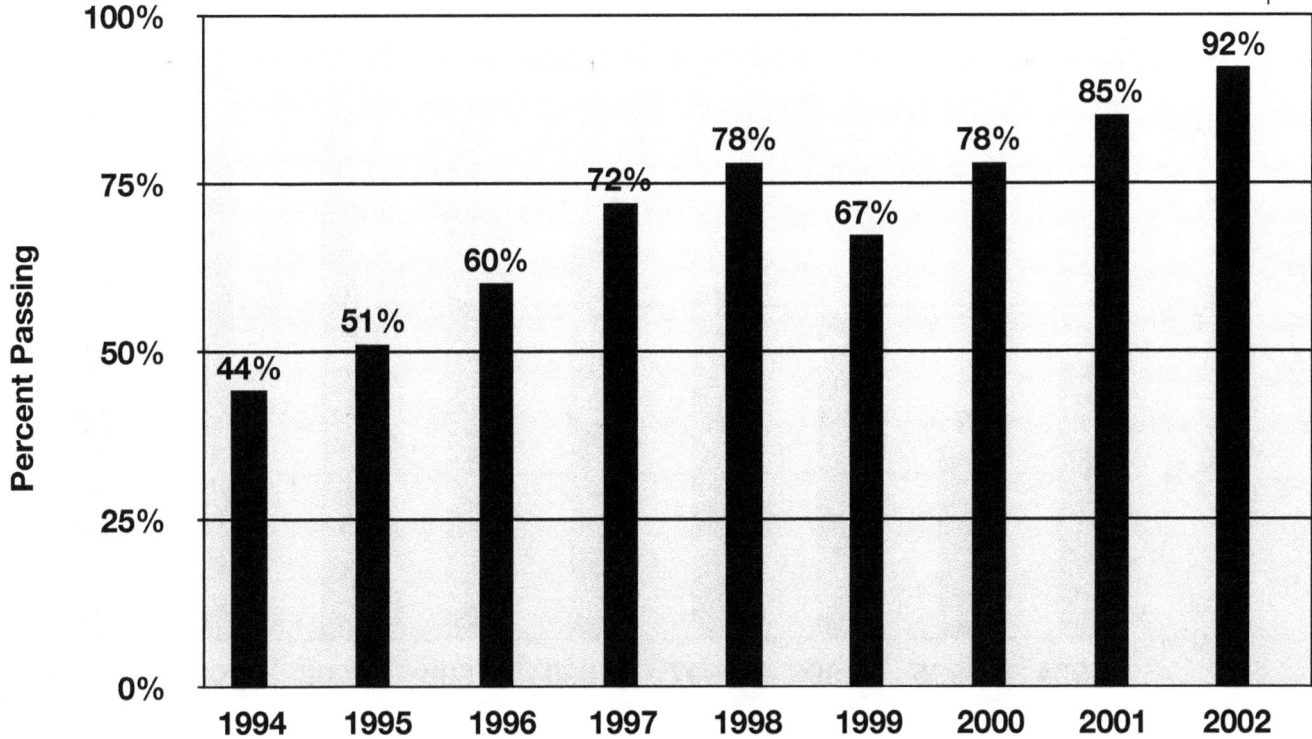

Davis High School Feeder School System
Marshall Middle School (Grades 6, 7, 8)
TAAS Math Results for All Students Not in Special Education
1994-2002

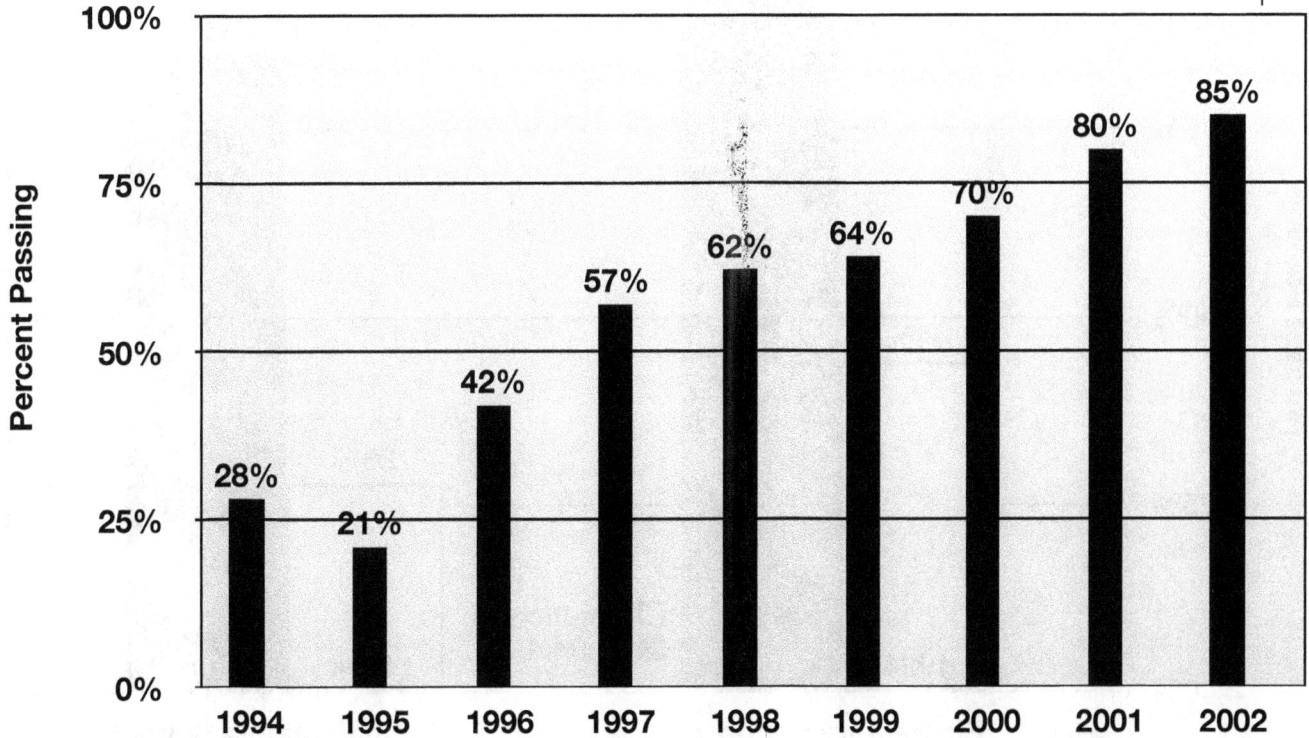

TAAS Math Results · Davis High School
Exit Level (Grade 10)
1994-2002

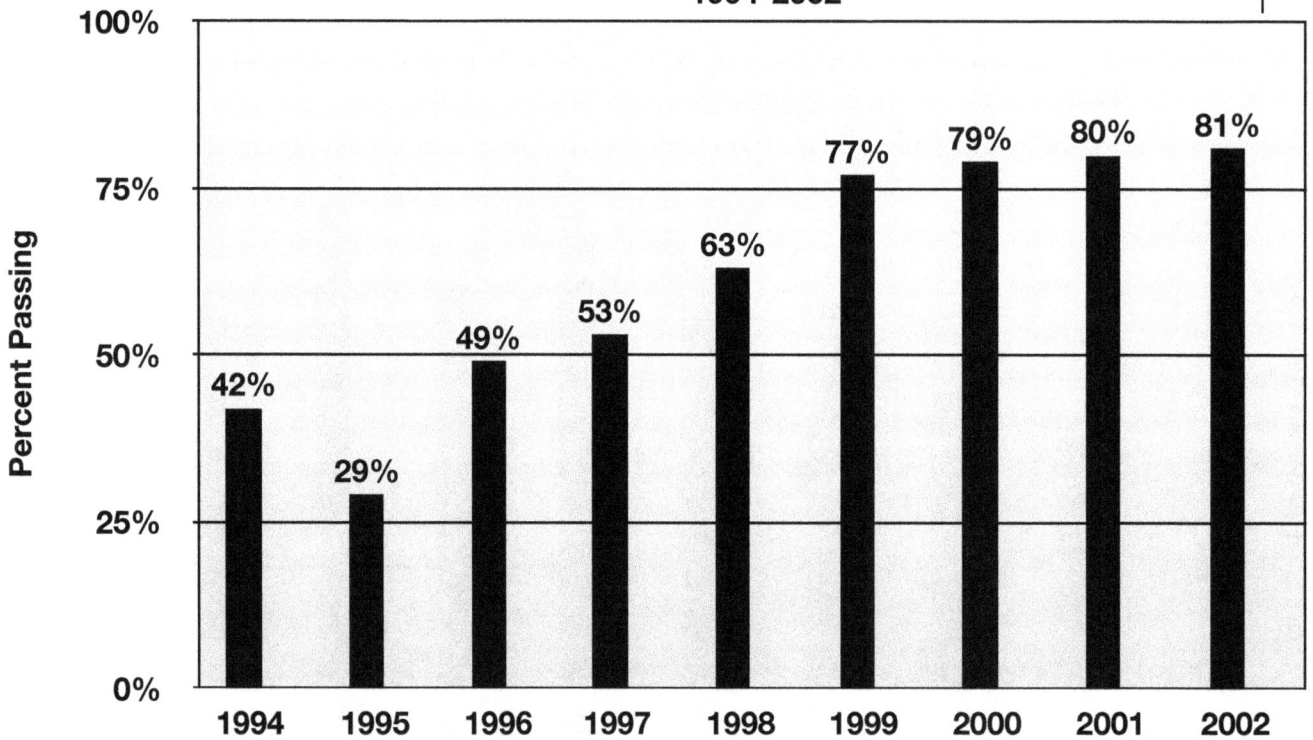

Davis High School Feeder Schools, TAAS Math Results, 1994 and 2002

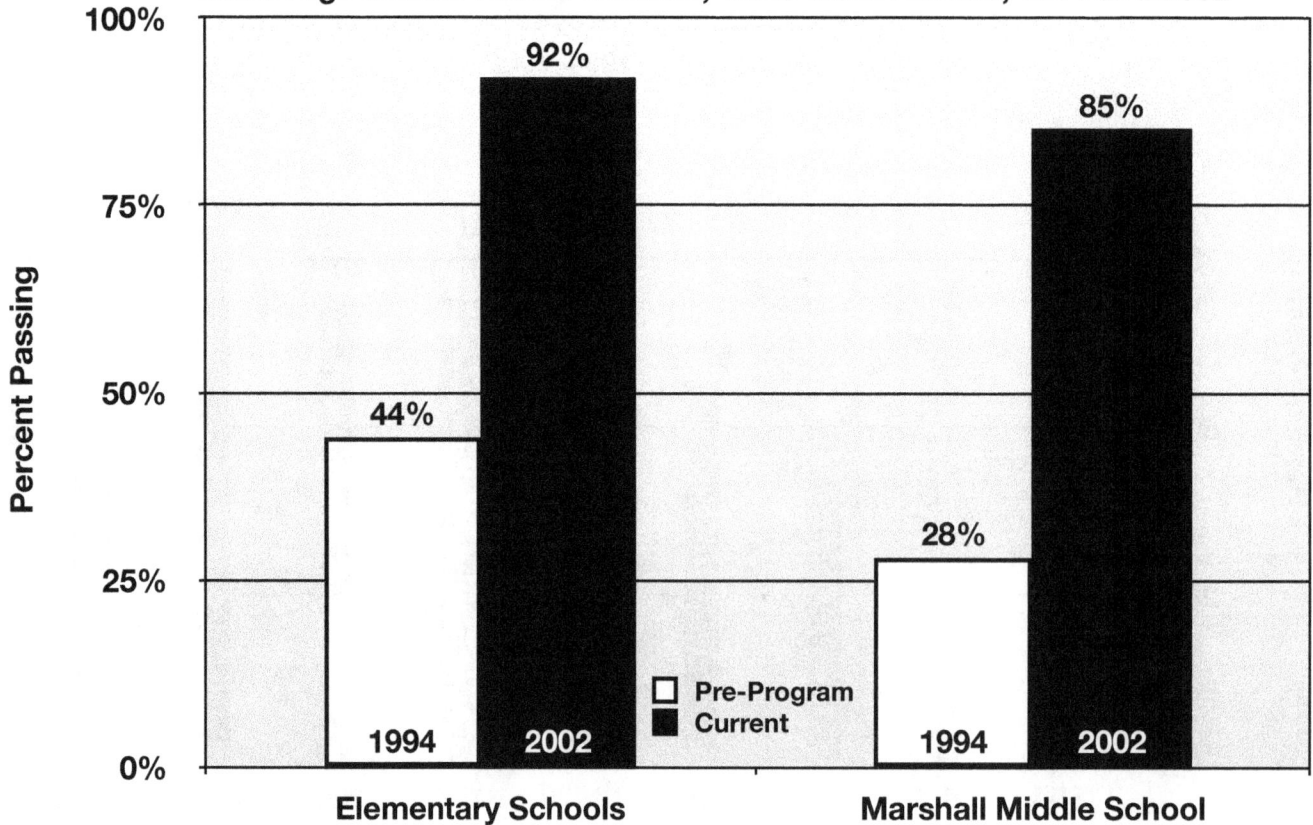

Texas State Criterion Referenced Test
(TAAS)
Jefferson Davis Feeder Pattern
All Students Not In Special Education
MATHEMATICS
Total Feeder Pattern Weighted Average, Percent Passing

Grade	Pre-K	K	1st	2nd	3rd	4th	5th	6th	7th	8th	9th	10th
Year					1994	1995	1996	1997	1998	1999	2000	2001
Class of 2003					34%	48%	65%	66%	57%	69%	NT	80%
Year					1995	1996	1997	1998	1999	2000	2001	2002
Class of 2004					51%	58%	83%	65%	62%	76%	NT	81%
Year				1995	1996	1997	1998	1999	2000	2001	2002	
Class of 2005				NT	54%	68%	87%	66%	65%	86%	NT	
Year			1995	1996	1997	1998	1999	2000	2001	2002		
Class of 2006			NT	NT	58%	79%	71%	68%	75%	84%		
Year		1995	1996	1997	1998	1999	2000	2001	2002			
Class of 2007		NT	NT	NT	66%	66%	84%	79%	82%			
Year	1995	1996	1997	1998	1999	2000	2001	2002				
Class of 2008	NT	NT	NT	NT	63%	84%	96%	88%				
Year	1996	1997	1998	1999	2000	2001	2002					
Class of 2009	NT	NT	NT	NT	71%	88%	97%					
Year	1997	1998	1999	2000	2001	2002						
Class of 2010	NT	NT	NT	NT	74%	90%						
Year	1998	1999	2000	2001	2002							
Class of 2011	NT	NT	NT	NT	88%							

Project GRAD USA / MOVE IT Math

Except for new hires, by the end of the 1998-1999 school year, most elementary school teachers in a Project GRAD school in Houston had taught MOVE IT Math 1 and 2 for at least one year, and, as always happens, some of them had made the switch to MOVE IT Math as if they had always taught math that way. From my perspective, these teachers were qualified to join the MOVE IT Math Guild, the group of 20 or so elementary school teachers who helped me disseminate MOVE IT Math in Texas and initiate it in Los Angeles, California; Yonkers, New York; and New York City, the latter for the United Federation of Teachers (UFT), an affiliate of the American Federation of Teachers (AFT).

Project GRAD no longer needed my services to inaugurate MOVE IT Math in a new site. It could employ its own "home grown" MOVE IT Math instructors for that purpose, the same as I had done to disseminate the program in Texas and other states.

In 1999, I licensed Project GRAD to launch MOVE IT Math nationwide with the elementary school teachers in Project GRAD schools who had demonstrated expertise with the program. *Free at last,* I thought, of the burden of making America aware of how its children are being held back in math in elementary school and what could be done about it.

By 2003, more than 135,000 students were enrolled in 217 Project GRAD USA/MOVE IT Math schools in Kenai, **Alaska**; Los Angeles, **California**; Atlanta, **Georgia**; Roosevelt, **New York**; Akron, Cincinnati, Columbus, and Lorain, **Ohio**; Knoxville, **Tennessee**; and Houston and Brownsville, **Texas**.

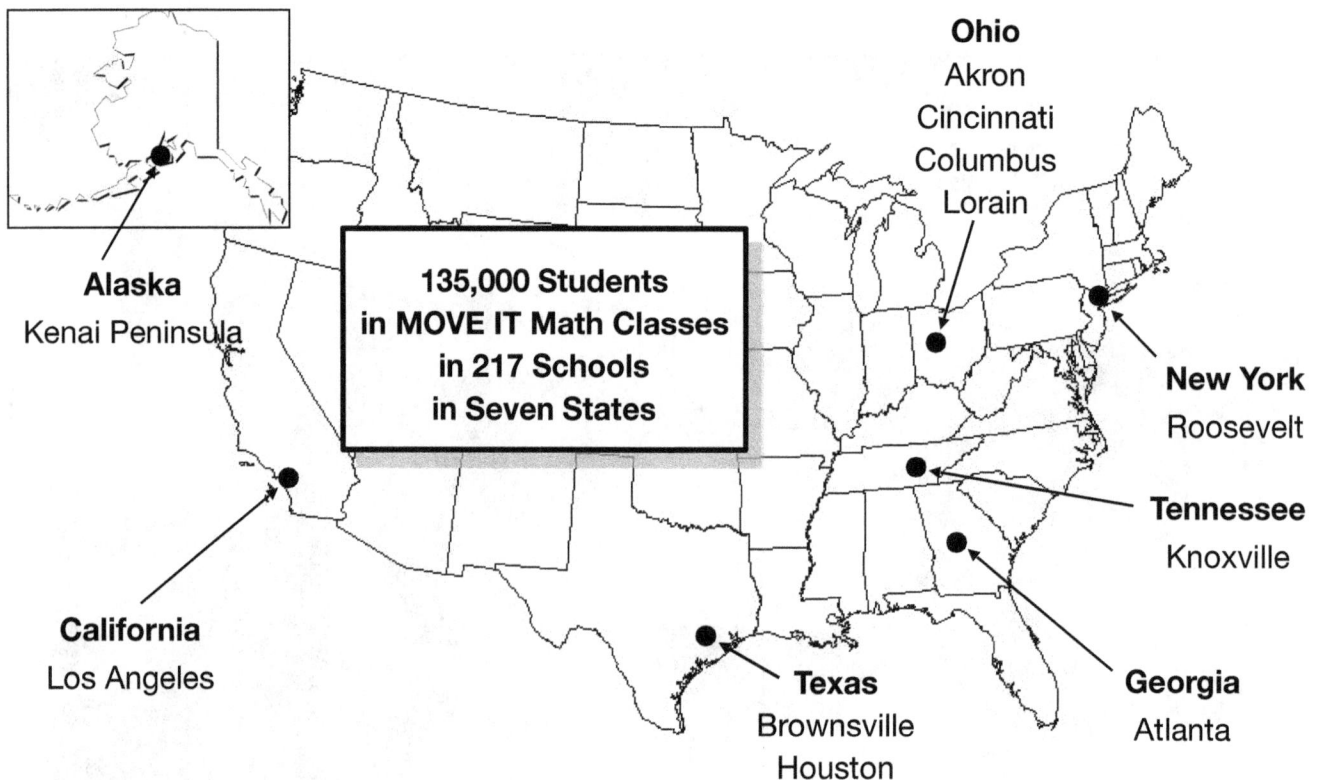

Ohio
Akron
Cincinnati
Columbus
Lorain

Alaska
Kenai Peninsula

135,000 Students
in MOVE IT Math Classes
in 217 Schools
in Seven States

New York
Roosevelt

Tennessee
Knoxville

California
Los Angeles

Texas
Brownsville
Houston

Georgia
Atlanta

Project GRAD USA / MOVE IT Math Results for Four Large American Cities

Project GRAD USA 2004: Overview, Results, and Indicators, a Project GRAD USA PowerPoint presentation. The following results from the presentation show that Project GRAD USA significantly increased high school graduation rates and college enrollment and completion rates in the high schools that partnered with it.

Houston, Texas

o The average number of Davis High School graduates in Houston who enrolled in college for years 1984-1988 was **174**. In 2004, after receiving financial incentives to graduate and enroll in college beginning in 1988 and support from Project GRAD beginning in 1994, the number of Davis graduates enrolling in college was **338**, nearly twice what it was before the incentives and Project GRAD.

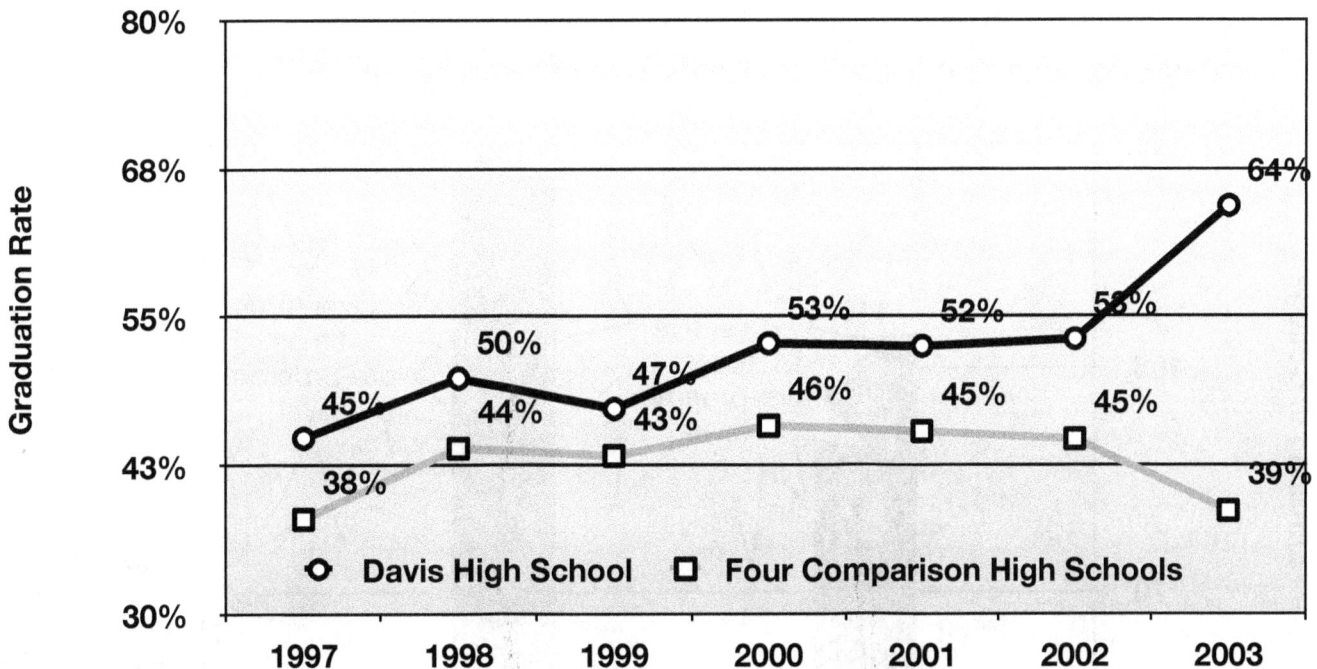

o By 2003, the graduation rate for Davis High School in Houston had significantly outpaced the average graduation rate for four comparable high schools in Houston.

Davis High School in Houston has shown steady improvement in the percentage of ninth graders that graduate four years later. In 1997, 45% of the ninth grade class from four years earlier graduated. In 2003, the graduation rate of ninth graders increased to 64%. These graduation rate increases are significantly higher than the historical trend line for high schools within the district that have student bodies with similar demographics and socio-economic status as Davis. We see that the average graduation rate for these comparison schools (Milby, Furr, Lee, Austin) stays relatively constant, and significantly lower than that of Davis over the past several years.

○ By 2004, Davis High School graduates in Houston were completing college at a rate 70% above the national average for low-income students.

Average number of low-income students nationally who completed college in 2004: **27%**

Average number of Davis High School graduates who completed college in 2004: **46%**

A study of the first eight classes of Project GRAD scholars shows that 46% of them completed college four to eight years after enrollment. In contrast, according to a national study by the U.S. Department of Education, only 27% of low income students across the country completed college four to eight years after enrollment. This indicates that Davis graduates, 91% of whom are low income, complete college at a rate 70% above the national low-income average.

○ By 2003, the graduates from the first three Project GRAD feeder school systems in Houston were enrolling in college in record numbers.

Davis: <u>from 20 to 110</u>, after partnering with Project GRAD in 1994

Yates: <u>from 40 to 135</u>, after partnering with Project GRAD in 1996

Wheatley: <u>from 15 to 28</u>, after partnering with Project GRAD in 1997

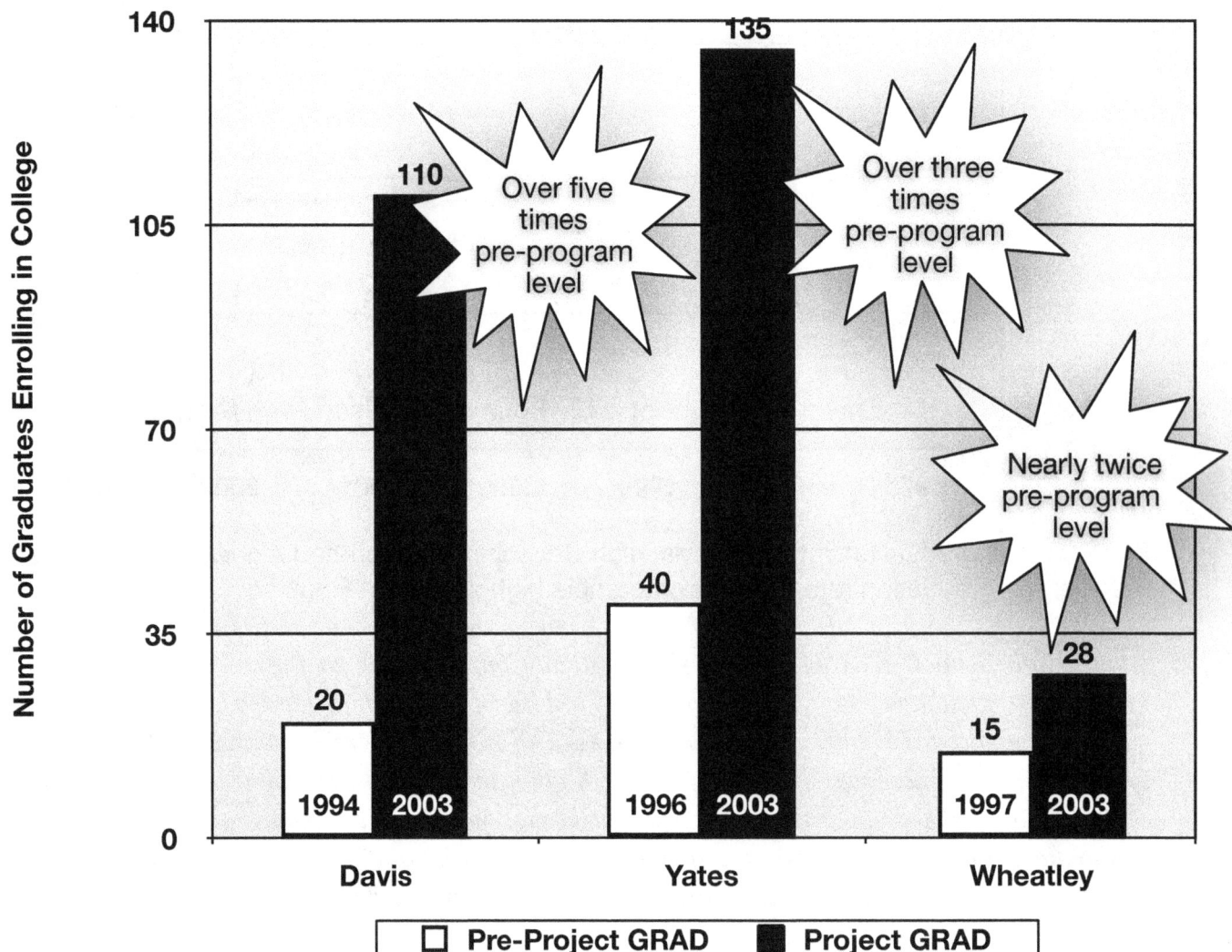

Columbus Ohio

○ Graduation rate for Linden-McKinley High School in Columbus, Ohio after partnering with Project GRAD USA in 1999 vs. other public high schools in Columbus. Said graduation rate was calculated as the percent of total ninth graders four years earlier.

	Linden – McKinley High School	Other Public High Schools
Grad Rate (1999)	21.4%	41.4%
Grad Rate (2003)	40.0%	50.0%
Gain in Grad Rate	**18.6%**	**8.6%**

○ In 2003, after partnering with Project GRAD USA in 1999, Linden McKinley High School in Columbus, Ohio outpaced the district in the number of graduates attending college.

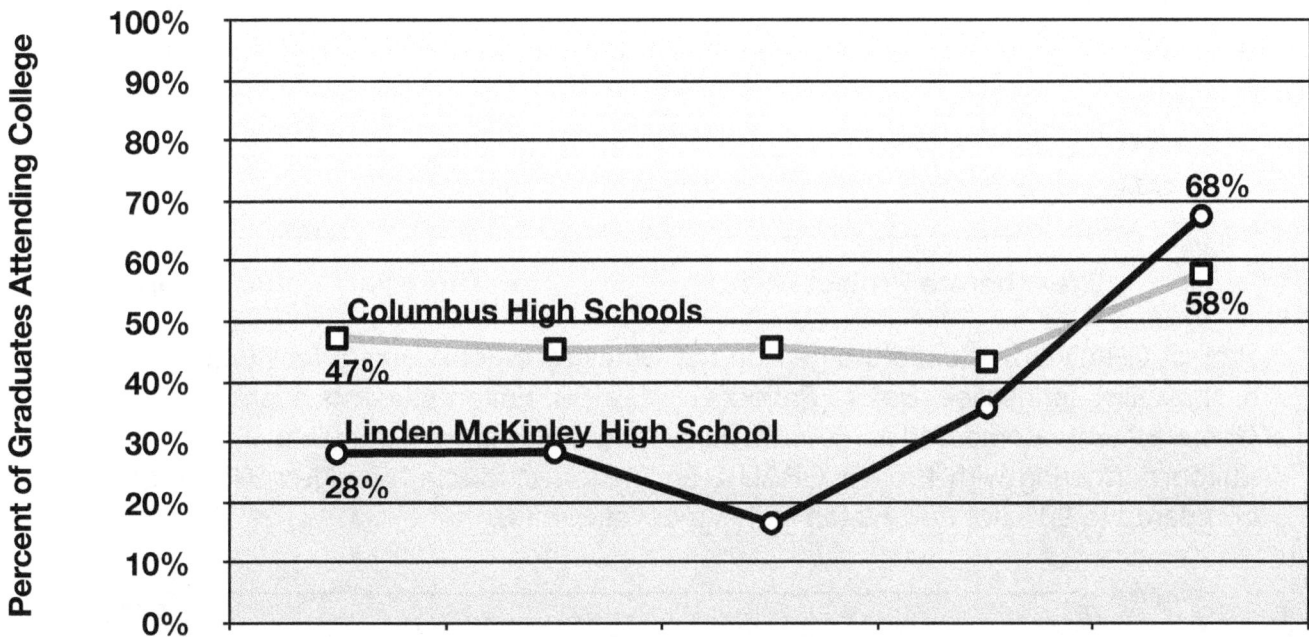

In 1999, the year before implementation, only 28% of Linden McKinley High School's graduating class went on to enroll in college. That same year, 47% of the district's total graduating class enrolled in college. By 2003, the first year of awarding Project GRAD USA scholarships, 68% of Linden McKinley's graduates entered college compared to 58% for the district's total graduating class.

Knoxville, Tennessee

○ Three years after partnering with Project GRAD USA in 2001, the number of graduates of Austin East High School in Knoxville, Tennessee increased by 45 percent, while that for the rest of the district's high schools increased by only two percent.

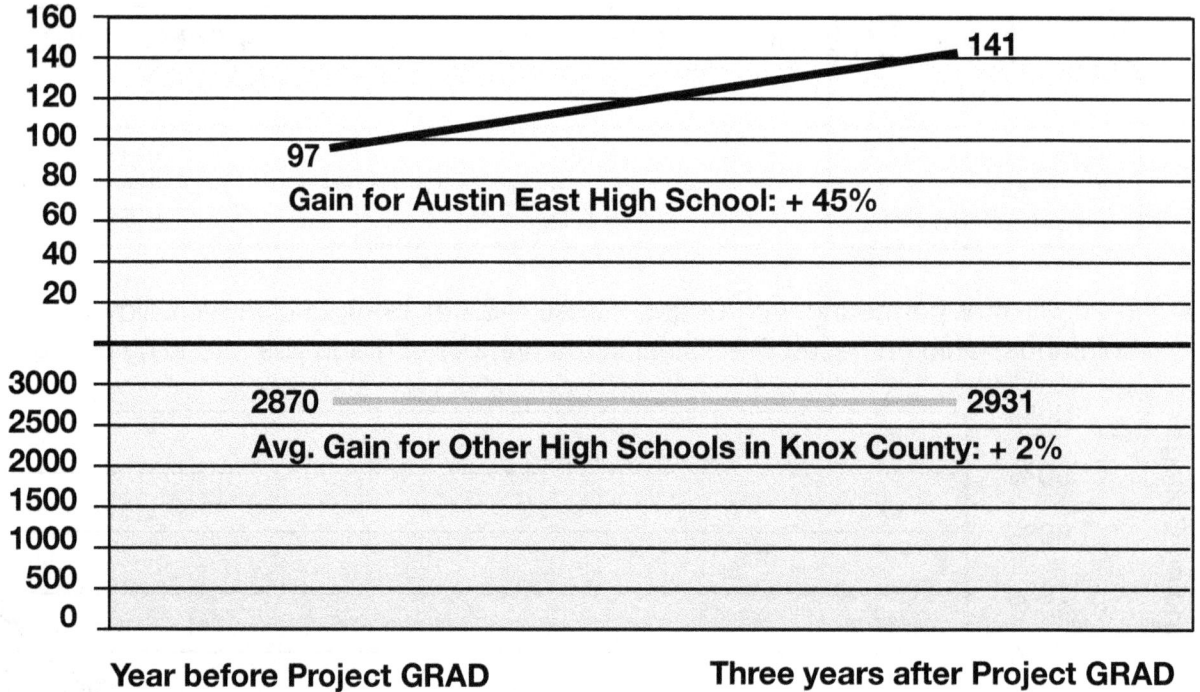

Gain for Austin East High School: + 45%
97 ... 141

Avg. Gain for Other High Schools in Knox County: + 2%
2870 ... 2931

Year before Project GRAD   **Three years after Project GRAD**

○ Project GRAD USA closed the gap on the Gateway **algebra** exam for Fulton High School in Knoxville, Tennessee. Before Project GRAD USA, Fulton students averaged 51% on the Gateway exam compared to 70% for non-Fulton high school students. After two years of Fulton partnering with Project GRAD USA, Fulton students averaged 94% on the exam compared to 90% for non-Fulton high school students.

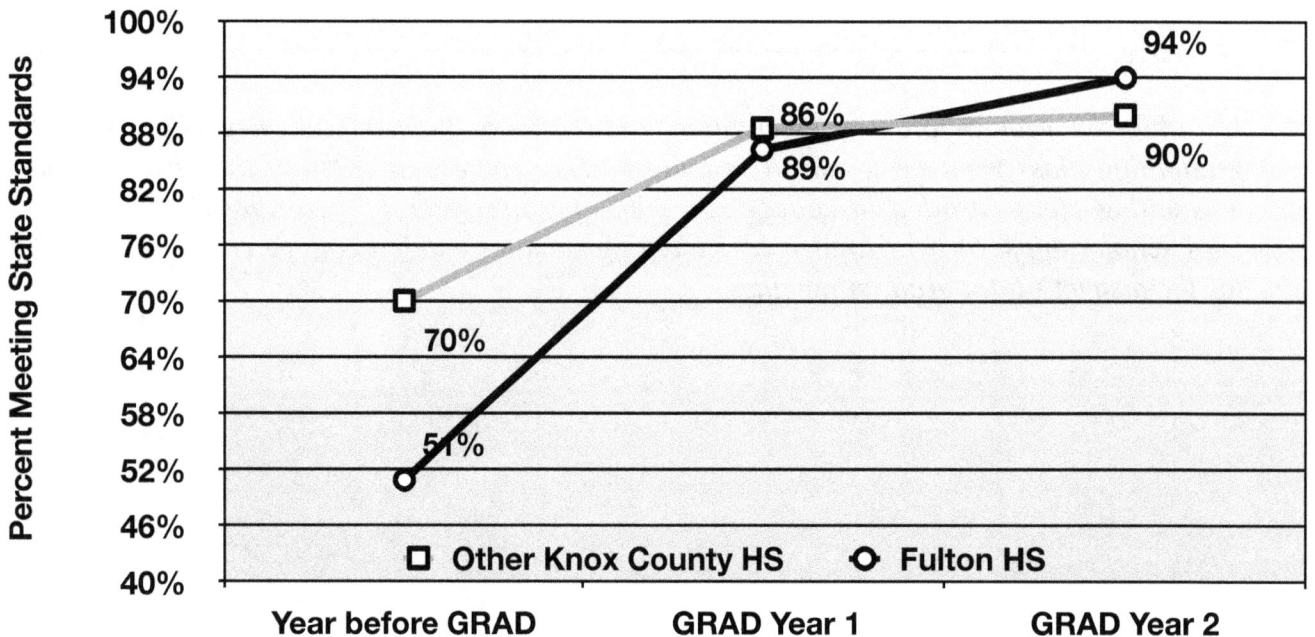

Percent Meeting State Standards

94%
86%
89%
90%
70%
51%

□ Other Knox County HS   O Fulton HS

Year before GRAD   **GRAD Year 1**   **GRAD Year 2**

Cincinnati, Ohio

○ Percent of Western Hills University High School students in Cincinnati, Ohio passing the Ohio Proficiency Test in 2004 after partnering with Project GRAD USA in 2002.

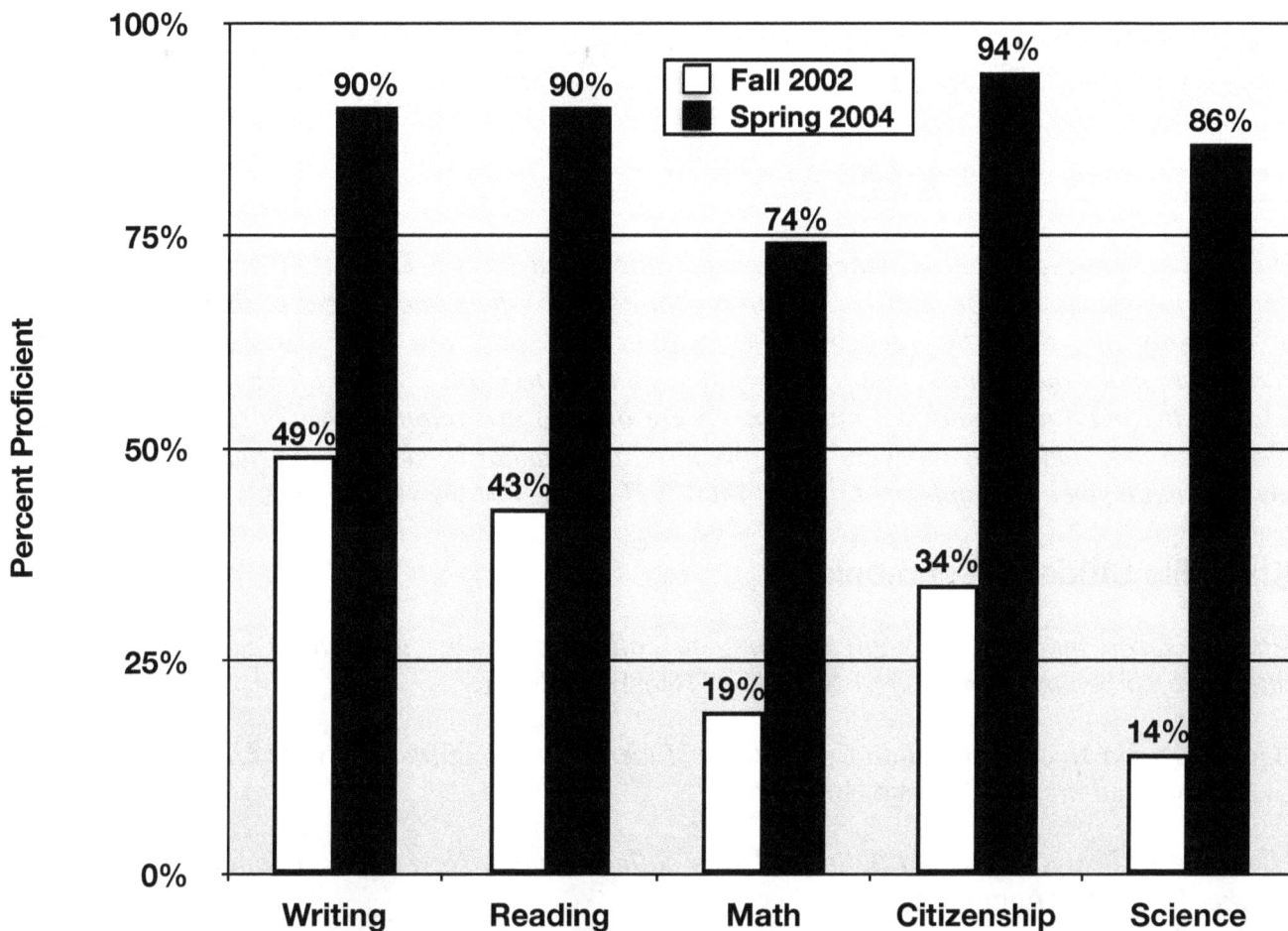

MORE Documented Results for Project GRAD USA / MOVE IT Math

Los Angeles, California

In Los Angeles, in only the third year of implementation, reading and math results have improved dramatically. On the nationally-normed Stanford 9 exam, the percentage of students scoring at or above the national median in reading and math increased in all Project GRAD USA elementary schools. In math, the increases ranged from 28% to 95%. — projectgradusa.org, c. 2002

Knoxville, Tennessee

Prior to partnering with Project GRAD USA in 2001, the average high school graduation rate for low-performing high schools in Knoxville, Tennessee was 50%. By 2019, it had surpassed 80%, and the percent of graduates from these schools who enrolled in college had increased from 30% to 56%. Moreover, a completion rate of 45% for those receiving Project GRAD USA Scholarships hugely exceeded the national average of 10% for similar high-poverty minority students (projectgradknoxville.org/about-us).

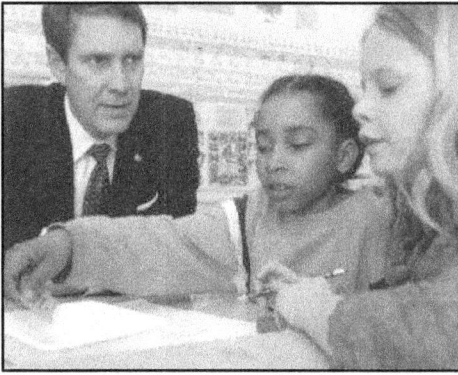

Students at Beaumont Elementary in Knoxville, Tennessee demonstrate MOVE IT Math to Bill Frist, U.S. Senate Majority Leader, 2003-2007.

[Knoxville elementary school] teachers are ecstatic about MOVE IT Math. One teacher explains that her favorite aspect of MOVE IT Math is that it "helps the students realize that math is a part of everyday life. Since they see that they will need it, they are eager to learn it. The math characters make the children laugh and relate even more to math." 100% of the teachers surveyed felt that the math manipulatives and games were okay to exceptional. Over 90% of the teachers responded that their students were more prepared for standardized tests and had better attitudes towards math since the implementation of MOVE IT Math. — Monica Harper, 2005

Knoxville Student Testimonials

"Project Grad made the thought of going to college a reality. Instead of being something I 'might do,' it became something I 'will do.'" Rilesha Holloway

"Going from a student at Fulton to a college student at East Tennessee State University, Project Grad never left my side." Blake Humphrey

"Because of Project GRAD, I believed I was college-bound. Now I am in Japan as an English teacher." Timothy Anderson

"Project GRAD not only gave me the opportunity to attend college but the blessing of finding myself." Courtney Winston

"Project GRAD pushed me through high school and undergrad just so I would be able to experience by far the best feeling ever — becoming." Jenae Antoinette Anderson

"Project GRAD pushed me to keep going no matter what over the years even after I had my son. I finally graduated with my Bachelors of Science in exercise science with a minor in dance." Dominique Sims

"Prior to Project Grad, college was simply a thought. After Project Grad, college was 'the next step.'" Ashford Collins

"When I think of Project GRAD, many words come to mind: friendship, providers, motivators, educators, mentors, and guides." Zachary Zialcita

"Because of Project GRAD, I was able to graduate with my bachelors in exercise science from Lincoln Memorial University. I am now one of the assistant basketball coaches for the women's team at Walters State Community College. Thank you Project GRAD!" Breona Hall

Closing Remarks

Results indicate that the Project GRAD model had consistent, predictable, positive impacts on student discipline, academic performance, time on task, and college attendance. — Kwame Opuni, 1999

The results for Project GRAD were amazing. They were music to the ears of parents who want assurance that their children will succeed in school. The results *proved* that America can significantly improve its schools, and Project GRAD showed how. What worked to significantly improve low-performing schools will work to significantly improve all schools.

Project GRAD partnered with low-performing high schools with students from mostly low-income families with the promise to increase graduation rates, up the number of students succeeding in college preparatory courses, and add to the number of graduates enrolling in college. Its brilliance was in accepting that to make good on its promise, it had to address the obvious, that high schools inherit problems that arose before their students got to high school. In particular, it had to grapple with too many ninth graders who lacked basic math, reading, and writing skills and were likely to quit school.

To resolve said problems, Project GRAD changed the curricula for math, reading, and writing in the elementary schools and middle schools whose students were slated to attend a Project GRAD high school. For math, it initiated MOVE IT Math; for reading and writing, it initiated Success for All; and to motivate students to stay in school, it created a supportive learning environment in kindergarten through grade 12 that *"insulated [students] from academic failure"* (Opuni, 1999).

It initiated Consistency Management and Cooperative Discipline in all grades, K-12, and provided counseling for all students with personal or family problems that inhibited their classroom performance. Additionally, it incentivized achievement from kindergarten on with a $1,000/year scholarship for each of four years for graduates who met Project GRAD's graduation requirements and enrolled in college.

What is now clear — to me, at least — is that trying to improve America's schools a subject at a time, as I had been doing with MOVE IT Math, is inadequate. It does not address the many factors that affect a student's likelihood of succeeding in school. What stands out in the results for Project GRAD is the importance of ensuring success in the three Rs in elementary school: Reading, 'Riting, and 'Rithmetic. Do that with the additional assists that Project GRAD provided, including scholarships to enroll in college, and achievement appears to soar in every subject, as shown in the results for Cincinnati, Ohio. There, the score for science rose from 14 percent to 86 percent and that for citizenship from 34 percent to 94 percent! Imagine having neighbors who believe in science with that much awareness of what it takes to be a good citizen.

References

American Federation of Teachers (AFT). *Consistency Management & Cooperative Discipline Overview: Improving Low-Performing High Schools. Ideas & Promising Programs for High Schools,* 1999.

Communities in Schools (CIS): communitiesinschools.org/about-us.

GRADcafé by Project GRAD, gradcafe.org.

Harper, Monica Balon. "What Is Move It Math and Is It Producing Improvements in the Test Scores of Dogwood Elementary?" University of Tennessee Honors Thesis Projects, 2005.

MOVE IT Math, moveitmath.com.

Opuni, Kwame A. *Project GRAD: Program Evaluation Report* (1994-95). Houston Independent School District, Houston, Texas, September, 1995.

Opuni, Kwame A. *Project GRAD: Program Evaluation Report* (1998-99). *Graduation Really Achieves Dreams.* Houston Independent School District, Houston, Texas, December 20, 1999. ERIC Number ED443779.

Project GRAD 2002: Math Success Closing the Gap. A PowerPoint presentation from Project GRAD's website when it was active.

Project GRAD 2004: Overview, Results, and Indicators. A PowerPoint presentation from Project GRAD's website when it was active.

Project GRAD Knoxville, projectgradknoxville.org/about-us.

Project GRAD USA, projectgradusa.org (defunct).

Success for All, successforall.org.

Texas Assessment of Academic Skills (**TAAS**), wikipedia.org.

Chapter 13: Demise of Project GRAD USA / MOVE IT Math

The results for Project GRAD USA were fantastic. They signaled full speed ahead for fixing America's schools, and Project GRAD USA showed the way. *But not so fast.*

In 1998, the Manpower Demonstration Research Corporation (MDRC), a nonprofit, nonpartisan research organization, began a four-year study to evaluate Project GRAD's results. In 2006, the MDRC released its findings in two documents that may be accessed online: *Striving for Student Success* (191 pages) about Project GRAD's results for elementary schools and *Charting a Path to Graduation* (131 pages) about Project GRAD's results for high schools. Summaries of the assessments follow.

Assessment of Project GRAD's Results for Elementary Schools

To evaluate Project GRAD's results for elementary schools, the MDRC examined student scores on <u>state achievement tests</u> and <u>national achievement tests</u> in Project GRAD elementary schools in Houston, Texas; Atlanta, Georgia; Newark, New Jersey; and Columbus, Ohio. Data was collected from the year when Project GRAD was initiated in these schools through the 2002-2003 school year. For Houston, the initiation year was 1993; for Atlanta and New Jersey, it was 1999; and for Columbus it was 2000. At Newark's insistence, Project GRAD Newark did not include MOVE IT Math in its initiation.

Implementing Project GRAD in selected elementary schools in Houston, Atlanta, and Columbus prompted comparison schools in those cities to implement reforms of their own. To determine the effectiveness of Project GRAD, the MDRC conducted a differential impact study. It compared the results for Project GRAD schools to those for comparison schools in the same city to determine the effectiveness of Project GRAD relative to the reforms the comparison schools initiated. It sought to determine if Project GRAD *"had an effect over and above whatever reforms would have been implemented in the absence of the program."*

Key Findings for Project GRAD Elementary Schools

The following are the MDRC's main findings for Project GRAD/MOVE IT Math elementary schools. Quotes from the MDRC documents are in italics. Underlining and boldface type in the quotes have been added.

o *In Houston and Atlanta, where Project GRAD implementation was strong, student scores on state achievement tests at the Project GRAD schools improved. During the same period, similar improvements on state tests also occurred at the comparison schools, which implemented other district- and school-level reforms (<u>often focused on boosting scores on state tests</u>).*

o *Scores on national achievement tests fell at comparison schools in Houston during the study period. Project GRAD frequently prevented or lessened a similar deterioration in performance on these tests, resulting in <u>significant positive effects on elementary student achievement relative to national norms</u>.*

o *In Columbus, the implementation of Project GRAD was initially weaker than in the other sites, and this appears to have lowered test scores — both absolutely and relative to comparison schools — in the early years of the initiative.*

The MDRC's main findings confirmed that the results for Project GRAD were fantastic, at least in Houston, and its findings for scores on *national* norm-referenced achievement tests asserted as much. Nonetheless, Project GRAD was abandoned after the MDRC assessment of Project GRAD's results.

I do not know why Project GRAD came to an end, but I suspect that the MDRC finding that comparison schools scored as well as Project GRAD schools on *state* achievement tests had something to do with it. The implication from that finding was that the reforms initiated by the comparison schools were just as effective as Project GRAD, which may have induced Project GRAD's benefactors, such as the Ford Foundation, to stop funding it, even though the MDRC admitted in its main findings that many of the so-called reforms in the comparison schools amounted to *"boosting scores on state tests."*

There are only two ways to quickly "boost" scores on state tests. One way is for the state education agencies that make the tests to make them easier to make themselves look good. The other way is to teach to the test. Kudos for the MDRC for its honesty. It admitted that improving scores on state achievement tests to match those obtained by Project GRAD did not achieve the same student outcomes as those obtained by Project GRAD.

> *Researchers have questioned whether improvements in state measures — absent progress on norm-referenced tests [like the SAT-9[57]] — represent genuine improvement in students' academic skills. Thus, in Houston, there is evidence that Project GRAD substantially improved overall measures of elementary student achievement relative to the levels that would have occurred without the program.*

In drawing attention to comparison schools achieving as well as Project GRAD schools on state achievement tests, the MDRC merely confirmed the obvious. State achievement tests sample gains in arithmetic with respect to the Common Core State Standards for math — the redundant, little-numbers-for-little-kids curriculum for elementary school math that was addressed in Chapter 3. State achievement tests are too easy for children in MOVE IT Math classes. *What they know by the end of the second grade will not show up on state achievement tests until the fifth grade.*

In MDCR's own words, …

> *Though Project GRAD did not have a differential impact on elementary students' TAAS [Texas Assessment of Academic Skills] scores, the program did have consistently positive, statistically significant, and substantively important effects on students' performance on the SAT-9. …*

> *This finding underscores the possibility … that the progress on the TAAS reflects improvements on different dimensions of skill than those measured by the SAT-9 and that the progress on the specific material germane to the TAAS may have come at the expense of the dimensions of academic skill measured by nationally norm-referenced tests such as the SAT-9. The findings also reflect the possibility that **ceiling effects** on the TAAS obscure meaningful variation across schools in student performance and school progress.*

---

[57] The SAT (Stanford Achievement Test) compares student achievement on the test to that of students nationally.

The net result is a consistent set of statistically significant positive effects on elementary-level SAT-9 achievement in both reading and math. For example, the analysis suggests that, in the absence of Project GRAD, third-grade SAT-9 math achievement throughout the Davis feeder pattern would have fallen to the 25th percentile; with Project GRAD, math achievement reached the 38th percentile. ...

In the context of a reform-rich environment focused on meeting state standards, Project GRAD had substantial positive effects on elementary students' achievement relative to national norms in two out of three feeder patterns [in Houston]. These effects may reflect Project GRAD's emphasis on relatively well-articulated curricular and instructional reforms at the elementary level that emphasize fundamental academic skills rather than specific state competencies. In particular, it is possible that ... the presence of Success for All and MOVE IT Math may have encouraged enough of a focus on core academic skills to support the current levels of achievement on the more general set of academic skills measured by the SAT-9.

The MDRC's findings confirmed that the results for Project GRAD were, indeed, fantastic, as stated at the outset of this chapter. Moreover, the 2004 accomplishments for Project GRAD that were reported in Chapter 12 suggest that given more time, the results for feeder systems in cities besides Houston would have confirmed Project GRADs efficacy to significantly improve America's schools. As noted by the MDRC, *"it takes at least five years for effective educational reforms to take hold and show results, which highlights the potential that results in Columbus and Atlanta might improve in the future."*

Assessment of Project GRAD's Results for High Schools

Unlike its intervention in elementary schools, Project GRAD did not change the curricula in high schools. Nor did it provide professional development for teachers, modify instruction, or address skill deficits of entering students. Instead, it offered the following components that encouraged planning for college and created an environment that was conducive to learning:

o *College scholarships are provided to students who have a cumulative 2.5 grade point average, graduate within a four-year time period, complete a recommended college preparatory curriculum, and participate in two summer institutes. Scholarship amounts and criteria vary slightly by site but usually average $1,000 to $1,500 each year during the four years of college. Each Project GRAD high school has a scholarship coordinator who provides counseling, tutoring, and college admission preparation.*

o *Summer institutes provide an opportunity for qualifying Project GRAD students to experience a college campus-based program taught by college faculty and to enhance their academic skills. The activities vary by site but typically include reading, writing, math, science, enrichment, and remedial activities. The institutes usually consist of four to six hours of instruction and related activities per day for four to six weeks.*

o *Parental and community involvement to engage parents and the community in the work of the schools, build awareness of the opportunity to attend college, and support the learning of students. At the high school level, annual Walks for Success are conducted, in which principals, teachers, Project GRAD staff, and community leaders visit students' homes to explain the program and encourage parents and students to participate.*

o ***Social services and academic enrichment*** *through one of two programs — Communities In Schools (CIS) or the Campus Family Support (CFS) Plan (developed by Project GRAD) — which bring additional social services, academic activities, and volunteers into Project GRAD schools to address issues that students and their families face and to build commitment to academic success.*

o ***Classroom management programs*** *developed by Consistency Management & Cooperative Discipline (CMCD) that are designed to produce orderly classrooms focused on learning, by promoting student responsibility and self-discipline and positive relationships among students, teachers, and other adults in the school.*

Key Findings for Project GRAD High Schools

To evaluate Project GRAD's results for high schools, the MDRC examined test scores, attendance records, and graduation rates in Project GRAD high schools in Houston, Texas; Columbus, Ohio; and Atlanta, Georgia. Data was collected from the year when the first component of Project GRAD was initiated in these schools through the 2002-2003 school year. For Davis, Yates, and Wheatley high schools in Houston, the initiation years were 1994, 1996, and 1997, respectively. For Linden-McKinley High School in Columbus and Booker T. Washington High School in Atlanta, the initiation year was 1999.

The following are the MDRC's main findings for Project GRAD high schools.

o *At Davis High School in Houston, the initiative's flagship school, Project GRAD had a statistically significant positive impact on the proportion of students who completed a core academic curriculum on time — that is, received an average grade of 75 out of 100 in their core courses; earned four credits in English, three in math, two in science, and two in social studies; and graduated from high school within four years.*

o *As Project GRAD expanded into two other Houston high schools [Yates and Reagan], these positive effects on students' academic preparation were not evident. Student outcomes at the newer Project GRAD high schools improved, but generally this progress was matched by progress at the comparison high schools.*

o *Improvements in graduation rates at the three Project GRAD Houston high schools were generally matched by improvements in graduation rates at the comparison schools.*

o *Looking at early indicators of student success, the initial Project GRAD high schools in Columbus and Atlanta showed improvements in attendance and promotion to tenth grade that appear to have outpaced improvements at the comparison schools, although the differences are only sometimes statistically significant.*

Key Accomplishments of Project GRAD

o *Through activities like the Walk for Success and outreach efforts that are part of specific Project GRAD components, the initiative engaged parents and community members in school improvement efforts in meaningful ways. For parents and key stakeholders, Project GRAD engagement strategies have helped build a constituency for school reform.*

o *Project GRAD's scholarship offer became a galvanizing force for teachers and parents of students in lower grades, fostering a greater awareness of college-going requirements and the importance of higher education.*

o *In each site, Project GRAD scholars have now graduated, and each year other students sign scholarship contracts attesting to their commitment to meet the criteria. To date, at least 4,300 young people from Project GRAD schools nationwide have been able to meet the scholarship requirements and go on to college — one of Project GRAD's priority goals.*

Closing Remarks

In spite of Project GRAD elementary schools performing grade levels above comparison schools on SAT-9 achievement tests in both reading and math, Project GRAD is now defunct, except for GRADcafé, a **free** online counseling service to advise minority students on how to secure financing to attend college.

> *GRADcafé by Project GRAD is a service of the Tejano Center for Community Concerns. We offer FREE services to help you decide on a career, apply to college, and find the financial aid you need. Whether you want to attend college to get a degree or work on a certificate in a professional trade, we are here to guide you through every step. You're never too young or old to get started!*

I do not know why Project GRAD was abandoned, but it probably occurred in response to MDRC's finding that schools comparable to Project GRAD schools performed as well on state achievement tests as Project GRAD schools: *If it ain't broke, it don't need fixin.* Never mind that the MDRC cautioned that the apparent "sameness" between the schools on state achievement tests was possibly due to how little achievement they measured.

Aside from all that, it is noteworthy that hardly any elementary school teachers or middle school math teachers balked at MOVE IT Math. The vast majority of those who taught it embraced its methodology and philosophy and declared that it enhanced their teaching skills. They valued how it put *them* in charge of teaching math instead of the pages in their textbooks.

References

GRADcafé by Project GRAD, gradcafe.org.

Manpower Demonstration Research Corporation (MDRC), *Charting a Path to Graduation*, mdrc.org/project/project-grad#overview, 2006.

Manpower Demonstration Research Corporation (MDRC), *Striving for Student Success*, mdrc.org/project/project-grad#overview, 2006.

National Council of Teachers of Mathematics (NCTM). *Principles to Actions: Ensuring Mathematical Success for All*, 2014.

Chapter 14: The Automaticity Standard Revisited

Yes, I lost — but not because nothing had changed, nor because things were worse, nor because I changed. I lost because I had shown what was possible, and because progress comes at a price. ... I was effective. I was bold. I was in love with possibility — and I remain those things. — Michael Tubbs, youngest Mayor (2016-2020) and ex-Mayor of Stockton, California, 2021

I used to believe the free-market maxim that building a better mousetrap would prompt the world to beat a path to the builder's door. That may be true for tangible items, like shoes and cars, but it is decidedly *not* true for certain ideas, like a better way to teach arithmetic. Those with a stake in the status quo beat the world to my door and turned it away.

First among them was the Texas Education Agency (TEA). After the publication of the NCTM Standards in 1989, the agency secured a government grant to produce a number of modules on standards-based math that it disseminated Texas-wide in workshops for teachers. In doing so, it collided head-on with the Multimodality Math 1 & 2/MOVE IT Math workshops that I and the teachers assisting me were conducting throughout Texas.

TEA saw MOVE IT Math as more than competition. It viewed it as The Enemy. It denigrated the program because it showed how to teach computation instead of handing out calculators to solve problems with numbers surrounded by words (e.g., 2 chickens plus 3 chickens = 5 chickens instead of 2 + 3 = 5). Suffice it to say that TEA wanted to extinguish MOVE IT Math.

TEA's capacity to squelch MOVE IT Math became evident when some of their math consultants visited a small Texas school that was in its second year of implementing the program. When the consultants asked the grade 2 teachers at the school about the program, they were proudly told that their students were already multiplying and dividing. In response, the consultants asked *"Why?"* and added that the state achievement tests did not include multiplication and division items until grade 4. That was all it took for the school to return to the TEA approved textbook-based math curriculum it had been using.

Today, some 30 years after the conversation just described, there is little evidence that the school that once bragged about the results it was getting with MOVE IT Math had ever used it. Modest reform that tweaks the status quo in mathematics education is readily accepted by those with a stake in the Common Core K-12 math curriculum. *Reform that outdoes it is suppressed.*

Nonetheless, the Project GRAD/MOVE IT Math partnership achieved results that cannot be denied. It showed that MOVE IT Math hugely accelerates the actual learning of arithmetic. Moreover, it demonstrated that children who understand arithmetic and know how to read arithmetic problems are more likely to graduate from high school than those in the Common Core standards-based math curriculum for elementary school.

So who wants to load their sling to slay the Goliath that feeds on the status quo and stymied the Project GRAD/MOVE IT Math partnership? I do. *And did.* This book is my sling, and I loaded it with information about how America's children are being robbed and held back in math. Now I am loading it with more to say about the automaticity standard and its authority to compel elementary school children to memorize arithmetic.

The justification for the automaticity standard is the belief that having to think or work out a number fact while computing are cognitive distractions from the task at hand that will cause mistakes and interfere with learning new mathematics. To debunk the belief, I showed that even if true, *so what?* The belief could not be used to justify memorizing the number facts because even if memorized, cognitive distractions will still arise while computing because they are built into all four algorithms.

The Automaticity Standard Is Two Standards in One

In debunking the justification for the automaticity standard, it was noted that doing so did not debunk the standard itself. It did not reject the standard. It just exposed the shallowness of the reason (excuse) for imposing it. This chapter debunks the standard itself.·

Children are taught arithmetic because it is the people's math. It is a useful mental tool. When grown, children will use it to determine the expenditures they can afford and the life style they can maintain. However, should they forget even one of the 390 number facts when using it and have to work it out, the automaticity standard warns of dire consequences. So the automaticity standard assumes that once children memorize the number facts, they will never forget them. If they did, their arithmetic tool would be unreliable when they are adults. Thus the automaticity standard is both a memorization standard and a retention standard: All children must 1) memorize all 390 number facts and 2) never forget even one of them.

Without acknowledging the "no forgetting" requirement of the automaticity standard, there would be no point in having children memorize the number facts except to prepare them to do well on timed tests while in school. *To be clear*, children are not taught the number facts because they need to know them to do well in school. They are taught the number facts because they will be expected to know them when they are adults. Thus to debunk the automaticity standard, it is enough to show that it is an unreasonable standard by showing that most children cannot remember the number facts indefinitely, which was done without fanfare in Chapter 7 in reporting on the DeMaioribus (2011) study. Now, however, I want to dig into the study to show that it reveals that the automaticity standard is not only unrealistic but harmful..

DeMaioribus Student Survey and Multiplication Facts Test

DeMaioribus administered a timed test on the multiplication facts for the 2s through the 9s to 125 college students enrolled in a general education course at the University of Minnesota in Duluth. Her intent was to assess the "fluency" (automaticity) level of college students with the multiplication facts and compare it to their achievement in math in high school and their current attitude toward the subject. Students who answered 40 or more items on the test correctly were deemed to have achieved automaticity with the multiplication facts.

Prior to administering the test, DeMaioribus acquired the following information from the students with a student survey: high school math classes taken, grades in those classes, attitude toward math, and their college major. Of the 125 students, 98% had completed three or more years of high school math, 75% had completed Algebra I, Algebra II, and pre-calculus or higher, 36% had completed calculus, AP (advanced placement) calculus, or AP statistics, 77% had received all As and Bs in the math courses they took in high school, and 64% reported that they liked math or were at least neutral in their attitude toward the subject. Derivatives of the student survey and multiplication facts test that DeMaioribus used are shown on the next page and included full size in the appendix with permission to copy to facilitate replicating her study.

Student Survey

STOP

Do not turn this paper over until instructed to do so.

Please answer the questions below by filling in the blanks or circling the answers. Then, when told to begin, turn the paper over and, working from left to right and top to bottom, answer as many of the single-digit multiplication problems as you can in 60 seconds.

1. Where did you attend elementary school (city and state or country)?
_____

2. Are you taking a math class now?   Yes   No   If no, when did you last take a math class?
   Last semester   About 1 2 3 4 5 year(s) ago   More than 5 years ago

3. Which word best completes this sentence for you? Math is _____.
   Easy   Fairly easy   Difficult   Very difficult

4. Which phrase best describes how much you like math?
   A lot   A little   Neutral   Not much   Not at all

5. How many years of math did you have in high school?   1   2   3   4

6. What was the name of your last high school math class (e.g., Algebra I, Honors, Geometry, etc.)? _____

7. What were your usual grades in high school math?
   As   As and Bs   Bs and Cs   Cs   Cs and Below

8. What is your current college level?
   1st Year   2nd Year   3rd Year   4th Year   Above 4th Year   N/A

9. What is your current/intended major? _____

10. How much math do you anticipate using in your future career?
    A lot   A little   None

Multiplication Facts

7 ×8	9 ×3	7 ×5	6 ×4	9 ×6	3 ×5	9 ×8	7 ×5	5 ×8	7 ×3
8 ×4	7 ×9	5 ×9	8 ×3	8 ×6	2 ×5	8 ×6	3 ×6	7 ×4	5 ×9
5 ×5	6 ×2	9 ×7	7 ×6	6 ×4	3 ×7	5 ×4	9 ×6	7 ×8	5 ×6
6 ×5	7 ×6	6 ×3	8 ×7	9 ×5	4 ×8	3 ×9	2 ×8	3 ×4	9 ×4
8 ×9	4 ×5	6 ×7	4 ×6	3 ×8	3 ×9	4 ×9	7 ×4	5 ×3	3 ×4
9 ×3	6 ×9	5 ×2	7 ×8	4 ×5	2 ×5	6 ×3	4 ×3	5 ×7	8 ×3
6 ×9	7 ×7	4 ×7	3 ×7	5 ×8	3 ×6	4 ×4	3 ×5	6 ×8	9 ×7
3 ×3	4 ×8	4 ×9	6 ×7	7 ×9	9 ×9	6 ×5	9 ×8	8 ×7	5 ×6
8 ×8	4 ×3	3 ×5	4 ×5	4 ×7	7 ×2	4 ×7	5 ×8	9 ×4	7 ×3
4 ×6	5 ×7	2 ×2	4 ×7	5 ×3	8 ×9	9 ×5	3 ×8	7 ×6	8 ×4
5 ×4	3 ×8	6 ×6	2 ×7	5 ×3	9 ×2	3 ×5	6 ×8	2 ×3	7 ×4
6 ×8	8 ×5	7 ×5	4 ×8	5 ×3	6 ×7	9 ×4	3 ×9	7 ×6	9 ×3

To analyze the student data and scores on the multiplication facts test, DeMaioribus sorted the students into "fluency" tiers based on their score on the test. Students who scored 40 or more and thereby demonstrated automaticity were assigned to Tier 1. Those who scored 30-39, 20-29, and less than 20 were assigned to Tiers 2, 3, and 4, respectively. Of the 125 students, 13 were assigned to Tier 1, 27 to Tier 2, 39 to Tier 3, and 46 to Tier 4, as shown below as percentages of 125.

⬤ Tier 1 (≥ 40)   ⬤ Tier 2 (30-39)   ⬤ Tier 3 (20-29)   ○ Tier 4 (< 20)

Percent of Students in Each Fluency Tier

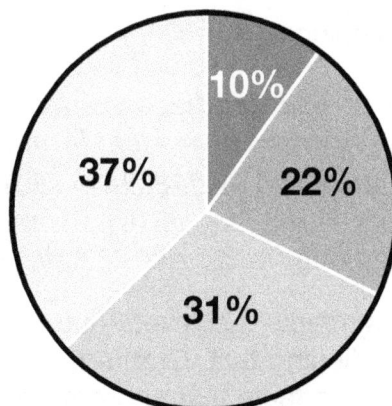

10%   22%   31%   37%

DeMaioribus then sorted the students in each tier into 1) four groups based on their major and 2) two groups based on their attitude toward math. The pie charts and bar graphs that follow portray her findings for those groups.

Fluency Levels by Major

Liberal Arts, Education, Human Services

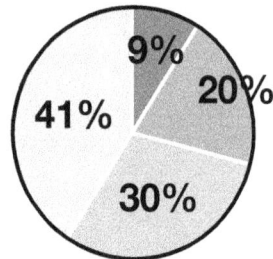

9%
20%
41%
30%

Science, Engineering, Math

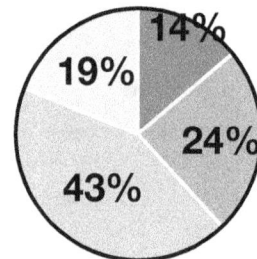

14%
19%
24%
43%

Business

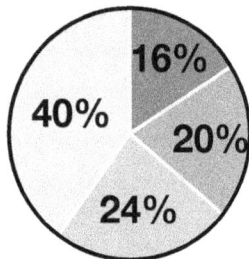

16%
20%
40%
24%

Misc. & Undecided

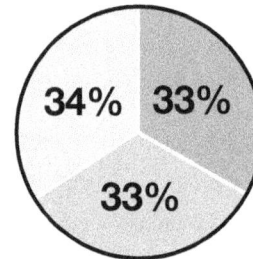

34% 33%
33%

Math Attitude by Fluency Tier

Liked Math a Lot, a Little, or Neutral

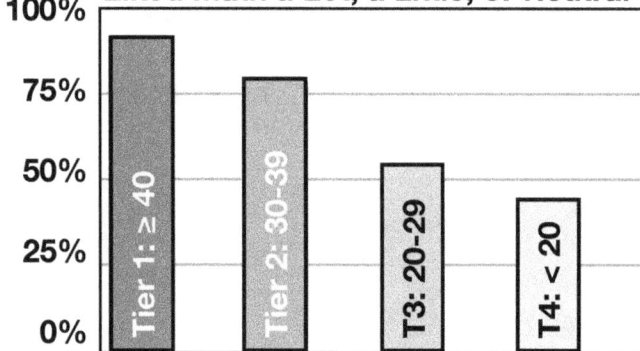

Tier 1: ≥ 40
Tier 2: 30-39
T3: 20-29
T4: < 20

Found Math Difficult or Very Difficult

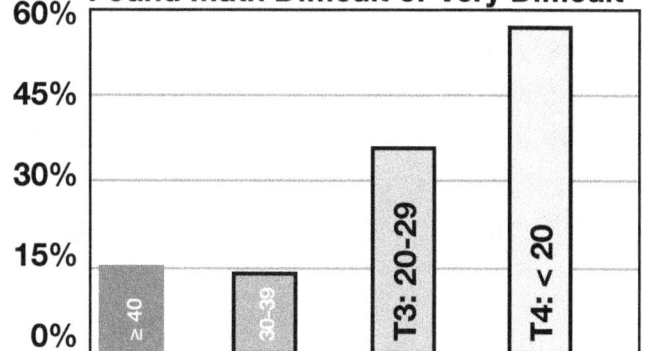

≥ 40
30-39
T3: 20-29
T4: < 20

DeMaioribus was surprised by her findings:

Students of all interests, attitudes, and abilities were represented at almost all levels of fluency and accuracy (number of mistakes made). In other words, <u>there were science majors as well as liberal arts majors in all four tiers of fluency. There were those who liked math who were represented in the bottom tier as well as in the top tier, and there were those who found math difficult in the top tier as well as in the bottom tier</u>.

DeMaioribus had assumed that students achieved automaticity with the number facts in elementary school, but she was not finding that. Granted, there were students in her study who scored 40 or more on her multiplication facts test and got better grades in math in high school

236

and liked math more than those who scored less than 40, which she expected, but other than that, her expectations were quashed. *Her main finding was that 90% of the students had not achieved automaticity with the multiplication facts:*

> *Because it is generally assumed that all students master their multiplication tables in elementary school, the surprising finding of this study was that 90% of the college students tested did not demonstrate automaticity (unconscious recall with speed and accuracy) of basic multiplication facts.*

Moreover, there were 225 students in the general education course she selected for her study, but only 137 of them chose to participate in it, and 12 of them were eliminated from it for not following instructions on the multiplication facts test. So imagine her surprise if she could have tested all 225 students. Knowing that 88 of them did not want to be tested on how well they still remembered the multiplication facts, it is plausible that they opted out of her study because they knew they had forgotten some of them.

> *The question now remains: 1) How does this lack of automaticity affect math education? 2) Is automaticity truly necessary for learning higher math?*

DeMaioribus answered her questions in her findings. She answered the first one when she found that students in tiers 3 and 4 with low fluency scores tended to dislike math and think it was difficult, whereas students in tiers 1 and 2 with relatively high fluency scores tended to like math and think it was easy. (See findings for math attitude by fluency tier.) And she answered *"No!"* to the second one when she found that even students in tier 4 with the lowest fluency scores had completed calculus or AP classes in high school.

> *Students at all levels of fluency reported having taken calculus or AP [Advanced Placement] classes. Even those students with very low fluency levels reported having completed the highest math classes in high school.*

Please note that DeMaioribus' findings are not factual information. *Not yet.* Not until her study is replicated multiple times, which would be easy to do. The data for the study can be acquired in about 15 minutes, and the only math required to analyze it is counting and turning fractions into percents with a calculator. Virtually anyone could replicate it with any willing group of adults.

From Chapter 2, ...

> *I see little hope for any further substantial improvements in mathematics education until we turn mathematics education into an experimental science, until we abandon our reliance on philosophical discussion based on dubious assumptions, and instead follow a carefully constructed pattern of observation and speculation, the pattern so successfully employed by the physical and natural sciences.* — Edward G. Begle, 1977

Unless the DeMaioribus study is an outlier, virtually every replication of it will confirm that the automaticity standard is delusional — that only a small percent of students not taking a math class when they take a multiplication facts test will demonstrate automaticity with the multiplication facts. If so, *the automaticity standard is inhumane.* Inhumane? Yes! It is inhumane to set a standard that *all* children are told they must meet knowing that most of them will not meet it.

Take into account that kids should not compete with each other unfairly in a way that makes some feel they are stupid. — Albert Shanker, President (1974-1997), American Federation of Teachers (AFT), 1989

People Do Not Forget the Multiplication Facts on Purpose

The following are conversations about knowing the multiplication facts that were on two blogs that were online in 2023 as I wrote this chapter.

○ **Is it normal to be an adult and not know your multiplication table?** (Reddit, 2023)

I hope so because I don't know them.

I can only multiply [by] a few numbers. Easiest for me is 2, 3, 4, 5, 6, and 10. Big [number] multiplication is rough for me, and I still sometimes need to use my fingers. I feel like a second grader. Is this normal?

The only reason I know my 9 times table is because I can sing it in the tune of take me out to the ball game.

○ **Is it bad to be an adult and not know multiplication?** (Quora, 2023)

Why can't my 13 year old daughter memorize her multiplication [facts]? I swear she tries her hardest, but she forgets them.

I'm 15 years old, in the 10th grade, and I still don't have my multiplication facts memorized. What should I do?

I'm years behind in math and don't have a solid foundation. How do I get back on track? I'm too embarrassed to tell anyone.

Gardner's Multiple Intelligences

In 1983, Howard Gardner, a Harvard psychologist, published *Frames of Mind: The Theory of Multiple Intelligences*, a bestseller that challenged the notion of "smartness" as one's score on an IQ test: the higher the score, the smarter the person. In Frames of Mind, he listed seven different kinds of intelligences, that is, *seven different kinds of smartness*. According to Anglia (2022), he has since added two more kinds of intelligence or smartness, the last two in the list below:

1. **Logical-Mathematical Intelligence:** Ability to spot trends and patterns, understand relationships, and think conceptually and abstractly, like a mathematician, economist, or scientist. *Like Albert Einstein.*

2. **Linguistic Intelligence:** Ability to use words effectively, like a journalist, lawyer, or politician. *Like Sir Winston Churchill.*

3. **Interpersonal Intelligence:** Ability to sense other people's emotions, like a counsellor, psychologist, or teacher. *Like Cesar Millan*, the Dog Whisperer.

4. **Intrapersonal Intelligence:** Ability to understand themselves and be self-aware, like a therapist, theologian, or poet. *Like Carl Sandburg.*

5. **Musical Intelligence:** Ability to sense rhythm and sound and create music, like a conductor, musician, or composer. *Like Taylor Swift.*

6. **Visual-Spatial Intelligence:** Ability to conjure up images and structures never seen before, like an artist, engineer, or architect. *Like Frank Lloyd Wright.*

7. **Bodily-Kinaesthetic Intelligence:** Ability to coordinate mind and body, like a dancer, carpenter, or athlete. *Like Michael Jordan.*

8. **Naturalist Intelligence:** Ability to read and understand nature, like a botanist, geologist, or cosmologist. *Like Carl Sagan.*

9. **Existential Intelligence:** Ability to deal with questions of fundamental importance, such as the meaning of existence, like an author, philosopher, or preacher. *Like Martin Luther King.*

Intelligence is the ability to acquire and apply knowledge and skills. Gardner's breakout of intelligence reveals the obvious: People have "gifts." We all do. In fact, we all have all nine of Gardner's gifts. We just have them in varying degrees, and no one person has all of them to the highest degree. We can also improve on our gifts. We can become a better listener, for instance, and thereby add IQ points to our interpersonal intelligence.

With Gardner's list of multiple intelligences in mind, can we agree that intelligence comes in different flavors? That people can be smart in different ways? That the factual information that some smart people remember whether they use it or not, other smart people forget once they quit using or reviewing it?

What about the nine people listed after Gardner's intelligences? Assuming that all of them memorized the number facts in elementary school, do you think all of them could still recall all 390 number facts automatically when they became famous? Were there a way to find out, would you bet $1,000 that all of them could do that?

Closing Remarks

The problem is never how to get new, innovative thoughts into your mind, but how to get old ones out. — Dee Hock, founder and former CEO of VISA

In replicating DeMaioribus' study, the following inferences from her findings may be denied or confirmed as factual: *After graduating from high school, …*

o Most students forget some of the multiplication facts after they quit using or practicing them.

o How quickly and accurately college students can recall the multiplication facts has little bearing on their long-term interests as indicated by their choice of a major in college.

o Students who struggle to quickly and accurately recall the multiplication facts tend to dislike math and think it is hard.

o Students who are reasonably adept at recalling the multiplication facts tend to like math and think it is easy.

o How quickly and easily college students can recall the multiplication facts is not indicative of how well they did in advanced math courses in high school.

o Even students who got As and Bs in advanced math courses in high school will forget some of the multiplication facts after they graduate.

References

Anglia, Nord. "What Are the Nine Types of Intelligence that should be Considered in All School Curricula?" Prague British International School (PBIS), December, 21, 2022.

Begle, Edward G. Director of the School Mathematics Study Group (SMSG), a major contributor to the New Math in the 1960s. See Rami, 1995.

DeMaioribus, Carmel E. (2011). "Automaticity of Basic Math Facts: The Key to Math Success." Retrieved from the University of Minnesota Digital Conservancy, https://hdl.handle.net/11299/187488.

Gardner, Howard. *Frames of Mind: The Theory of Multiple Intelligences*, 1983.

Hock, Dee. brainyquote.com/quotes/dee_hock.

quora.com, 2023.

reddit.com/CasualConversation/comments, 2023.

Shanker, Albert. Education Reform: "Quick Fixes or Enduring Cures?" Union Carbide Corporate Forum on Emerging Education Issues, April 3-5, 1989.

Tubbs, Michael. *The Deeper the Roots, A Memoire of Hope and Home*, Amazon.com, 2021.

Chapter 15: The Numerals 1, 2, 3, 4, 5, 6, 7, 8, 9 Were Made to Count On

The Arabic Numerals aka Hindu-Arabic Numerals for 1 through 9

The Arabic numerals aka Hindu-Arabic numerals 1, 2, 3, 4, 5, 6, 7, 8, 9 were introduced in Europe in the 12th century through the writings of Arabian mathematicians, especially al-Khwarizmi (c. 780-850) and al-Kindi (c. 801-873 AD),[58] who got them from two Indian mathematicians, Aryabhata (c. 476-50 CE) and Brahmagupta (c. 598-668 CE),[59] who got them, called Brahmi numerals, from where they originated in the Indian subcontinent in the third century BCE.[60]

The Brahmi Numerals for 1 through 9

–	=	≡	+	ʰ	Ɛ	ꝰ	ꞃ	ʔ
1	2	3	4	5	6	7	8	9

I learned that about the numerals 1 through 9 in a math history course I took while working on my doctorate in mathematics education at the University of Michigan. My initial reaction was disbelief. Except for the tally mark for the numeral 1, I could not imagine how the characters in the top row ever became the familiar characters in the bottom row; however, my wonderment about the Brahmi numerals for 2 and 3 dissipated when I learned that they were stylized renditions of the tally marks being connected in cursive writing, as shown below.

The Brahmi numerals for 4 through 9, however, were a different matter. They looked like random scribbles to me, but what if they were like the Brahmi numerals for 1, 2, and 3 that originated from tally marks that indicated their count?

Please see if you can answer the question just posed for the numerals 4, 5, 6, and 8 before turning the page. *No peeking!* See if you can make the numeral 4 with four tally marks, the numeral 5 with five tally marks, the numeral 6 with six tally marks, and the numeral 8 with eight tally marks. Better yet, do it hands-on. See if you can make them with toothpicks, one toothpick per tally mark. Then imagine how what you made would appear if written cursively.

---

[58] Encyclopedia Britannica/Arabic Numerals.

[59] Aryabhata developed the notion of place value in a numeration system with a dot as a place holder in the fifth century, and Brahmagupta introduced the symbol for zero a century later and was the first to compute with zero as if it were a number like 1, 2, 3, 4, 5, 6, 7, 8, 9 (Boyer, 1968).

[60] Wikipedia/Brahmi Numerals.

If you did as requested, you may have found that you could easily make the numerals 4, 5, and 6 with four, five, and six tally marks or toothpicks, respectively, as shown below. If you gave up on the numeral 8, that is understandable. It is so easy to make with seven tally marks by drawing two squares, one on top of the other with a shared side, that your mind may have gotten stuck on that. To make the numeral 8, draw a square, but rotate it 45 degrees to where it looks like a baseball diamond. Then, beginning at "second base" on the square, draw another square like it on top of it, as shown below.

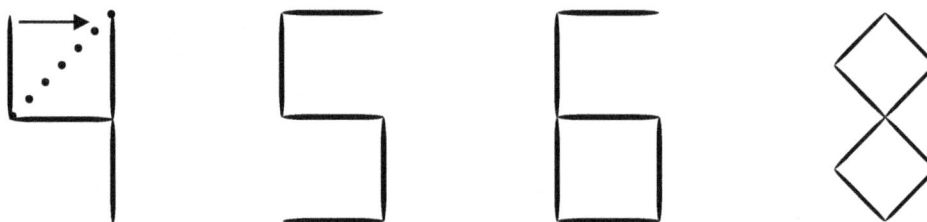

In looking at the numerals 4, 5, 6, and 8 made with tally marks, it is easy to see how they turned out the way they did. Cursive writing connected the tally marks at the top of the numeral 4, as shown, and rounded out most of the sharp corners on the numerals 5, 6, and 8. Thus a defensible play on words for the numerals 4, 5, 6, and 8 would be to say that they were "cursed" by cursive writing. Because of it, they became abstract symbols whose meaning had to be memorized.

As for the numerals 7 and 9, they, too, could have originated from seven and nine tally marks, respectively. As shown below, if the stem of each one had once been crossed with a line to stand for five tally marks, the same as the line commonly drawn across four tally marks to indicate a count of five, the numerals 7 and 9 displayed the amounts they represented.

Five tally marks on a chalkboard

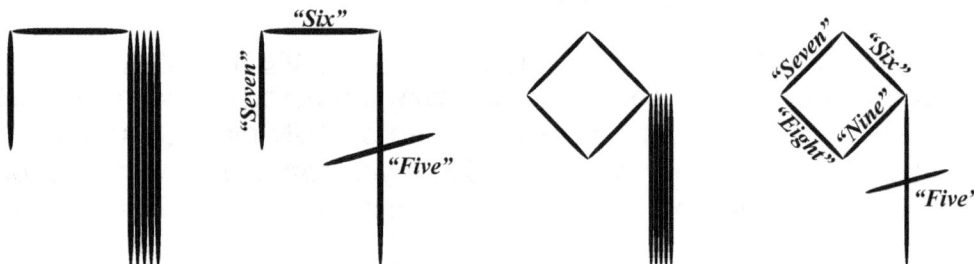

If so, the lines across the stems of the numerals 7 and 9 lost their original meaning. Now, a line across the stem of the numeral 7 is drawn to distinguish the numeral from the numeral 1, and the line that might have been drawn across the stem of the numeral 9 is drawn beneath the numeral (9) to distinguish it from the numeral 6 viewed upside down. So it appears that at some point in passing on the Brahmi numerals from Indian mathematicians to Arabian mathematicians to European Mathematicians, the numerals for 4, 5, 6, 7, 8, 9 that were passed on were seen as meaningless scribbles and were made anew with tally marks that showed their count before they were smoothed out with cursive writing.

To my knowledge, the tally mark connection with the numerals 4, 5, 6, 7, 8, 9 has not been publicized until this publication. I discovered it a long time ago and wrote about it in the *Mad Minute Primer*, the book Pearson Education published in 2005 but killed. In the book were three activity sheets that directed children to make the numerals 1 through 9 with toothpicks that they might make the connection between the numerals and their count. Copies of these activity sheets, entitled Toothpick Numerals, are included in this book on pages 322-324 in the appendix.

Closing Remarks

The probability that the numerals 4, 5, 6, 7, 8, 9 just happen to be connected to their count, as revealed with tally marks (but hidden with cursive writing), is astonishingly small. To figure the probability of that "coincidence," an approximation of it may be obtained by referencing Unicode.

Unicode is an encoding standard for digitizing textual information in all 294 of the world's known writing systems, both modern, like those for Arabic, English, and Mandarin text, and historical, like cuneiform text, wedge-shaped characters impressed on clay tablets. As such, version 16.0 of the standard digitally defines 154,998 characters (letters, numerals, punctuation marks, and other symbols) that are used or have been used historically in ordinary, literary, academic, and technical communications worldwide (Wikipedia/Unicode). Thus, assuming that all four tally mark renditions of the numerals 4, 5, 6, and 8 are among the 154,998 characters that have been digitized for Unicode, the probability of drawing all four of them in any order out of a box filled with all 154,998 Unicode characters on stamp-sized pieces of paper, one character per piece of paper, is (4/154998)(3/154997)(2/154996)(1/154995) = 24/577148492455751100000, which is approximately 1/24047853852322960000, which is virtually zero.

As further evidence of the numerals 4, 5, 6, and 8 having evolved from tally mark representations that showed their count, consider the numerals that are used to express numbers that are greater than ten, like the numerals 2 and 5 in 25. Such numerals are called "digits," *the same as the fingers on your hands*. How can that be unless people once counted on the numerals 1, 2, 3, 4, 5, 6, 7, 8, 9 with their fingers?

The first rung of the Math Ladder is for counting to 100. Thus arithmetic, algebra, calculus and so on up the Math Ladder are all based on counting. Had humans never counted, the Math Ladder would not exist.

The ultimate root of "calculus" is the Latin "calx," meaning "stone." ... "Calculus" is a diminutive of "calx," meaning small stone or pebble, and here comes the math. "Calculi" (the plural form) were the small stones or pebbles used in counting before the advent of more advanced methods. It was from this use of "calculi" that eventually the whole process of adding, subtracting and otherwise manipulating numbers came to be called "calculation." — Word Detective, 2005

References

Boyer, Carl B. *A History of Mathematics*, John Wiley & Sons, 1968.

Encyclopedia Britannica (britannica.com/Hindu-Arabic Numerals).

Unicode.org

Wikipedia.com/Brahmi Numerals.

Wikipedia.com/Unicode.

Word Detective (word-detective.com), December 5, 2005.

Chapter 16: America's Choice

From the Introduction, page i:

No problem is too big to run away from. — Charles M. Schulz, creator of Peanuts

From Chapter 2, page 30:

Our nation is at risk. The educational foundations of our society are presently being eroded by a rising tide of mediocrity that threatens our very future as a Nation and a people. If an unfriendly foreign power had attempted to impose on America the mediocre educational performance that exists today, we might well have viewed it as an act of war.
— National Commission on Excellence in Education, 1983

Forty years and counting have passed since America was warned of the underperformance of its schools in math and science. Yet, as shown below from the Introduction, international testing reveals that America is still at risk. It shows that America's children are still being held back in math in elementary school, the same as my classmates and I were held back in math in the 1940s.

United States Rank in TIMSS Results for Math and Science, 4th Grade, 1995-2019

4th Grade	1995	1999	2003	2007	2011	2015	2019
Math	12th	Not tested	12th	11th	11th	14th	15th
Science	3rd	Not tested	6th	8th	7th	10th	8th

United States Rank in TIMSS Results for Math and Science, 8th Grade, 1995-2019

8th Grade	1995	1999	2003	2007	2011	2015	2019
Math	28th	19th	15th	9th	9th	10th	12th
Science	17th	18th	9th	11th	10th	10th	11th

In reading this book, you at least know why the math results for America's fourth graders are so dismal. The Common Core arithmetic curriculum for kindergarten through most of grade 3 prescribes just adding and subtracting small numbers while memorizing the number facts, so fourth graders in other nations do not have to know much more than how to add and subtract to make America's fourth graders look like math dummies. *But do "we, the people," care?*

Public acceptance of deficient standards contributes significantly to poor performance in mathematics education. — National Research Council, 1989

Too many Americans seem to believe that it does not really matter whether or not one learns mathematics. Only in America do adults openly proclaim their ignorance ("I never was very good at math") as if it were some sort of merit badge. — National Research Council, 1989

Few adults would say, "I'm not so good at reading." But many [Americans] say, "I'm not so good at math," and somehow that's socially acceptable. Very intelligent people brag about not being good at math. — Eric Westervelt, 2015

People always say, like, don't do math, not on live TV, not even two and two. Don't even try to add unless you've got somebody to walk you through it. — Rachael Maddow, 2021

I can attest to Americans volunteering their ineptness in math. When people I have just met learn that I used to teach math, some of them make a point of telling me that they were never good in math. Invariably, they smile as they do, as if to say *"See, I'm normal."*

This Is Embarrassing!

U.S. fourth- and eighth-grade students are not the only ones being compared to their international peers, as in the TIMSS results in the Introduction. U.S. adults are as well.

In 2011-12 and again in 2013-14, the Program for the International Assessment of Adult Competencies[61] (PIAAC) surveyed the proficiency of working-age adults (16-65) in the U.S. and other countries with three cognitive skills that are essential to socially and economically participate in the 21st century: literacy, numeracy, and problem solving in technology-rich environments. The following italicized paragraphs are quotes from Goodman (2013) and others who outlined the overall goal of PIAAC and its subgoals for numeracy and problem solving in technology-rich environments.

The goal of PIAAC is to assess and compare the basic skills and the broad range of competencies of adults around the world. The assessment focuses on cognitive and workplace skills necessary for successful participation in 21st-century society and the global economy. Specifically, PIAAC measures relationships between individuals' educational background, workplace experiences and skills, occupational attainment, use of information and communication technology, and cognitive skills in the areas of literacy, numeracy, and problem solving in technology-rich environments.

Numeracy

The primary goal of PIAAC's numeracy assessment is to evaluate basic mathematical and computational skills that are considered fundamental for functioning in everyday work and social life. Numeracy in the PIAAC framework is defined as the ability to access, use, interpret, and communicate mathematical information and ideas, to engage in and manage mathematical demands of a range of situations in adult life.

Problem Solving in Technology-Rich Environments

PIAAC represents the first attempt to assess problem solving in technology-rich environments on a large scale and as a single dimension in an international context. PIAAC defines problem solving in technology-rich environments as using digital technology, communication tools, and networks to acquire and evaluate information, communicate with others, and perform practical tasks.

Skill Use and Background Questionnaire

In addition to the skills assessment, PIAAC's background questionnaire surveys adults about their educational background; work history; their intrapersonal, interpersonal, and professional skills; and their use of those skills on the job and at home.

---

[61] PIAAC is a cyclical, large-scale international survey of adult skills that was developed by the Organization for Economic Cooperation and Development (OECD, 2012).

Proficiency Levels for Numeracy[62]

Below Level 1: Count, sort, and perform basic arithmetic with whole numbers and money in concrete, familiar contexts where the mathematical content is explicit with little or no text or distractors. Understand common <u>spatial representations</u> (e.g., maps).

Level 1: Perform one- or two-step processes involving basic arithmetic in concrete, familiar contexts with little text and minimal distractors. Understand simple percents, like 50%. Locate, identify, and use elements of common graphical or spatial representations.

Level 2: Apply two or more steps or processes involving calculations with whole numbers and common decimals, percents, and fractions. Perform simple measurements. Understand spatial representations. Interpret data and statistics in texts, tables, and graphs.

> **Nearly two-thirds (61%) of Americans tested
> no better than Level 2 on numeracy.**

Level 3: Apply two or more steps that may involve choosing problem-solving strategies that require <u>number sense</u> (an understanding of numbers and ability to manipulate them) and <u>spatial sense</u> (an understanding of shape, size, position, direction, and movement). Recognize and work with mathematical relationships, patterns, and proportions expressed in verbal or numerical form. Interpret and analyze data and statistics in texts, tables and graphs.

Level 4: Apply multiple steps that may involve choosing relevant problem-solving strategies requiring analysis and complex reasoning about quantities and data, statistics and chance, spatial relationships, proportions, and formulas. Understand and communicate well-reasoned explanations for answers or choices.

Level 5: Integrate multiple types of mathematical information where translation or interpretation is required. Draw inferences. Develop or work with mathematical arguments or models. Evaluate, justify, and critically reflect upon solutions or choices. Understand and communicate well-reasoned explanations for answers or choices.

Proficiency Levels for Problem Solving in Technology-Rich Environments[63]

Level 1: Perform common tasks such as sending an email, texting, or browsing the Internet with a computer or mobile phone.

> **Nearly two-thirds (64%) of Americans tested no better than
> Level 1 on problem solving in technology-rich environments.**

Level 2: Perform tasks that involve multiple steps that require the use of both generic and specific technology applications that may require navigating across pages and applications (as in searching the Internet to acquire information to include in a word-processing document). In terms of cognitive processing, a task may require inferential reasoning and have to be defined by the problem solver even though the goal of the task is explicit. Additionally, the task may include distractors that must be discarded and outcomes and impasses that are unexpected.

Level 3: Same as Level 2 except more demanding.

---

[62] Rampey, 2016.

[63] Rampey, 2016.

For the PIAAC surveys in the United States, the data for the 2011-12 survey was collected from a nationally representative sample of 5,010 adults, age 16-65. For the 2013-14 survey, it was collected from 3660 adults from targeted subgroups: adults, age 16-34, unemployed adults, age 16-65, and adults, age 66-74. The following tables dated 2012/2014 are from Rampey and others, as are those dated 2012/2014 *and* 2017. Those dated 2012 are from Goodman and others. The results for the current PIAAC survey will be published in December, 2024.

Average PIAAC Scores for Adults, Age 16-65, in 22 Countries on Numeracy: 2012/2014

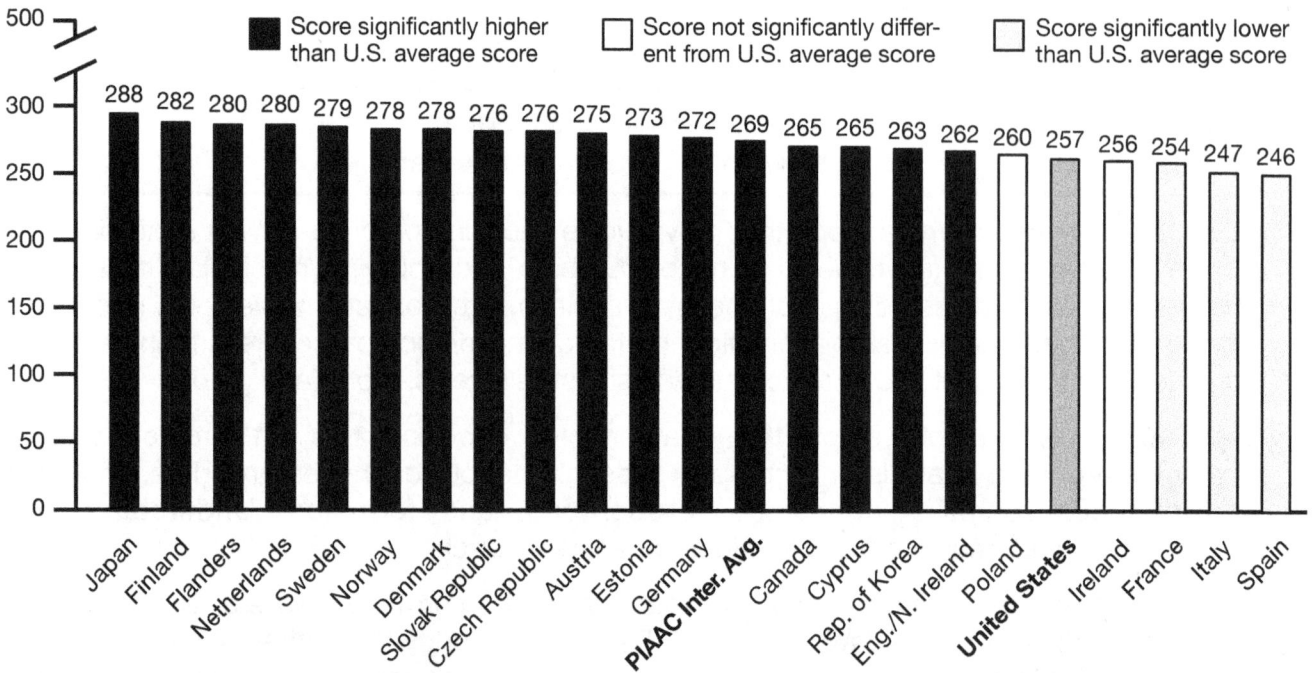

Average PIAAC Scores for Adults, Age 16-65, in 18 Countries on Problem Solving in Technology-Rich Environments: 2012/2014

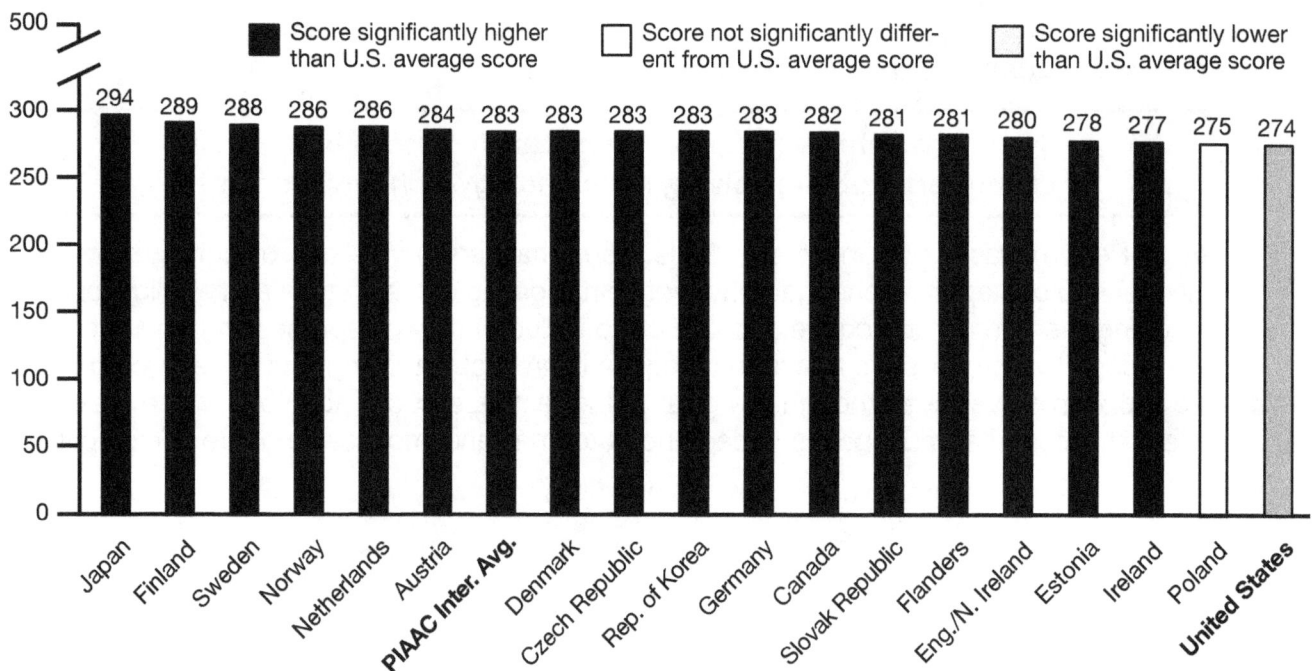

Percent of Adults, Age 16-65, in PIAAC Proficiency Levels for Numeracy: 2012

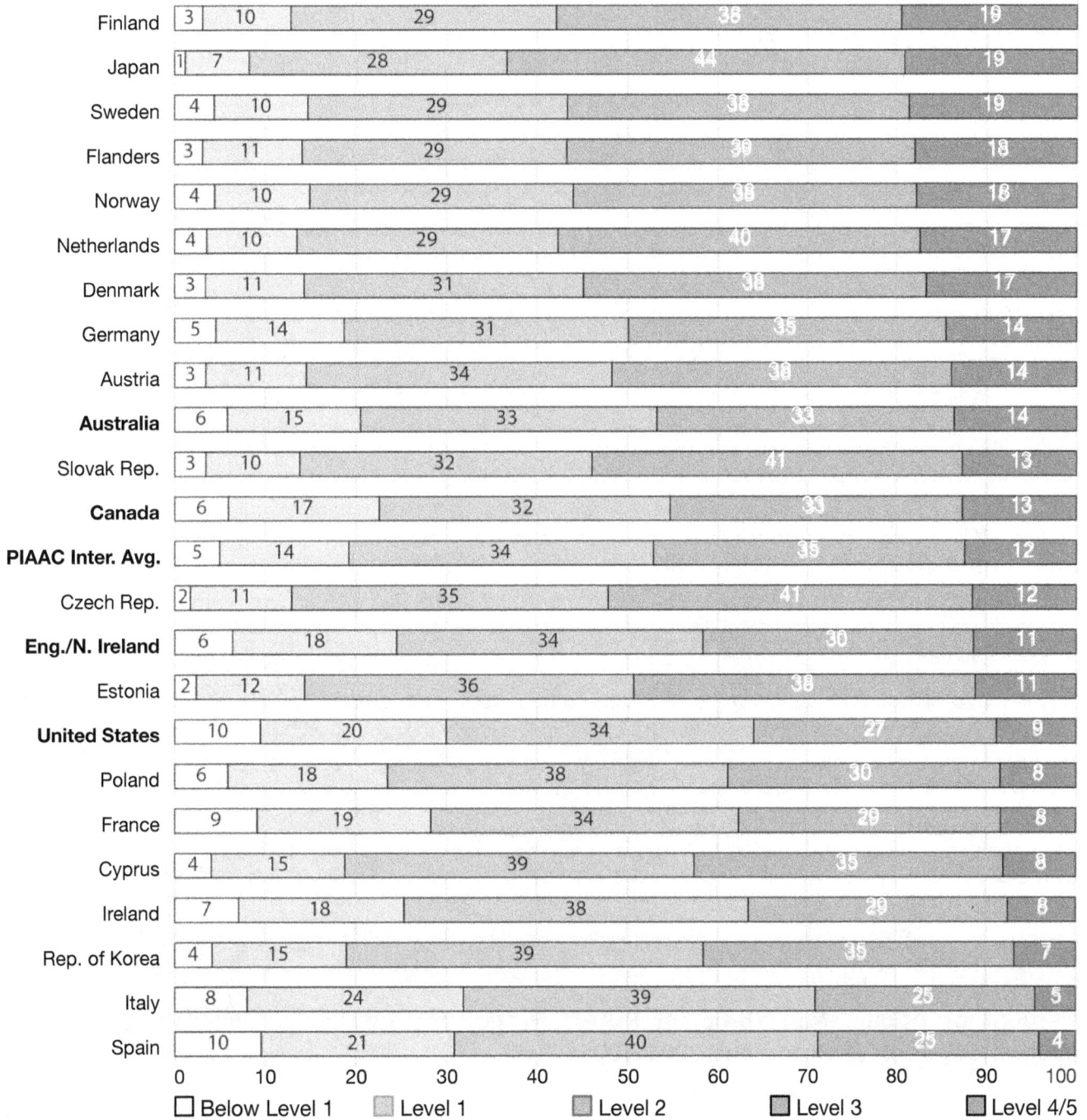

Country	Below Level 1	Level 1	Level 2	Level 3	Level 4/5
Finland	3	10	29	38	19
Japan	1	7	28	44	19
Sweden	4	10	29	38	19
Flanders	3	11	29	39	18
Norway	4	10	29	38	18
Netherlands	4	10	29	40	17
Denmark	3	11	31	38	17
Germany	5	14	31	35	14
Austria	3	11	34	38	14
Australia	6	15	33	33	14
Slovak Rep.	3	10	32	41	13
Canada	6	17	32	33	13
PIAAC Inter. Avg.	5	14	34	35	12
Czech Rep.	2	11	35	41	12
Eng./N. Ireland	6	18	34	30	11
Estonia	2	12	36	38	11
United States	10	20	34	27	9
Poland	6	18	38	30	8
France	9	19	34	29	8
Cyprus	4	15	39	35	8
Ireland	7	18	38	29	8
Rep. of Korea	4	15	39	35	7
Italy	8	24	39	25	5
Spain	10	21	40	25	4

□ Below Level 1   ▨ Level 1   ▨ Level 2   ▨ Level 3   ▨ Level 4/5

Percent of Adults. Age 16-65. in PIAAC Proficiency Levels for Numeracy: 2012/2014

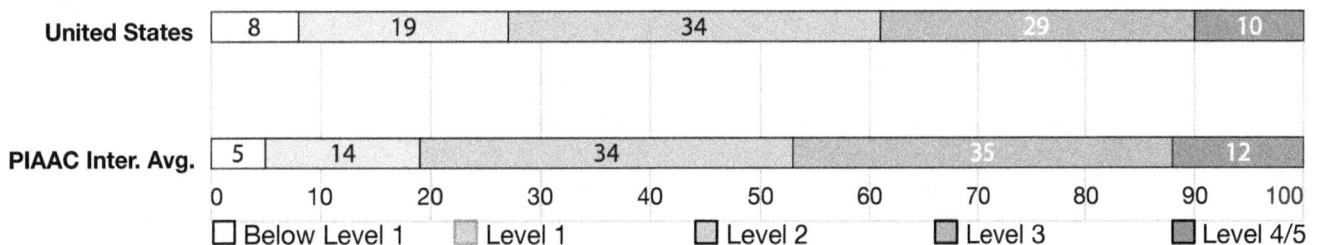

Country	Below Level 1	Level 1	Level 2	Level 3	Level 4/5
United States	8	19	34	29	10
PIAAC Inter. Avg.	5	14	34	35	12

□ Below Level 1   ▨ Level 1   ▨ Level 2   ▨ Level 3   ▨ Level 4/5

Percent of Adults, Age 16-65, in PIAAC Proficiency Levels for Problem Solving in Technology-Rich Environments: 2012

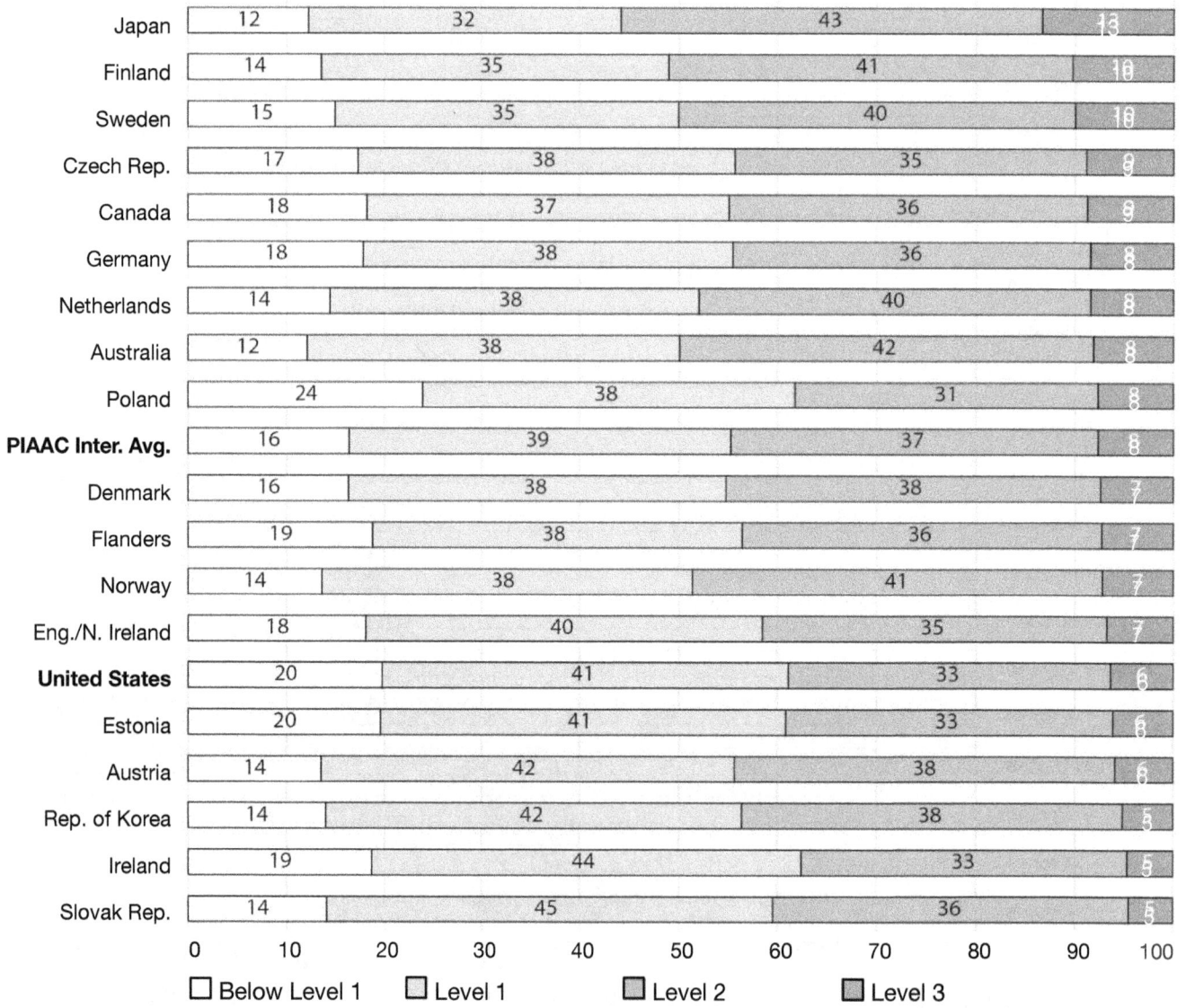

Country	Below Level 1	Level 1	Level 2	Level 3
Japan	12	32	43	13
Finland	14	35	41	10
Sweden	15	35	40	10
Czech Rep.	17	38	35	9
Canada	18	37	36	9
Germany	18	38	36	8
Netherlands	14	38	40	8
Australia	12	38	42	8
Poland	24	38	31	8
PIAAC Inter. Avg.	16	39	37	8
Denmark	16	38	38	7
Flanders	19	38	36	7
Norway	14	38	41	7
Eng./N. Ireland	18	40	35	7
United States	20	41	33	6
Estonia	20	41	33	6
Austria	14	42	38	6
Rep. of Korea	14	42	38	5
Ireland	19	44	33	5
Slovak Rep.	14	45	36	5

☐ Below Level 1 ☐ Level 1 ☐ Level 2 ☐ Level 3

Percent of Adults, Age 16-65, in PIAAC Proficiency Levels for Problem Solving in Technology-Rich Environments: 2012/2014

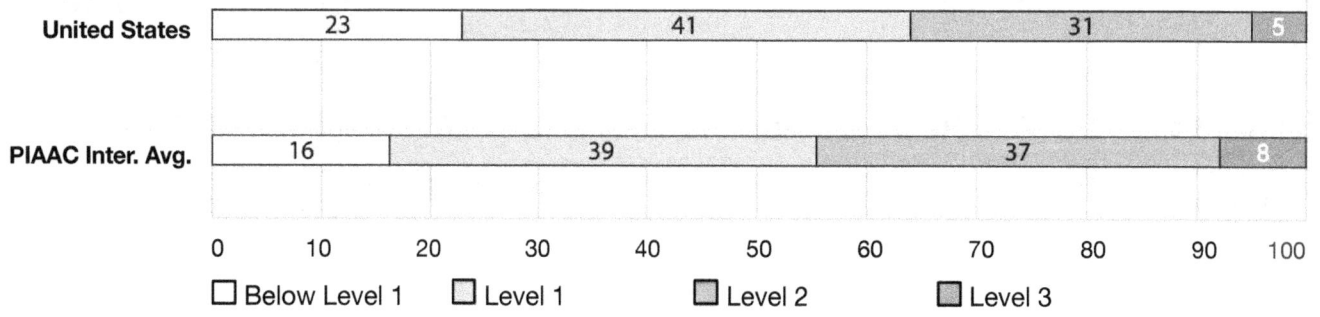

Country	Below Level 1	Level 1	Level 2	Level 3
United States	23	41	31	5
PIAAC Inter. Avg.	16	39	37	8

☐ Below Level 1 ☐ Level 1 ☐ Level 2 ☐ Level 3

Average Scores for U.S. Adults, Age 16-65, on the PIAAC Scales for Numeracy and Digital Problem Solving: 2012/2014 and 2017

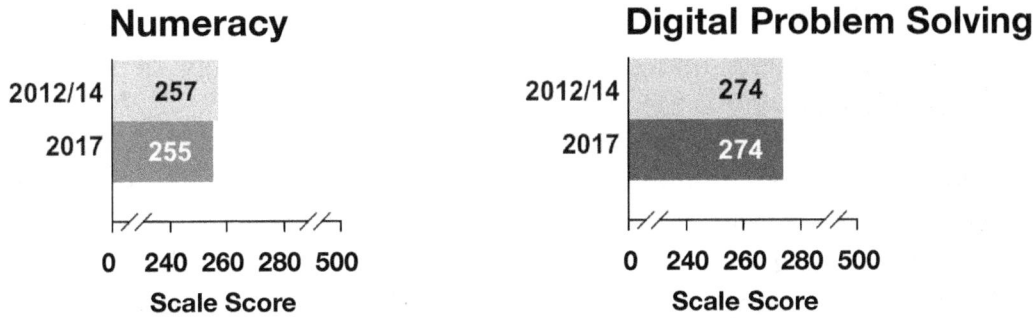

Numeracy

	Scale Score
2012/14	257
2017	255

0 240 260 280 500
Scale Score

Digital Problem Solving

	Scale Score
2012/14	274
2017	274

0 240 260 280 500
Scale Score

Percent of U.S. Adults, Age 16-65, in PIAAC Proficiency Levels for Numeracy and Digital Problem Solving: 2012/2014 and 2017

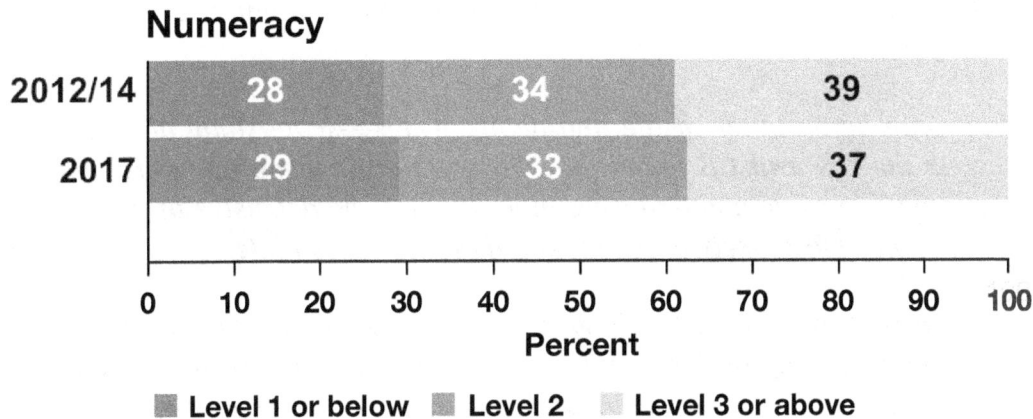

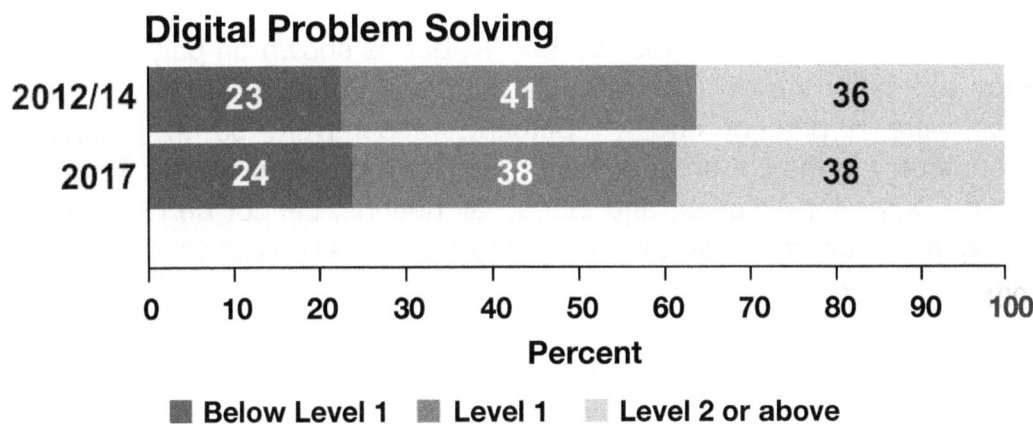

Numeracy

	Level 1 or below	Level 2	Level 3 or above
2012/14	28	34	39
2017	29	33	37

0 10 20 30 40 50 60 70 80 90 100
Percent

◼ Level 1 or below   ◼ Level 2   Level 3 or above

Digital Problem Solving

	Below Level 1	Level 1	Level 2 or above
2012/14	23	41	36
2017	24	38	38

0 10 20 30 40 50 60 70 80 90 100
Percent

◼ Below Level 1   ◼ Level 1   Level 2 or above

As you can see in comparing the PIAAC results over time, if the educational input for K-12 math is not changed, how Americans compare with their international peers in numeracy and problem solving in technology-rich environments will not change.

Do Low Scores for Numeracy Matter?

Low scores for numeracy absolutely matter! As explained by Butterworth, a cognitive psychologist, in his book *Can Fish Count?* (2022), adults who score below Level 1 on numeracy are severely disadvantaged, and one can only imagine the plight of those who score below Level 1 on problem solving in technology-rich environments.

> *"Poor understanding of numbers is a serious handicap for individuals. ... It makes individuals less employable, creates a risk of depression in adulthood, and lowers lifetime earnings significantly. In the UK about 25 per cent or 15 million adults ... have numeracy skills lower than those expected of an 11-year-old. Of these, 6.8 million have skills below the standard expected for a 9-year-old: These problems persist into adulthood: 74 per cent of 37-year-olds have problems with division, 57 per cent with subtraction, 15 per cent could not manage their household accounts and 8 per cent could only manage their household accounts with difficulty. Low numeracy causes poorer educational outcomes, lower earnings and trouble with the law, and impacts on mental and physical health. It is a cause of distress, low self-esteem, stigmatization and disruptive behavior in class. ...*
>
> *"Indeed, low numeracy can be a matter of life and death. A large-scale study of adults with colorectal cancer in the UK **and US** found that those with low numeracy were less likely to intend to participate in screening and were more likely to be defensive in getting cancer information, and hence more likely to be untreated or treated too late.*
>
> *"Low Numeracy is also a cost to nations. In 2009, it was calculated that the cost of the lowest 6 per cent of numeracy to England is about £2.4 billion a year in lost taxes due to lower earnings, higher rates of unemployment, and increased costs of crime, social security, education and health."*

Granted, Butterworth's statistics are for numeracy in the UK, but as shown on page 249, the below Level 1 numeracy scores for the UK (Australia, Canada, and England/Northern Ireland) and the United States were 6 percent and 10 percent, respectively, so the negative consequences of low numeracy in the UK are less than those for the United States. Thus the importance of *Arithmetic Counts* for America, and indeed all nations, cannot be overstated. Simply put, the book explains how to dispel low numeracy for virtually *all* children, thus all adults worldwide when they mature.

From Chapter 9, page 160:

> *One of the most prominent merits of [the methodology to teach arithmetic that is outlined in Arithmetic Counts] is its unparalleled capability of instilling in early primary students a math-related self-confidence and the notion that math is fun and easy. When a first grader correctly solves upper grade math problems such as subtracting 2,540,159 from 8,002,011; multiplying 7,586,423 by 3; and dividing 7,842,843,096 by 6 and ends with a shrug, a smile, and words of pride, 'It's simple,' one cannot resist appreciating the effectiveness of [said methodology].* — Kwame Opuni, 1999

So What?

"We, the people," can ignore the TIMSS results that portray America's fourth and eighth graders as math and science dummies and the PIAAC findings that depict America as a backward nation of technologically-challenged math dullards. We can stick with the Common Core State Standards curriculum for elementary school math that makes memorizing arithmetic the Law of the Land. We can adhere to the elitism that rejects counting out the number facts *"quickly and accurately."* We can cling to the false belief that success in advanced mathematics requires being able to recall the multiplication facts as quickly and unthinkingly as recalling one's name.

We can do all of the above, or we can say *"enough!"* to what is holding America's children back in math and branding them, when adults, as unfit to fully participate in 21st-century society and the global economy. We can nix the forced memorization of arithmetic and instead teach it with understanding. We can replace the hundreds of worksheets and years of drill and practice on the number facts with songs and dances and how to count them out and be done with them by the end of grade 2. *We can teach every child whole number arithmetic by then and can prove it with the Monster Math certificates they earn for all four operations.*

"We, the people," will eventually say *"enough"* to what is holding our children back in math because it is the sensible thing to do or because forces beyond America's control will require it, the same as they required adopting the NCTM Standards in 1989. The following illustrate the intensity of the forces that motivated the adoption of said standards as the *first-ever* national curriculum for K-12 math.

Business

The Sputnik of the 1980's is economic competition, and the United States is beginning to recognize that the educational attainment of its people is an intrinsic element of national economic well-being. — John Jennings, 1987

A 1989 poll of the FORTUNE 500 companies indicated that over half of the 400 respondents currently find it difficult to hire employees with good basic skills. One third of the companies offer their employees remedial courses to improve reading, writing and math abilities. — Union Carbide Corporate Task Force on Education, 1989

If you don't have workers with strong math and science backgrounds, you don't have the edge in technology. And if you don't have the edge in technology, you don't have the competitive edge. The fact that other nations are doing a better job educating their young people is reflected in those countries' gains in the world marketplace. — John Clendenin, Chief Executive Officer, BellSouth Corporation, 1989

In the old economy of Texas, there was little foreign competition. Texans had little to fear from other nations. In the new economy of Texas, we live with a climate of heavy foreign competition. ... We understand more than ever the economic interrelations and trade interrelations that bind us in a complex world where decisions made in Tokyo, as we prepare to sleep, reverberate around to the London stock market by the time we awake and are felt on the stock market in New York by the time we are at work at 9:00. We live in that kind of interconnected electronically wired world where foreign competition is a dimension of our lives. — Henry Cisneros, Mayor (1981-1989), San Antonio, Texas, 1989

Government

All industrialized countries have experienced a shift from an industrial to an information society, a shift that has transformed both the aspects of mathematics that need to be transmitted to students and the concepts and procedures they must master if they are to be self-fulfilled, productive citizens in the next century. — National Council of Teachers of Mathematics, 1989

Mathematics is the key to opportunity. No longer just the language of science, mathematics now contributes in direct and fundamental ways to business, finance, health, and defense. For students, it opens doors to careers. For citizens, it enables informed decisions. For nations, it provides knowledge to compete in a technological economy. To participate fully in the world of the future, America must tap the power of mathematics. — National Research Council, 1989

Mathematics is the foundation of science and technology. Increasingly, it plays a major role in determining the strength of the nation's work force. Yet, evidence all around us shows that American students are not fulfilling their potential in mathematics education. — National Research Council, 1989

When one compares the potential return on investment in education with the consequences of inaction, it becomes clear that we as a nation have no choice: we must improve the ways our children learn mathematics. — National Research Council, 1989

Education

We've inherited a woefully limited set of expectations of what schools can accomplish and what children can learn." — William R. Graham, 1987

Supposedly the [elementary school math] curriculum is sequential. Actually, it's circular. It goes back every year and starts a little bit further than it did the previous year, but not much. — Zalman Usiskin, Director, University of Chicago School Mathematics Project, 1989

Our school system wastes people by underestimating them; it is oriented toward finding out what kids can't do rather than what they can do. — Albert Shanker, President (1965-1986), American Federation of Teachers (AFT), 1989

We must completely transform American education. Not with incremental reform. Not with casual, cautious reform. It's too late for tinkering. We need massive, system wide restructuring. — Mary Hatwood Futrell, President (1983-1989), National Education Association (NEA),1989

We need to emphasize math and science early on in education before it's too late. We can't wait to interest students in entering high tech professions right before they graduate, because they can't go back and get the basics they may have missed up front. — Raymond A. Reed, Director of Community Relations, Rockwell International, 1989

The same forces are building again as America reckons with the failure of the NCTM Standards and its derivative, the Common Core State Standards for Mathematics, to meet expectations.

Based on national and international assessments, the United States math scores have stayed stagnant, while other countries have seen significant growth in their scores. — Denise Rawding, 2016

U.S. Students' Academic Achievement Still Lags That of Their Peers in Many Other Countries. — Drew Desilver, 2017

Math Scores for U.S. Students Hit All-Time Low on International Exam. — Donna St. George, 2023

U.S. Reading and Math Scores Drop to Lowest Level in Decades. — Sequoia Carrillo, 2023

Being Real

If *"we, the people,"* want our children to perform as well or better in math and science as their international peers, we have to teach arithmetic better and teach it in conjunction with hands-on algebra from kindergarten on. We have to change the arithmetic curriculum and create a K-6 algebra curriculum with concrete referents for the symbolics. Instead, we hear our politicians talk about paying teachers more, building more charter schools, and handing out tax supported vouchers that parents can "spend" on tuition to send their children to private schools.

Paying elementary school teachers more will not affect how they teach arithmetic. Their students will not even know they got a raise. Grade school teachers teach the curriculum they are given, which, for arithmetic, means to teach what is in their textbooks.

Likewise, more charter schools and issuing school vouchers will not affect how arithmetic is taught. Charter schools and private schools use the same textbooks for elementary school math as the public schools, so they hold their students back the same as the public schools. Besides, charter schools and school vouchers are only secondarily about improving education. Their main reason for being is to privatize public education with taxpayer dollars.

The education industry represents ... the final frontier of a number of sectors once under public control that ... have been <u>forced</u> to open up to private enterprise. Indeed, ... the education industry represents the largest market opportunity since health-care services were privatized during the 1970s. ... The larger developing opportunity is in the K-12 ... market led by private elementary school providers who are well positioned to exploit potential political reforms such as school vouchers. From the point of view of private profit, the K-12 market is the <u>Big Enchilada</u>. — Jonathan Kozal, 2007

To teach arithmetic better, the *curriculum* for elementary school math must be reformed, which is what this book is about. Until that occurs, publishers of textbooks for elementary school math will continue to publish textbooks that are in sync with the NCTM Standards/Common Core State Standards for Mathematics, as emblazoned with CCSSM on their covers. Said textbooks will *never* result in America's fourth and eighth graders catching up in math and science with their peers in most developed nations, *not even when they are grown,* <u>which is the problem.</u>

What Changes in How Arithmetic Is Taught Can Be Made Now?

Although individual teachers cannot teach *"too much"* math in grade school without risking being fired, schools can teach as much math as their teachers know how to teach. An elementary school that wishes to teach arithmetic with the methodology used in MOVE IT Math can use it for that purpose.[64] They can call it whatever they want and use their textbooks for everything else in the math curriculum, like geometry, graphing, and measurement until the textbooks are updated to reflect the methodology used to teach arithmetic in MOVE IT Math. All MOVE IT Math owns is the name itself, not its methodology. That aside, for a school to implement MOVE IT Math's methodology for teaching arithmetic would be a daring undertaking.

An elementary school teaching arithmetic with the methodology used in MOVE IT Math would be on a collision course with their state education agency, the National Council of Teachers of Mathematics, the U.S. Department of Education, and publishers of educational materials for elementary school math because they all have invested heavily in the status quo. The school would also invite the wrath of professors of early childhood education who, in the 1990s, insisted that the K-2 math curriculum adhere to developmental theories that focused on what children supposedly could not do based on their age, even when shown otherwise.

> *Piaget has been wrongly used as the guru to say you shouldn't do this until a certain age.*
> — Zalman Usiskin, Director, University of Chicago School Mathematics Project, 1989

Obstacles to Reforming the Elementary School Math Curriculum

> *"If you can't change your mind, you can't change anything."* — Joe Manchin, U.S. Senator

The biggest obstacle to changing how arithmetic is taught in America's elementary schools is the pervasive belief among Americans that it should be taught as they were taught:

> *Dominant cultural beliefs about the teaching and learning of mathematics continue to be obstacles to consistent implementation of effective teaching and learning in mathematics classrooms. Many parents and educators believe that students should be taught as they were taught, through memorizing facts, formulas, and procedures and then practicing skills over and over again. This view perpetuates the traditional lesson paradigm that features review, demonstration, and practice and is still pervasive in many classrooms. Teachers, as well as parents, are often not convinced that straying from these established beliefs and practices will be more effective for student learning.* — National Council of Teachers of Mathematics, 2014

The problem with Americans wanting arithmetic to be taught how it was taught to them is what they were taught *besides* arithmetic. They were taught that it should be memorized and that counting out a number fact on their fingers was BAD. However, I have never seen a chapter on no finger counting in any of the textbooks for elementary school math that I have examined since I began doing so in the 1990s. Why not is because a chapter about no finger counting would not be about arithmetic. It would be about how to *behave* in math class.

---

[64] The particulars on how to teach arithmetic with MOVE IT Math is **free** @ moveitmath.com.

How counting out the number facts became unacceptable in teaching math is a mystery, but the NCTM's snide attitude about counting out a number fact is unprofessional and deceitful. They know that its negativity about counting out the number facts is personal. They know it has nothing to do with arithmetic, per se, or they would just say it: *"No counting allowed!"* Instead, they hide their intolerance of any behavior that is reminiscent of children counting on their fingers by complaining that such behavior lacks *"fluency,"* which they define as *"skill in carrying out procedures flexibly, accurately, efficiently, and appropriately."*

The only word in the NCTM's definition of *"fluency"* that has to do with arithmetic is the word *"accurately."* *"Flexibly"* and *"efficiently"* are code for *"If you forget a number fact while computing, don't count it out. Figure it out with mental math and tough luck if you can't."* And *"appropriately"* means *"Stop counting out the number facts. We don't care if you can count them out 'quickly and accurately.' We don't like how it looks when you do that."*

What America CAN Do for Its Children?

What America needs is for the U.S. Congress to enact another bill like the National Defense Education Act to fund another round of New Math, *except this time for elementary school teachers*. Said teachers need to be teaching arithmetic instead of covering pages in textbooks, and the ones that I have worked with have been eager to do that. But to do that, all 1.9 million of them need to know the truth about the automaticity standard and how it is holding children back in math.

One way for every elementary school teacher in America to know that about the automaticity standard is for the U.S. Department of Education to replicate the DeMaioribus study nationwide. To do so, they could task their ten regional service centers to have the study replicated in the states they serve. Having done that, each regional service center could then ask the education departments of the universities in the states they serve to replicate the study and submit their findings to the USDOE to analyze and summarize in a report for the American public with a title like The Automaticity Standard, a Pump or a Sieve (TASPS).

USDOE Regional Service Centers

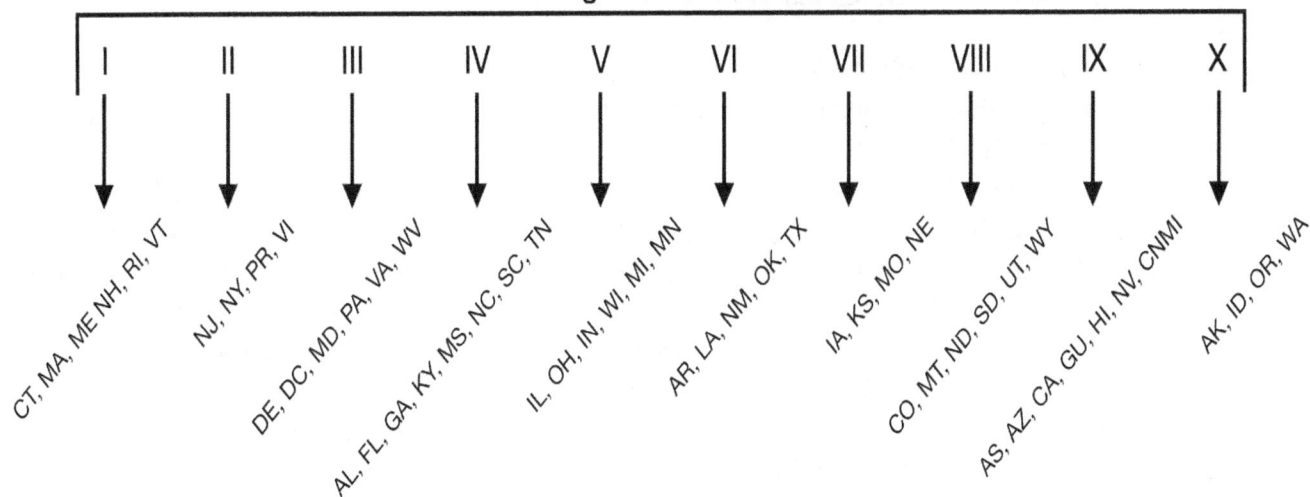

I	II	III	IV	V	VI	VII	VIII	IX	X
CT, MA, ME NH, RI, VT	NJ, NY, PR, VI	DE, DC, MD, PA, VA, WV	AL, FL, GA, KY, MS, NC, SC, TN	IL, OH, IN, WI, MI, MN	AR, LA, NM, OK, TX	IA, KS, MO, NE	CO, MT, ND, SD, UT, WY	AS, AZ, CA, GU, HI, NV, CNMI	AK, ID, OR, WA

We know what the imagined TASPS report is going to say about the automaticity standard. It will second DeMaioribus' findings and show that the automaticity standard is not a pump that causes students to rise up the Math Ladder. Instead, it will show that it is a finely-meshed sieve that filters out the vast majority of students, leaving them to overcome doubts about their mental capacity.

I do not remember ever struggling to memorize the number facts. Nor do I remember ever forgetting one, but I remember forgetting lots of names, events, and dates in history class in high school. Importantly, though, I never thought I was stupid when I forgot something in history class. *I thought history was stupid*. Although I do not think that about history now, I cannot imagine how I would have ended up in life if my high school history teacher had intimated that there was something wrong with me because I kept forgetting what he wanted me to remember.

Americans need to debate around dinner tables the findings in a report like the imagined TASPS. To reject the automaticity standard, they need to grapple with the fact that most students will forget some of the multiplication facts when they no longer practice them or use them in a math class. In rejecting it, they need to question the age-old practice of devoting the first three years of school — *all of K-2!* — to just adding and subtracting small numbers while striving to memorize the addition and subtraction facts. They need to see for themselves that first graders can learn how to count out all 200 of them in a matter of *hours* — not years — the same as I realized that with my daughter Jessica when she was in the first grade, as told on page 122.

Now for the fun part — *really!* — that of teaching all 1.9 million elementary school teachers in America how to actually/finally *teach* arithmetic. Assuming that America has come to grips with the crippling effect of the automaticity standard on the arithmetic curriculum and the toll it exacts on children's perception of their cognitive ability, it is full steam ahead to scrap the Common Core humdrum for arithmetic with the smorgasbord of games and hands-on activities that characterize teaching arithmetic the MOVE IT Math way, hereafter referred to as the new methodology for teaching arithmetic.

Teaching the New Methodology for Teaching Arithmetic to Every Elementary School Teacher in America

To ensure that all of America's children excel in arithmetic and thereby dispel the effects of low numeracy among them when they are grown, the U.S. Congress must enact a bill to fund teaching the new methodology for teaching arithmetic to every elementary school teacher in America and equipping their classrooms with the means to implement it. Specifically, said funding must provide for the creation of instructional videos for the five "golden keys" in Chapter 11 and a teacher's manual keyed to the videos. Additionally, it must provide for follow-up conferences and compensation for teachers who attend them plus compensation for teachers who receive instruction on the new methodology outside normal school hours. By putting the videos on YouTube and making the teacher's manual downloadable on a website for the endeavor, all 1.9 million elementary school teachers could be taught the new methodology for teaching arithmetic in one year.

Closing Remarks

The burden of citizenship is accepting that what is neither your fault nor your responsibility may be your problem. — Anand Giridharadas

Skip count by two to count the letters: *If-it-is-to-be-it-is-up-to-us.*

It is up to us — you, me, and our fellow Americans — to speak on behalf of our children and grandchildren to halt the century-old practice of intimidating them in elementary school with the forced memorization of arithmetic and the threat of bad things to follow if they forget a number fact while computing. It is up to us to end the 35-year drought of intellectual stimulation brought on by the NCTM Standards/Common Core State Standards curriculum for elementary school math. Said curriculum is holding our children and grandchildren back in math and will do so indefinitely.

Only a small percentage of children can memorize arithmetic and remember it for the rest of their lives without practicing it periodically if they do not use it in their work or around the home. For those that can remember it, they can dream big about their future because they can expect to do well in algebra and advanced mathematics, *but not really.*

Many classrooms focus on math facts in unproductive ways [e.g., endless repetition, timed testing], giving students the impression that math facts are the essence of mathematics, and, even worse, that the fast recall of math facts is what it means to be a strong mathematics student. <u>Both of these ideas are wrong</u>. — Jo Boaler, 2015

What is not a good thing is the many children who keep forgetting some of the number facts, especially the multiplication facts, and take that as a sign that they lack the innate ability to pursue a dream that requires a knowledge of advanced mathematics.

Schools should be teaching children — *all children* — to dream big. They should not be limiting their dreams with a math curriculum that is guaranteed to make them look stupid when compared to children from other developed nations.

If taught how to count out the number facts, all 200 addition and subtraction facts can be taught to every child in a matter of *weeks* in kindergarten, and all 190 multiplication and division facts can be taught to every child with songs and dances by the end of grade 2 by teaching them how to skip count by 2, 5, and 10 in kindergarten, 3 and 4 in grade 1, and 6, 7, 8, and 9 in grade 2.

The MOVE IT Math methodology for arithmetic provides a clear choice for America: *"We, the people,"* can cling to the automaticity standard and the onus on counting while computing and accept the stifling affect that has on our children's ability to advance in math. Instead, we can teach arithmetic with understanding and let children count out the number facts and zip through whole number arithmetic by the end of grade 2, with achievement certificates to prove it.

All children can realize the self-esteem and self-confidence that comes with excelling in arithmetic. When children understand base ten numeration and are empowered *for life* with knowing how to count out all 390 number facts, they *embrace* arithmetic. Knowing they know arithmetic, they are eager and prepared to learn algebra and move up the Math Ladder, unlike the many today who do poorly in arithmetic and come to dread math.

References

Boaler, Jo. "Fluency without Fear," YouCubed (youcubed.org), 2015.

Carrillo, Sequoia. "Test Scores Drop Among U.S. Students in Reading and Math." NPR (National Public Radio), June 21, 2023.

CCSSM. See *Common Core State Standards for Mathematics.*

Cisneros, Henry. Mayor (1981-1989), San Antonio, Texas. Keynote address at the 4th annual Title II conference of the Education for Economic Security Act (EESA), San Antonio, January 4-6, 1989.

Clendenin, John. Union Carbide Corporate Task Force on Education, *Undereducated Uncompetitive USA.* Union Carbide Corporation, 1989.

Common Core State Standards for Mathematics, National Governors Association in collaboration with the Council of Chief State School Officers, 2010.

Curriculum and Evaluation Standards for School Mathematics, National Council of Teachers of Mathematics, 1989.

Desilver, Drew. "U.S. Students' Academic Achievement Still Lags That of Their Peers in Many Other Countries," Pew Research Center, February 15, 2017.

Futrell, Mary Hatwood. "Mission Not Accomplished: Education Reform in Retrospect." *Phi Delta Kappan*, Vol. 71, No. 1, 1989.

Giridharadas, Anand. American journalist and political pundit. Former columnist for The New York Times.

Goodman, M. and others. *Literacy, Numeracy, and Problem Solving in Technology-Rich Environments Among U.S. Adults: Results from the Program for the International Assessment of Adult Competencies* (PIAAC), *First Look*. National Center for Education Statistics, U.S. Dept. of Education, 2013.

Graham, William R. "Challenges to the Mathematics Community." Notices of the American Mathematical Society, 34 (February), 1987.

Jennings, John F. "The Sputnik of the Eighties." *Phi Delta Kappan*, Vol. 69, No. 2, 1987.

Kozal, Jonathan, "The Big Enchilada," Harper's Magazine Notebook, August, 2007.

Maddow, Rachael. The Rachael Maddow Show, MSNBC, June 29, 2021.

Manchin, Joe. U.S. Senator, 2010-2024.

National Commission on Excellence in Education. *A Nation at Risk: The Imperative for Educational Reform*, U.S. Department of Education, 1983.

National Council of Teachers of Mathematics, *Principles to Actions: Ensuring Mathematical Success for All*, 2014.

National Research Council. *Everybody Counts: A Report to the Nation on the Future of Mathematics Education.* National Academy Press, Washington, D.C., 1989.

NCTM Standards. See *Curriculum and Evaluation Standards for School Mathematics.*

OECD (Organization for Economic Cooperation and Development). *Literacy, Numeracy, and Problem Solving in Technology-Rich Environments: Framework for the OECD Survey of Adult Skills*, OECD Publishing, 2012.

Rampey, B.D. and others. "Skills of U.S. Unemployed, Young, and Older Adults in Sharper Focus: Results From the Program for the International Assessment of Adult Competencies (PIAAC) 2012/2014, First Look," National Center for Education Statistics, U.S. Department of Education, 2016.

Rampey, B.D. and others. "Highlights of the 2017 U.S. PIAAC Results Web Report," National Center for Education Statistics, U.S. Department of Education, 2019.

Rawding, Denise M. "Common Core State Standards for Mathematics: How Well do the Textbook and Instructional Methods Align?" College of Saint Elizabeth, ProQuest Dissertations Pub., 2016.

Reed, Raymond A. "Establishing/Enhancing Industry/Education Partnerships." Address at the Texas Education for Economic Security Act (EESA), Title II 4th Annual Conference, San Antonio, Texas, January 4-6, 1989.

Schulz, Charles M. BrainyQuote.com.

Shanker, Albert. President, American Federation of Teachers (1964-1985), Union Carbide Corporate Forum, 1989.

St. George, Donna. "Math Scores for U.S. Students Hit All-Time Low on International Exam," *The Washington Post*, December 5, 2023.

Union Carbide Corporate Task Force on Education. *Under-Educated Uncompetitive USA*, 1989.

Usiskin, Zalman. University of Chicago. The University of Chicago School Mathematics Project (UCSMP), "Preparing Teachers and Students for the Present." Address at the Texas Education for Economic Security Act (EESA), Title II 4th Annual Conference, January 4-6, 1989.

Westervelt, Eric. Broadcast on All Things Considered, October 8, 2015.

Appendix 263

Common Core State Standards for Arithmetic, K-6

Copied from the Common Core State Standards website: corestandards.org/Other Resources/Key Shifts in Mathematics.

Greater focus on fewer topics: The Common Core calls for greater focus in mathematics. Rather than racing to cover many topics in a mile-wide, inch-deep curriculum, the standards ask math teachers to significantly narrow and deepen [drag out] the way time and energy are spent in the classroom. This means focusing deeply [reducing the curriculum to just a few topics that are carried over from one year to the next for years] on the major work of each grade as follows:

○ Kindergarten through Grade 2

  Concepts, skills, and problem solving related to addition and subtraction of whole numbers.

○ Grades 3 through 5

  Concepts, skills, and problem solving related to multiplication and division of whole numbers and fractions.

○ Grade 6

  Ratios and proportional relationships, and early algebraic expressions and equations.

Common Core State Standards for Arithmetic, K-3

Copied from *Common Core State Standards for Mathematics*, National Governors Association Center for Best Practices and Council of Chief State School Officers, 2010, except for italicized examples and comments in brackets.

Kindergarten

In kindergarten, instructional time [for arithmetic] should focus on [one] critical area: representing, relating, and operating on whole numbers, initially with sets of objects. Students use numbers, including written numerals, to represent quantities and to solve quantitative problems, such as counting objects in a set; counting out a given number of objects; comparing sets or numerals; and modeling simple joining and separating situations with sets of objects, or eventually with equations such as 5 + 2 = 7 and 7 – 2 = 5.

Counting & Cardinality

Know number names and the count sequence.

○ **K.CC.1**

  Count to 100 by ones and by tens.

  [Count 1, 2, 3 ... 100 and 10, 20, 30 ... 100.]

- **K.CC.2**

   Count forward beginning from a given number within the known sequence (instead of having to begin at 1).

   [For example, to count on forward 3 from 9, count 10, 11, 12.]

- **K.CC.3**

   Write numbers from 0 to 20. Represent a number of objects with a written numeral 0-20 (with 0 representing a count of no objects).

   [Write 0, 1, 2, 3 ... 20.]

Count to tell the number of objects.

- **K.CC.4**

   Understand the relationship between numbers and quantities; connect counting to cardinality.

   [The cardinal number of a set is the number of elements in the set. For example, the cardinal number of {a, b, c, d, e} is five.]

- **K.CC.4a**

   When counting objects, say the number names in the standard order, pairing each object with one and only one number name and each number name with one and only one object.

   [When counting objects, say the number names in the standard order and count all of the objects to be counted and count each one just once.]

- **K.CC.4b**

   Understand that the last number name said tells the number of objects counted. The number of objects is the same regardless of their arrangement or the order in which they were counted.

- **K.CC.4c**

   Understand that each successive number name refers to a quantity that is one larger.

- **K.CC.5**

   Count to answer "how many?" questions about as many as 20 things arranged in a line, a rectangular array, or a circle, or as many as 10 things in a scattered configuration. Given a number from 1-20, count out that many things.

   [For example, count things (like eggs in a carton) by counting each thing (egg) just once and assigning a numeral (12) to the number of things (eggs) counted.]

Compare numbers.

o **K.CC.6**

Identify whether the number of objects in one group is greater than, less than, or equal to the number of objects in another group, e.g., by using matching and counting strategies. **Note:** Include groups with up to ten objects.

o **K.CC.7**

Compare two numbers between 1 and 10 presented as written numerals.

Operations & Algebraic Thinking

Understand addition as putting together and adding to, and understand subtraction as taking apart and taking from.

o **K.OA.1**

Represent addition and subtraction with objects, fingers, mental images, drawings, sounds (e.g., claps), acting out situations, verbal explanations, expressions, or equations. **Note:** Drawings need not show details, but should show the mathematics in the problem. (This applies wherever drawings are mentioned in the Standards.)

o **K.OA.2**

Solve addition and subtraction word problems, and add and subtract within 10, e.g., by using objects or drawings to represent the problem.

o **K.OA.3**

Decompose numbers less than or equal to 10 into pairs in more than one way, e.g., by using objects or drawings, and record each decomposition by a drawing or equation (e.g., $5 = 2 + 3$ and $5 = 4 + 1$).

o **K.OA.4**

For any number from 1 to 9, find the number that makes 10 when added to the given number, e.g., by using objects or drawings, and record the answer with a drawing or equation.

■ **K.OA.5**

Fluently add and subtract **within 5**.

*[**Kindergarteners** must be able to quickly and accurately work out the addition and subtraction facts with sums and differences less than or equal to **five** (e.g., $2 + 3 = 5$ and $5 - 2 = 3$).*

***Note:** Unless the numerals 6, 7, 8, and 9 are not allowed in kindergarten, "fluently" subtracting within five means kindergarteners must be able to quickly and accurately work out a subtraction fact like $6 - 2 = 4$, but not its related addition facts, $2 + 4 = 6$ and $4 + 2 = 6$, because their sums are not within (less than or equal to) five, which is nonsensical.]*

Number & Operations in Base Ten

Work with numbers 11–19 to gain foundations for place value.

○ **K.NBT.1**

Compose and decompose numbers from 11 to 19 into ten ones and some further ones, e.g., by using objects or drawings, and record each composition or decomposition by a drawing or equation (e.g., 18 = 10 + 8); understand that these numbers are composed of ten ones and one, two, three, four, five, six, seven, eight, or nine ones.

Measurement and Data

Classify objects and count the number of objects in each category.

○ **K.MD.3**

Classify objects into given categories; count the numbers of objects in each category and sort the categories by count. **Note:** Limit category counts to be less than or equal to 10.

Grade 1

In Grade 1, instructional time [for arithmetic] should focus on [two] critical areas: (1) developing understanding of addition, subtraction, and strategies for addition and subtraction within 20 and (2) developing understanding of whole number relationships and place value, including grouping in tens and ones.

Operations & Algebraic Thinking

Represent and solve problems involving addition and subtraction.

○ **1.OA.1**

Use addition and subtraction within 20 to solve word problems involving situations of adding to, taking from, putting together, taking apart, and comparing, with unknowns in all positions, e.g., by using objects, drawings, and equations with a symbol for the unknown number to represent the problem. **Note:** See Glossary, Table 1.

○ **1.OA.2**

Solve word problems that call for addition of three whole numbers whose sum is less than or equal to 20, e.g., by using objects, drawings, and equations with a symbol for the unknown number to represent the problem.

[In limiting the sum of the digits in a column to 20 or less, this content standard reinforces "always carrying the 1" when adding a column of three or more numbers. As shown in Chapter 4, in adding the first column in 46 + 18 + 27 (written vertically) and getting 21, many children "carry the 1" instead of the 2. They have been conditioned to do that because almost all of the addition problems in the textbooks for K-3 are two-addend problems.]

Understand and apply properties of operations and the relationship between addition and subtraction.

o **1.OA.3**

Apply properties of operations as strategies to add and subtract. Examples: If 8 + 3 = 11 is known, then 3 + 8 = 11 is also known. (Commutative property of addition.) To add 2 + 6 + 4, the second two numbers can be added to make a ten, so 2 + 6 + 4 = 2 + 10 = 12. (Associative property of addition.) **Note:** Students need not use formal terms for these properties.

o **1.OA.4**

Understand subtraction as an unknown-addend problem. For example, subtract 10 − 8 by finding the number that makes 10 when added to 8.

[Unknown-addend problems are also called missing-addend problems because the number to add to 8 to get 10 is "missing." That is, 8 + ? = 10.]

Add and subtract within 20.

o **1.OA.5**

Relate counting to addition and subtraction (e.g., by counting on 2 to add 2).

[Also, by counting back 2 to subtract 2.]

■ **1.OA.6**

Add and subtract within 20, demonstrating **fluency** for addition and subtraction **within 10**. Use strategies such as counting on;[65] making ten (e.g., 8 + 6 = 8 + 2 + 4 = 10 + 4 = 14); decomposing a number leading to a ten (e.g., 13 − 4 = 13 − 3 − 1 = 10 − 1 = 9); using the relationship between addition and subtraction (e.g., knowing that 8 + 4 = 12, one knows 12 − 8 = 4); and creating equivalent but easier or known sums (e.g., adding 6 + 7 by creating the known equivalent 6 + 6 + 1 = 12 + 1 = 13).

*[**First graders** must be able to use the counting and mental strategies that are listed in the standard to work out all 100 addition facts from 0 + 0 = 0 to 9 + 9 = 18 and all 100 subtraction facts from 0 − 0 = 0 to 18 − 9 = 9. In particular, they must be able to quickly and accurately work out the addition and subtraction facts with sums and differences less than or equal to **ten** (e.g., 4 + 6 = 10 and 10 − 4 = 6).]*

---

[65] *Counting on is the only strategy* listed *without an example*. Counting on means to count forward a certain number of counts after some number. For example, to count on 3 from 9, count 10, 11, 12, which yields the addition fact 9 + 3 = 12, which is why counting on is *not* exemplified. Counting on may be used to work out all 100 addition facts, and counting back may be used to work out all 100 subtraction facts (e.g., to count back 3 from 9, count 8, 7, 6, which yields the subtraction fact 9 − 3 = 6), but the leadership of the National Council of Teachers of Mathematics abhors counting in arithmetic.

Work with addition and subtraction equations.

o **1.OA.7**

Understand the meaning of the equal sign, and determine if equations involving addition and subtraction are true or false. For example, which of the following equations are true and which are false? $6 = 6$, $7 = 8 - 1$, $5 + 2 = 2 + 5$, $4 + 1 = 5 + 2$.

o **1.OA.8**

Determine the unknown whole number in an addition or subtraction equation relating three whole numbers. For example, determine the unknown number that makes the equation true in each of the equations $8 + ? = 11$, $5 = \_\_ - 3$, $6 + 6 = \_\_$.

[Note the avoidance of literal numbers like x, y, z to represent the unknown numbers.]

Number & Operations in Base Ten

Extend the counting sequence.

o **1.NBT.1**

Count to 120, starting at any number less than 120. In this range, read and write numerals and represent a number of objects with a written numeral.

Understand place value.

o **1.NBT.2**

Understand that the two digits of a two-digit number represent amounts of tens and ones. Understand the following as special cases:

o **1.NBT.2a**

[The number] 10 can be thought of as a bundle of ten ones — called a "ten."

o **1.NBT.2b**

The numbers from 11 to 19 are composed of a ten and one, two, three, four, five, six, seven, eight, or nine ones.

o **1.NBT.2c**

The numbers 10, 20, 30, 40, 50, 60, 70, 80, 90 refer to one, two, three, four, five, six, seven, eight, or nine tens (and 0 ones).

o **1.NBT.3**

Compare two two-digit numbers based on meanings of the tens and ones digits, recording the results of comparisons with the symbols >, =, and <.

Use place value understanding and properties of operations to add and subtract.

o **1.NBT.4**

Add within 100, including adding a two-digit number and a one-digit number, and adding a two-digit number and a multiple of 10, using concrete models or drawings and strategies based on place value, properties of operations, and/or the relationship between addition and subtraction; relate the strategy to a written method and explain the reasoning used. Understand that in adding two-digit numbers, one adds tens and tens, ones and ones; and sometimes it is necessary to compose a ten.

o **1.NBT.5**

Given a two-digit number, mentally find 10 more or 10 less than the number, **without having to count**; explain the reasoning used.

[Note the negativity toward counting.]

o **1.NBT.6**

Subtract multiples of 10 in the range 10-90 from multiples of 10 in the range 10-90 (positive or zero differences), using concrete models or drawings and strategies based on place value, properties of operations, and/or the relationship between addition and subtraction; relate the strategy to a written method and explain the reasoning used.

Geometry

Reason with shapes and their attributes.

o **1.G.3**

Partition circles and rectangles into two and four equal shares, describe the shares using the words halves, fourths, and quarters, and use the phrases half of, fourth of, and quarter of. Describe the whole as two of, or four of the shares. Understand for these examples that decomposing into more equal shares creates smaller shares.

Grade 2

In Grade 2, instructional time [for arithmetic] focus on [two] critical areas: (1) extending understanding of base-ten notation and (2) building **fluency** with addition and subtraction [memorizing the algorithms for the operations].

Operations & Algebraic Thinking

Represent and solve problems involving addition and subtraction.

o **2.OA.1**

Use addition and subtraction within 100 to solve one- and two-step word problems involving situations of adding to, taking from, putting together, taking apart, and comparing, with unknowns in all positions, e.g., by using drawings and equations with a symbol for the unknown number to represent the problem. **Note:** See Glossary, Table 1.

Add and subtract within 20.

■ **2.OA.2**

Fluently add and subtract **within 20** using mental strategies. (See standard 1.OA.C.6 for a list of mental strategies.) By [the] end of Grade 2, **know from memory** all sums of two one-digit numbers.

*[**Second graders** must be able to use mental strategies like the ones listed in 1.OA.C.6 to quickly and accurately figure out all 100 addition facts from 0 + 0 = 0 to 9 + 9 = 18 and all 100 subtraction facts from 0 − 0 = 0 to 18 − 9 = 9. Additionally, by the end of grade 2, they must know from memory all 100 addition facts. They must be able to recall them automatically without conscious thought, the same as they can recall their own names.]*

Work with equal groups of objects to gain foundations for multiplication.

o **2.OA.3**

Determine whether a group of objects (up to 20) has an odd or even number of members, e.g., by pairing objects or counting them by 2s; write an equation to express an even number as a sum of two equal addends.

o **2.OA.4**

Use addition to find the total number of objects arranged in rectangular arrays with up to 5 rows and up to 5 columns; write an equation to express the total as a sum of equal addends.

Number & Operations in Base Ten

Understand place value.

o **2.NBT.1**

Understand that the three digits of a three-digit number represent amounts of hundreds, tens, and ones; e.g., 706 equals 7 hundreds, 0 tens, and 6 ones. Understand the following as special cases:

o **2.NBT.1a**

100 can be thought of as a bundle of ten tens — called a "hundred."

o **2.NBT.1b**

The numbers 100, 200, 300, 400, 500, 600, 700, 800, 900 refer to one, two, three, four, five, six, seven, eight, or nine hundreds (and 0 tens and 0 ones).

o **2.NBT.2**

Count within 1000; skip-count by 5s, 10s, and 100s.

[5-10-15-20-25-30-35-40-45-50, 10-20-30-40-50-60-70-80-90-100, 100-200-300-400-500-600-700-800-900-1000.]

o **2.NBT.3**

Read and write numbers to 1000 using base-ten numerals, number names, and expanded form.

o **2.NBT.4**

Compare two three-digit numbers based on meanings of the hundreds, tens, and ones digits, using >, =, and < symbols to record the results of comparisons.

Use place value understanding and properties of operations to add and subtract.

■ **2.NBT.5**

Fluently add and subtract **within 100** using strategies based on place value, properties of operations, and/or the relationship between addition and subtraction.

*[**Second graders** must be able to use mental strategies based on place value, properties of operations (e.g., a + b = b + a), and/or addition and subtraction being inverses (opposites) of one another (e.g., 36 + 64 = 100 implies 100 – 36 = 64 and 100 – 64 = 36, and vice versa) to quickly and accurately add and subtract **two-digit numbers** with sums and differences less than or equal to **100**.]*

o **2.NBT.6**

Add up to four two-digit numbers using strategies based on place value and properties of operations.

o **2.NBT.7**

Add and subtract within 1000 using concrete models or drawings and strategies based on place value, properties of operations, and/or the relationship between addition and subtraction; relate the strategy to a written method. Understand that in adding or subtracting three-digit numbers, one adds or subtracts hundreds and hundreds, tens and tens, ones and ones; and sometimes it is necessary to compose or decompose tens or hundreds.

[Add and subtract three-digit numbers with sums and differences less than or equal to 1000 using the strategies listed for this standard.]

o **2.NBT.8**

Mentally add 10 or 100 to a given number 100-900, and mentally subtract 10 or 100 from a given number 100-900.

o **2.NBT.9**

Explain why addition and subtraction strategies work using place value and the properties of operations. **Note:** Explanations may be supported by drawings or objects.

Measurement and Data

Relate addition and subtraction to length.

○ **2.MD.5**

Use addition and subtraction within 100 to solve word problems involving lengths that are given in the same units, e.g., by using drawings (such as drawings of rulers) and equations with a symbol for the unknown number to represent the problem.

○ **2.MD.6**

Represent whole numbers as lengths from 0 on a number line diagram with equally spaced points corresponding to the numbers 0, 1, 2, ... and represent whole-number sums and differences within 100 on a number line diagram.

Work with time and money.

○ **2.MD.8**

Solve word problems involving dollar bills, quarters, dimes, nickels, and pennies, using $ and ¢ symbols appropriately. Example: If you have 2 dimes and 3 pennies, how many cents do you have?

Geometry

Reason with shapes and their attributes.

○ **2.G.2**

Partition a rectangle into rows and columns of same-size squares and count to find the total number of them.

○ **2.G.3**

Partition circles and rectangles into two, three, or four equal shares, describe the shares using the words halves, thirds, half of, a third of, etc., and describe the whole as two halves, three thirds, four fourths. Recognize that equal shares of identical wholes need not have the same shape.

Grade 3

In Grade 3, instructional time [for arithmetic] should focus on [three] critical areas: (1) developing understanding of multiplication and division and strategies for multiplication and division within 100, (2) developing understanding of fractions, especially unit fractions (fractions whose numerators are 1), and (3) developing understanding of the structure of rectangular arrays and of area.

Operations & Algebraic Thinking

Represent and solve problems involving multiplication and division.

○ **3.OA.1**

Interpret products of whole numbers, e.g., interpret 5 × 7 as the total number of objects in 5 groups of 7 objects each. For example, describe a context in which a total number of objects can be expressed as 5 × 7.

○ **3.OA.2**

Interpret whole-number quotients of whole numbers, e.g., interpret 56 ÷ 8 as the number of objects in each share when 56 objects are partitioned equally into 8 shares, or as a number of shares when 56 objects are partitioned into equal shares of 8 objects each. For example, describe a context in which a number of shares or a number of groups can be expressed as 56 ÷ 8.

○ **3.OA.3**

Use multiplication and division within 100 to solve word problems in situations involving equal groups, arrays, and measurement quantities, e.g., by using drawings and equations with a symbol for the unknown number to represent the problem. **Note:** See Glossary, Table 2.

○ **3.OA.4**

Determine the unknown whole number in a multiplication or division equation relating three whole numbers. For example, determine the unknown number that makes the equation true in each of the equations 8 × ? = 48, 5 = __ ÷ 3, 6 × 6 = ?

[Note the avoidance of literal (letter) numbers like x, y, z.]

Understand properties of multiplication and the relationship between multiplication and division.

○ **3.OA.5**

Apply properties of operations as strategies to multiply and divide. (Students need not use formal terms for these properties.) Examples: If 6 × 4 = 24 is known, then 4 × 6 = 24 is also known. (Commutative property of multiplication.) 3 × 5 × 2 can be found by 3 × 5 = 15, then 15 × 2 = 30, or by 5 × 2 = 10, then 3 × 10 = 30. (Associative property of multiplication.) Knowing that 8 × 5 = 40 and 8 × 2 = 16, one can find 8 × 7 as 8 × (5 + 2) = (8 × 5) + (8 × 2) = 40 + 16 = 56. (Distributive property.)

○ **3.OA.6**

Understand division as an unknown-factor problem. For example, find 32 ÷ 8 by finding the number that makes 32 when multiplied by 8.

[In other words, solve 32 ÷ 8 by guess and check.]

Multiply and divide within 100.

■ **3.OA.7**

Fluently multiply and divide **within 100** using strategies such as the relationship between multiplication and division (e.g., knowing that 8 × 5 = 40, one knows 40 ÷ 5 = 8 and 40 ÷ 8 = 5) or properties of operations. By the end of Grade 3, **know from memory** all products of two one-digit numbers.

*[**Third graders** must be able to use mental strategies based on multiplication and division being the inverse (opposite) of one another (e.g., 10 × 5 = 50 implies 50 ÷ 10 = 5 and 50 ÷ 5 = 10, and vice versa) and multiplication being commutative (e.g., 10 × 5 = 5 × 10) to quickly and accurately figure out all 100 multiplication facts from 0 × 0 = 0 to 9 × 9 = 81 and all 90 division facts from 0 ÷ 1 = 0 to 81 ÷ 9 = 9.[66] Additionally, by the end of grade 3, they must know from memory all 100 multiplication facts. They must be able to recall them automatically without conscious thought, the same as they can recall their own names.]*

Solve problems involving the four operations, and identify and explain patterns in arithmetic.

○ **3.OA.8**

Solve two-step word problems using the four operations. Represent these problems using equations with a letter standing for the unknown quantity. Assess the reasonableness of answers using mental computation and estimation strategies including rounding. **Note:** This standard is limited to problems posed with whole numbers and having whole-number answers; students should know how to perform operations in the conventional order when there are no parentheses to specify a particular order (Order of Operations).

○ **3.OA.9**

Identify arithmetic patterns (including patterns in the addition table or multiplication table), and explain them using properties of operations. For example, observe that 4 times a number is always even, and explain why 4 times a number can be decomposed into two equal addends.

Number & Operations in Base Ten

Use place value understanding and properties of operations to perform multi-digit arithmetic. Note: A range of algorithms may be used.

○ **3.NBT.1**

Use place value understanding to round whole numbers to the nearest 10 or 100.

---

[66] There are only 90 division facts because division by zero is not allowed. To understand why it is not allowed, consider what happens when a number like 10 is divided by numbers close to zero: 10 ÷ 1 = 10, 10 ÷ 0.1 = 100, 10 ÷ 0.01 = 1,000, 10 ÷ 0.001 = 10,000, and so on. As you can see, the quotient is getting bigger and bigger (becoming infinitely large) as the divisor is getting closer and closer to zero (becoming infinitely small). Thus dividing 10 by numbers that are super close to zero would yield quotients so big that it would take practically forever to record them by hand.

■ **3.NBT.2**

Fluently add and subtract **within 1000** using strategies and algorithms based on place value, properties of operations, and/or the relationship between addition and subtraction.

*[**Third graders** must be able to use mental strategies based on place value, properties of operations (e.g., a + b = b + a), and/or addition and subtraction being inverses (opposites) of one another (e.g., 45 + 50 = 95 implies 95 − 45 = 50 and 95 − 50 = 45, and vice versa) to quickly and accurately add and subtract **three-digit numbers** with sums and differences less than or equal to **1000**.]*

○ **3.NBT.3**

Multiply one-digit whole numbers by multiples of 10 in the range 10-90 (e.g., 9 × 80, 5 × 60) using strategies based on place value and properties of oerations.

Number & Operations — Fractions

Note: Grade 3 expectations in this domain are limited to fractions with denominators 2, 3, 4, 6, and 8.

Develop understanding of fractions as numbers.

○ **3.NF.1**

Understand a fraction 1/b as the quantity formed by 1 part when a whole is partitioned into b equal parts; understand a fraction a/b as the quantity formed by a parts of size 1/b.

○ **3.NF.2**

Understand a fraction as a number on the number line; represent fractions on a number line diagram.

[Choosing a number line as a referent for fractions is a poor choice because the fractional length of a line segment relative to some fixed length cannot be determined unless the fixed length is present. For instance, to determine if a line segment is half a yard or half a meter, one would need to compare the line segment to a yard stick or a meter stick.

The best referent for fractions is a circle, because the unit is always the circle itself, regardless of the size of the circle. For instance, half of a small circle illustrates 1/2 the same as half of a large circle.]

○ **3.NF.2a**

Represent a fraction 1/b on a number line diagram by defining the interval from 0 to 1 as the whole and partitioning it into b equal parts. Recognize that each part has size 1/b and that the endpoint of the part based at 0 locates the number 1/b on the number line.

○ **3.NF.2b**

Represent a fraction a/b on a number line diagram by marking off a lengths 1/b from 0. Recognize that the resulting interval has size a/b and that its endpoint locates the number a/b on the number line.

- ### 3.NF.3

  Explain equivalence of fractions in special cases, and compare fractions by reasoning about their size.

- ### 3.NF.3a

  Understand two fractions as equivalent (equal) if they are the same size, or the same point on a number line.

- ### 3.NF.3b

  Recognize and generate simple equivalent fractions, e.g., 1/2 = 2/4, 4/6 = 2/3. Explain why the fractions are equivalent, e.g., by using a visual fraction model.

  [To illustrate, use large circles printed on cardstock and cut into halves, thirds, fourths, fifths, sixths, eighths, ninths, tenths, twelfths, fifteenths, and sixteenths.]

- ### 3.NF.3c

  Express whole numbers as fractions, and recognize fractions that are equivalent to whole numbers. Examples: Express 3 in the form 3 = 3/1; recognize that 6/1 = 6; locate 4/4 and 1 at the same point of a number line diagram.

- ### 3.NF.3d

  Compare two fractions with the same numerator or the same denominator by reasoning about their size. Recognize that comparisons are valid only when the two fractions refer to the same whole. Record the results of comparisons with the symbols >, =, or <, and justify the conclusions, e.g., by using a visual fraction model.

Measurement and Data

Solve problems involving measurement and estimation of intervals of time, liquid volumes, and masses of objects.

- ### 3.MD.1

  Tell and write time to the nearest minute and measure time intervals in minutes. Solve word problems involving addition and subtraction of time intervals in minutes, e.g., by representing the problem on a number line diagram.

- ### 3.MD.2

  Measure and estimate liquid volumes and masses of objects using standard units of grams (g), kilograms (kg), and liters (l). **Note:** Excludes compound units such as cm^3 and finding the geometric volume of a container. Add, subtract, multiply, or divide to solve one-step word problems involving masses or volumes that are given in the same units, e.g., by using drawings (such as a beaker with a measurement scale) to represent the problem. **Note:** Excludes multiplicative comparison problems (problems involving notions of "times as much"; see Glossary, Table 2).

Geometric measurement: understand concepts of area and relate area to multiplication and addition.

o   **3.MD.7**

   Relate area to the operations of multiplication and addition.

Hands-on Algebra in K-6

If algebra is just for big kids, then it must be hard. Why else can kids in middle school and high school do it when kids in grade school cannot? The answer to that question is that it is too abstract for children to comprehend. That is true if you chalk-and-talk algebra to a class of 5-year-olds, but it is not true if you make it real to them, just like touchpoints make whole numbers real and fraction cakes make fractions real to them.

The Equabeam makes equality real. It balances when what is on one side of an equation "is the same as" what is on the other side of the equation. Colored chips with, say, yellow for positive numbers on one side and red for negative numbers on the other side make real working with signed numbers, like adding $+5$ and -7. Five yellows cancel five of the seven reds with 2 reds or -2 remaining. Thus $+5 + -7 = -2$. Geoboards make real the Pythagorean theorem ($a^2 + b^2 = c^2$). The sum of the areas of the squares built with rubber bands on the small sides of a right triangle on a geoboard equal the area of the square built with a rubber band on the biggest side of the triangle. And so on with concrete materials as referents for the symbolics.

See page 112 for how to create a K-6 algebra curriculum to run side by side with the new methodology for teaching arithmetic.

Hands-on Materials for Modeling Algebraic Concepts and Equations

○ Two-colored counters for positive and negative numbers.

○ Algeblocks for adding, subtracting, multiplying and dividing polynomials.

○ Algebra Tiles for adding and subtracting first degree and second degree polynomials and factoring second degree trinomials, such as $X^2 + 5X + 6 = (X + 2)(X + 3)$.

○ Square geoboards for understanding square roots and discovering the Pythagorean theorem, as sequenced in *Math Games & Activities*, Vol. 2. Paul Shoecraft, Dale Seymour Pub., 1983.

○ Plastic hand mirrors for learning mirror (line) symmetry.

○ Circular geoboards for discovering formulas pertaining to circles and central and inscribed angles (e.g., the degree measure of a central angle that subtends the same arc as an inscribed angle is twice that of the inscribed angle), *Math Games & Activities*, Vol. 2. Paul Shoecraft, Dale Seymour Pub., 1983.

○ GeoPieces for discovering formulas pertaining to geometric shapes (e.g., the angle-sum formula for triangles), *Math Games & Activities*, Vol. 2. Paul Shoecraft, Dale Seymour Pub., 1983.

○ Math balances for understanding equality and the need to treat both sides of an equation the same to maintain equality.

○ VersaTiles Algebra Starter Set. ETA Cuisenaire Algebra, 2003.

○ Spinners, dice, and playing cards for learning probability.

Resources for Algebra Lessons that Are Suitable or Can Be Made Suitable for K-6

- Algebraic Expressions & Equations Dominoes, gr. 5-6. Children's Mathematical Learning Aids, Didax.

- Algebra Thinking, First Experiences, 112 hands-on activities, gr. 5-8. Linda H. Charles, Creative Pub., 1990. Algebra Tiles Workbook. gr. 6 and up. Renee Burgdorf, 2002.

- Algebra Warm-Ups, gr. 8 and up, 96 pages. Scott McFadden, Dale Seymour Pub., 1987.

- Amazing Algebra Book: 20 Engaging Tricks. Fleron and Edwards, Didax, 2007.

- Developing Algebraic Thinking with Number Tiles. Don Balka, Didax, 2005.

- Dinah Zike's Teaching Mathematics with Foldables. McGraw-Hill, 2002.

- EquaBeam™ Tasks. Paul Shoecraft, MOVE IT Math, 2013. **Free** e-book. Download @ moveitmath.com.

- Exploring Algebra and Pre-Algebra with Manipulatives. Don Balka. Didax, 1994.

- GeoGebra. **Free** software for graphing equations and much more. geogebra.org.

- Hands-On Algebra: Ready-to-Use Games & Activities, gr. 7-12. Frances McBroom Thompson, 1998.

- Hands-On Equations: Making Algebra Child's Play, gr. 3-9. Henry Borenson, 1994.

- Intermediate MathLink Cubes Book: Fractions, probability, and algebra-readiness, gr. 3-6. Judith K. Wells & Carol A. Thornton, 1991.

- Key to Algebra, gr. 5-12. Peter Rasmussen, Key Curriculum Press, 1971.

- Making Math Accessible to Students with Special Needs: Practical Tips and Suggestions, gr. 3-5. Educated Solutions, 2010.

- Making Math Accessible to Students with Special Needs: Practical Tips and Suggestions, gr. 6-8. Educated Solutions, 2010.

- Math Games & Activities, Vols. 1 & 2. Paul Shoecraft, Dale Seymour Pub., 1983.

- Teaching Pre-Algebra with Manipulatives. Glencoe Mathematics Pre-Algebra, 2005.

- Using Cuisenaire Rods: Patterns & Algebra. Barbara Berman & Fredda Friederwitzer, Learning Resources, 2002.

- Working with Algebra Tiles, gr. 6-12. Don Balka and Laurie Boswell, Didax, 2006.

Number of Pages on Arithmetic in Seven Textbook Series

GO MATH!
Houghton Mifflin Harcourt, 2012

Table of Contents for the Go Math! Student Edition (SE)

Kindergarten	Grade 1	Grade 2	Grade 3
1. Represent, Count, Write 0 to 5 2. Compare 0 to 5 3. Represent, Count, Write 6 to 9 4. Represent, Compare 0 to 10 5. **Addition** 6. **Subtraction** 7. Represent, Count, Write 11-19 8. Represent, Count, Write to 100 9. Identify, Describe 2D Shapes 10. Identify, Describe 3D Shapes 11. Measurement 12. Classify, Sort Data	1. Addition Concepts 2. Subtraction Concepts 3. Addition Strategies 4. Sub. Strategies 5. Addition, Subtraction Relationships 6. Count, Model Numbers 7. Compare Numbers 8. **2-Digit Addition, Subtraction** 9. Measurement 10. Represent Data 11. 3D Geometry 12. 2D Geometry	1. Number Sense, Place Value 2. Numbers to 1,000 3. Basic Facts, Relationships 4. **2-Digit Addition** 5. **2-Digit Subtraction** 6. **3-Digit Addition, Subtraction** 7. Money, Time 8. Length in Customary Units 9. Length in Metric Units 10. Data 11. Geometry, Fraction Concepts	1. **Add, Subtract [3-digit numbers] within 1,000** 2. Represent, Interpret Data 3. Understand Mult. 4. **Mult. Facts,** Strategies 5. Use **Mult. Facts** 6. Understand Division 7. **Division Facts,** Strategies 8. Understand Fractions 9. **Compare Fractions** 10. Time, Length, Volume, Mass 11. Perimeter, Area 12. 2D Shapes
Total Pages: 548	**Total Pages: 556**	**Total Pages: 578**	**Total Pages: 554**
Pages on Arith: 336 (61%)	**Pages on Arith: 382 (69%)**	**Pages on Arith: 341 (59%)**	**Pages on Arith: 347 (63%)**

Kindergarten: Number of Pages on Whole Number Arithmetic

	Whole Numbers		TOTAL number of pages dedicated to Arithmetic: 336 of 548 pages or 61 percent of the student textbook. No multiplication and division. No fractions.
	Represent, Count, Write Numbers to 100	Add. and Subt. Facts[1]	
Pages	244	92	
Percent	45% (244/548)	17% (92/548)	

[1]Includes problems like 10 = ___ + 1 with the sum on the left side of the equal sign. That is good because problems like that depict the equal sign as meaning "is the same as" instead of "get the answer," like on a calculator. <u>What is *not* good is referring to such problems as *algebra*</u>. Doing so is a shameful marketing ploy based on research that correlates success in Algebra II in high school with success in college.

Algebra involves literal (letter) numbers, like x, y, z. For 10 = ___ + 1 to be an algebra problem, it should be presented as 10 = x + 1. Underlining just shows where answers go.

Grade 1: Number of Pages on Whole Number Arithmetic

	Whole Numbers		Fractions[3]	TOTAL number of pages dedicated to Arithmetic: 382 of 556 pages or 69 percent of the student textbook. No multiplication and division.
	Numeration and Place Value to 100[1]	2-Digit Add. and Subt.[2]		
Pages	312	52	20	
Percent	56% (312/556)	9% (52/556)	4% (20/556)	

[1]Only after 99 in a Hundred Chart.

[2]Max 2-digit numbers! Examples: 20 + 13 and 60 − 30. Only a few addition problems with three addends, all with single digits that are amenable to "making a ten" in order to avoid a sum that's not a number fact (e.g., 2 + 3 + 4 = 5 + 4 = 9).

[3]The last 20 pages of the last chapter show what is meant by the words "half," "halves," "fourths," and "quarters" as equal parts of circles, squares, and rectangles. Just the *words*, not the symbols. Example: Shade one-fourth of a rectangle instead of ¼ of a rectangle.

Grade 2: Number of Pages on Whole Number Arithmetic

	Whole Numbers		Fractions[2]	TOTAL number of pages dedicated to Arithmetic: 341 of 578 pages or 59 percent of the student textbook. No multiplication and division.
	Numeration and Place Value to 1000	2- and 3-Digit Add. and Subt.[1]		
Pages	108	217	16	[1]Max 3-digit numbers! Examples: 766 + 125 and 706 − 681. Few addition problems with three or four addends, all of which reinforce putting a 1 at the top of the tens column.
Percent	19% (108/578)	38% (217/578)	3% (16/578)	[2]No addition, subtract, multiplication, or division of fractions and no decimals and percent. *No fraction notation!* Just the concept of fraction as equal parts of a whole and only for halves, thirds, and fourths. Example: Identify the shapes that show fourths. No decimals and percent.

Grade 3: Number of Pages on Whole Number Arithmetic

	Whole Numbers			Fractions[4]	TOTAL number of pages dedicated to Arithmetic: 347 of 554 pages or 63 percent of the student textbook. No multiplication and division except for the number facts for those operations.
	3-Digit Add. and Subt.[1]	Mult. Facts[2]	Div. Facts[3]		
Pages	58	112	96	81	[1]Max 3-digit numbers *again*! Examples: 218 + 342 and 261 − 150. Few addition problems with three addends, all of which reinforce putting a 1 at the top of the tens column.
Percent	10 % (58/554)	20% (112/554)	17% (96/554)	15% (81/554)	[2]Including developing the concept of multiplication and multiplying the numerals 1 through 9 by 10,

[3]Including developing the concept of division and dividing 10, 20, 30 ... 90 by 10.

[4]No addition, subtract, multiplication, or division of fractions. No decimals and percent. Only understanding fractions followed by a section on equivalent fractions. Examples: Compare 2/6 and 2/8. Order 4/8, 4/4, 4/6. Equivalent fractions: 3/4 = ?/8.

Saxon Mathematics

Harcourt Achieve Inc., 2008

Number of Lessons

	K	Gr. 1	Gr. 2	Gr. 3	Gr. 4	Gr. 5	Total
Lessons	146	158	160	110	120	120	814

• Approximately three pages per lesson for K-2.

Number of Pages in Lessons on Whole Number Arithmetic

W. Numbers	K	Gr. 1	Gr. 2	Gr. 3	Gr. 4	Gr. 5	Total
Number Sense					1.5	3	4.5
Numbers	31.5	17.5	10.5				59.5
No., Numeration	73	32.5	23	11	16	14	169.5
Operations	20	21.5	8.5				50
Addition Facts		23.5	9	2.5			35
Addition		8	14.5	5.5	4.5	2	34.5
Subt. Facts		11	8.5	2.5			22
Subtraction		2	7	7.5	11	7	34.5
Mult. Facts			6	8.5	5	2.5	22
Multiplication			4	12.5	13.5	8.5	38.5
Division Facts			5.5	1.5	1		8
Division				4	13.5	9	26.5
Total	124.5	116	96.5	55.5	66	46	504.5

Number of Pages in Lessons on Fractions, Decimals, and Percent

Fractions	K	Gr. 1	Gr. 2	Gr. 3	Gr. 4	Gr. 5	Total
Concept	1.5	7.5	12	8	6	7.5	42.5
Equivalent				2	8	10	20
Addition					1.5	1.5	3
Subtraction					1.5	5.5	7
Multiplication				1	1	3	5
Division						2	2
Total	1.5	7.5	12	11	18	29.5	79.5

Decimals	K	Gr. 1	Gr. 2	Gr. 3	Gr. 4	Gr. 5	Total
Concept					1	7.5	8.5
Addition				1	2	1	4
Subtraction				1	1.5	2	4.5
Multiplication				1		3	4
Division						3	3
Total	0	0	0	3	4.5	16.5	24

Percent	K	Gr. 1	Gr. 2	Gr. 3	Gr. 4	Gr. 5	Total
Concept						2	2
Total	0	0	0	0	0	2	2

Number and Percent of Pages in Lessons on Arithmetic

Topic	K	Gr. 1	Gr. 2	Gr. 3	Gr. 4	Gr. 5	Total
W. Numbers	124.5	116	96.5	55.5	66	46	504.5
Fractions	1.5	7.5	12	11	18	29.5	79.5
Decimals				3	4.5	16.5	24
Percent						2	2
Total	126	123.5	108.5	69.5	88.5	94	610
Percent	49% (126/259.5)	57% (123.5/218.5)	53% (108.5/205.5)	51% (69.5/135)	61% (88.5/144)	63% (94/149)	55% (610/1111.5)

Number and Percent of Pages in Lessons on Other than Arithmetic

Topic	K	Gr. 1	Gr. 2	Gr. 3	Gr. 4	Gr. 5	Total
Number Sense					1.5	3	4.5
Numbers	31.5	17.5	10.5				59.5
Comparing, Sorting, Ordering	31	10.5	5.5	3	1	2	53
Patterns	10.5	1	2.5	2	2	2	20
Number Sense					1.5	3	4.5
Place Value		2.5	3.5	4	8	5	23
Rounding		1	1	2	3.5	2	9.5
Time	3	4	8.5	4	2	2	23.5
Money	13	11	7	1	1		33
Measurement	16.5	20	24	23	12	18	113.5
Metrics				2	1	1	4
Temperature	.5	1	2	1	1	2	7.5
Length	1	.5		2	1		4.5
Area		.5	3	6		1	10.5
Geometry	18.5	17.5	16	9	12.5	9	82.5
Graphs	7.5	6.5	11	3	1	1	30
Data/Probability	.5	1.5	1.5	3		1	7.5
Statistics			1		2	1	4
Algebra				0.5	4.5	2	7
Total	**133.5**	**95**	**97**	**65.5**	**55.5**	**55**	**501.5**
Percent	**51%** (133.5/259.5)	**43%** (95/218.5)	**47%** (97/205.5)	**49%** (65.5/135)	**39%** (55.5/144)	**37%** (55/149)	**45%** (501.5/1111.5)

Comments

Kindergarten: Numbers through 20 in last chapter.

Grade 1: Only instance of 3 addends is in Lesson 114.

Grade 2: Multiplication and division facts in last 2 lessons.

Grade 3: Column addition in Sec. 3 avoids "carrying" two or more. The multiplication facts are extended to the 12s! Money is used as a referent for adding and subtracting decimals. "Investigations" include word problems involving all four arithmetic operations, pictographs and bar graphs, working with money, scale maps, probability games, symmetry, geometric solids, and making designs on coordinate graph paper.

Texas enVisionMATH

Scott Foresman – Addison Wesley, 2009

Number of Instructional Pages

	K	Gr. 1	Gr. 2	Gr. 3	Gr. 4	Gr. 5	Total
Instruc. Pages	120			320	306	328	

• Misc. pages includes daily TEKS/TAKS (Texas Essential Knowledge & Skills/Texas Assessment of Knowledge & Skills) reviews, problems of the day, Quick Checks, center activities, chapter reviews, and enrichment.

Number of Instructional Pages on Whole Number Arithmetic

W. Numbers	K	Gr. 1	Gr. 2	Gr. 3	Gr. 4	Gr. 5	Total
No., Numeration	64			19	11	8	
Operations	19						
Addition Facts	3						
Addition				32	14	15	
Subt. Facts	3						
Subtraction				31	14	15	
Mult. Facts				42	19		
Multiplication				18	43	21	
Division Facts				24	14		
Division					31	42	
Total	89			166	146	101	

• Numbers and numeration includes comparing, sorting, ordering, patterns, and place value.

• Operations refers to acting out, illustrating, and guess and check problem solving for the four fundamental arithmetic operations of addition, subtraction, multiplication, and division.

Number of Instructional Pages on Fractions, Decimals, and Percent

Fractions	K	Gr. 1	Gr. 2	Gr. 3	Gr. 4	Gr. 5	Total
Concept	2			19	14		
Equivalent					18	48	
Addition						8	
Subtraction						8	
Multiplication							
Division							
Total	**2**			**19**	**32**	**64**	

Decimals	K	Gr. 1	Gr. 2	Gr. 3	Gr. 4	Gr. 5	Total
Concept					27	18	
Addition							
Subtraction							
Multiplication							
Division							
Total	**0**			**0**	**27**	**18**	

Percent	K	Gr. 1	Gr. 2	Gr. 3	Gr. 4	Gr. 5	Total
Concept							
Total	**0**			**0**	**0**	**0**	

Number and Percent of Instructional Pages on Arithmetic

Topic	K	Gr. 1	Gr. 2	Gr. 3	Gr. 4	Gr. 5	Total
W. Numbers	89			166	146	101	
Fractions	2			19	32	64	
Decimals					27	18	
Percent							
Total	**91**			**185**	**205**	**183**	
Percent	**76%** (91/120)			**58%** (185/320)	**67%** (205/306)	**56%** (183/328)	

Number and Percent of Instructional Pages on Other than Arithmetic

Topic	K	Gr. 1	Gr. 2	Gr. 3	Gr. 4	Gr. 5	Total
Time	10			11	3	11	
Measurement	8			45	43	45	
Geometry	7			43	30	36	
Graphs	4			17		8	
Graphing						13	
Data/Probability					8	11	
Statistics					8	8	
Algebra					9	13	
Total	**29**			**116**	**101**	**145**	
Percent	**24%** (29/120)			**39%** (116/301)	**33%** (101/306)	**44%** (145/328)	

- Time includes temperature.
- Measurement includes length, perimeter, area, volume (capacity), weight (mass), and metrics.

Comments (by grade)

Kindergarten: Numbers through 20 not until the last chapter.

Grade 3: Numbers through six digits. Multiplication facts through 12 x 12! Multiplication by one digit. Noteworthy is the inclusion of "order of operations," that in reading from left to right, multiply *or* divide, *whichever comes first*, then add *or* subtract, *whichever comes first*, unless instructed otherwise with parentheses (e.g., 3 + 2 x 9 = 21, *not 45*, p. 165). It should be in the main text, though, not in Algebra Links.

Order of operations is about the priority to give the operations in *arithmetic*, only secondarily in algebra. To put something in arithmetic in Algebra Links just because it occasionally shows up in algebra would be to put all of arithmetic in Algebra Links. Better would be to just get arithmetic done as in MOVE IT Math by grade 2 and make most of K-6 math a hands-on bonafide algebra course.

Grade 4: Numbers through nine digits. Multiplication by two digits. Division by one digit. Notable are the input/output tables to develop the concept of function (e.g., gr. 4, p. 270-79).

Grade 5: Numbers through 12 digits. Addition and subtraction of fractions only with like denominators!

Comments (general)

The word "algebra" has market value because success in algebra, especially Algebra II in high school, correlates highly with graduating from college (Final Report, National Mathematics

Advisory Panel). That being the case, serious algebra could and should be interwoven into the elementary school math curriculum. However, until that is done, implying that it has been done misleads parents and the community into falsely thinking that their children are on a path to success in algebra.

Instances of improperly labeling certain content as algebra in Algebra Links is of three types:

Type 1: Filling in blanks in number patterns

Type 2: Filling in blanks with <, =, or > to state relationships between quantities

Type 3: Filling in blanks with +, −, x, or ÷ to make quantities equal

Examples of the three types in grades 3-5 follow:

Grade 3:

Type 1: 5, 10, 15, 20, ___, 30 (p. 13, ans. 25)

Type 2: 3 + 4 ___ 2 + 7 (p. 39, ans. <)

Type 3: 9 ___ 36 = 45 (p. 149, ans. +)

Grade 4:

Type 1: 2, 4, 6, 8, ___ (p. 21, ans. 10)

Type 2: 5 x 71 ___ 5 x 70 (p. 139, ans. >)

Type 3: 4 ___ ___ = 8 (p. 351, ans. + 4 or x 2)

Grade 5:

Type 1: 18, 27, 36, 45, ___ (p. 33, ans. 6 x 9 = 54)

Type 2: 3 + 48 ___ 3 x 48 (p. 101, ans. <)

These are worthwhile exercises, but they should not be linked to algebra any more than the number facts should be linked to it just because they come up a lot in solving equations. All three types are arithmetic problems, nothing more. Granted, a lot of arithmetic is used in solving equations, but that does not mean arithmetic is algebra. Arithmetic is adding, subtracting, multiplying, and dividing whole numbers, fractions, and decimals and working with percent. Period.

My desktop dictionary defines algebra as that part of mathematics in which *"letters and other general symbols are used to represent numbers and quantities in formulae and equations."* In other words, algebra is the generalization of arithmetic with literal (letter) numbers, like x, y, z that are used like pronouns — like "him," "her," "it" — until their number names are known.

Type 1 problems are arithmetic problems, *not algebra problems*. Filling in the blank for a specific term in a number pattern involves nothing more than doing the computation suggested by the pattern. For 5, 10, 15, ___, for example, it would be realizing that 5 = 1 x 5, 10 = 2 x 5, and 15 = 3 x 5 and assuming that 4 x 5 = 20 would be the next number. For the sequence to be deemed algebra, it would need to be generalized to the Nth term instead of a specific term. For the example given, it would need to be presented as 5, 10, 15 ... ___N for N = 1, 2, 3, and so on. Filling in the blank with 5 in the example would identify the rule being used, that the Nth term is 5N for N = 1, 2, 3, and so on indefinitely.

Even remotely suggesting that Type 2 and 3 problems have to do with algebra is terribly wrong. The unknowns in algebra — the x, y, z nomenclature — refer to *numbers*, usually those represented by the points on a number line. They *never* stand for relational symbols like <, =, > or operations such as +, −, x, ÷. The Type 2 and 3 problems given in the examples are okay in the context of arithmetic but become sources of misunderstanding in algebra.

Missing addend problems, like 9 + ___ = 11 below, cry out to be expressed algebraically but to no avail. What, really, is so different about 9 + ___ = 11 vs. 9 + X = 11? Being somewhat facetious, here is what is different: An empty space "cries out" to be filled in, whereas an X is too scary for young children. Baloney. Young children have yet to hear that algebra is hard, so they do not balk at it. They think the X in an equation like 9 + X = 11 is just a number they do not know yet, which is a good way to think about it.

Grade 3:

Missing addends: 9 + ___ = 11 (p. 77, ans. 2)

Answers: 7 + 3 x 2 = ___ (p. 165, ans. 13, not 20)

Grade 4:

Missing addends: ___ + 6 = 21 (p. 33, ans. 15)

Answers: 12 x 0 = ___ (p. 95, ans. 0)

Whatever the reason for sheltering young children from the use of variables in arithmetic, this much is certain. Long ago, somebody decided that elementary school math should be that way, and it has been that way ever since. Worth considering is that it has never been that way for the "variables" that pervade ordinary speech, the pronouns his, him, her, hers, your, yours, we, they, us, them, their, theirs, our, ours, it, its, and so on. Most 5- and 6-year-olds are adept with them in speech, so why not in algebra as well?

Texas Everyday Mathematics

The University of Chicago School Mathematics Project, 2008

Number of Instructional Pages

	K	Gr. 1	Gr. 2	Gr. 3	Gr. 4	Gr. 5	Total
Instruc. Pages	264	407	328	318	✕	650	

Number of Instructional Pages on Whole Number Arithmetic and Fractions

W. Numbers	K	Gr. 1	Gr. 2	Gr. 3	Gr. 4	Gr. 5	Total
No., Numeration	128	129	14	18		42	
Operations	10	7					
Addition Facts	12	47.5	16	7		.5	
Addition		15	32	20		18.5	
Subt. Facts	7	15	10	6		.5	
Subtraction		11	24	23		17	
Mult. Facts			33	28		12.5	
Multiplication				43		64.5	
Division Facts		4	11	7		1.5	
Division				10		27	
Total	157	228.5	140	162		184	

Fractions	K	Gr. 1	Gr. 2	Gr. 3	Gr. 4	Gr. 5	Total
Concept	3	14	22	11		26.5	
Equivalent		4		10		25	
Addition						15	
Subtraction						9	
Multiplication						11	
Division						3	
Ratios						20	
Total	3	18	22	21		109.5	

• Number and numeration includes comparing, sorting, ordering, patterns, and place value.

Number of Instructional Pages on Decimals and Percent

Decimals	K	Gr. 1	Gr. 2	Gr. 3	Gr. 4	Gr. 5	Total
Concept				8		17.5	
Addition						4.5	
Subtraction						3.5	
Multiplication						1	
Division						3	
Total	0	0	0	8		**29.5**	

Percent	K	Gr. 1	Gr. 2	Gr. 3	Gr. 4	Gr. 5	Total
Concept				1		9.5	
Percent of Number						7.5	
Total	0	0	0	1		**17**	

Number and Percent of Instructional Pages on Arithmetic

Topic	K	Gr. 1	Gr. 2	Gr. 3	Gr. 4	Gr. 5	Total
Whole Nos.	157	228.5	140	162		184	
Fractions	3	18	22	21		109.5	
Decimals				8		29.5	
Percent				1		17	
Total	160	246.5	162	192		340	
Percent	61% (160/264)	61% (246.5/407)	49% (162/328)	60% (192/318)		52% (340/650)	

Number and Percent of Instructional Pages on Other than Arithmetic

Topic	K	Gr. 1	Gr. 2	Gr. 3	Gr. 4	Gr. 5	Total
Time	16	44.5	25	10		8.5	
Money	21	38.5	37	7			
Measurement	26	36	55	50		78.5	
Geometry	21	25	28	34		81.5	
Graphs	16	6.5	21	6		42	
Data/Probability	4	9.5		19		46	
Algebra						53	
Total	**104**	**160**	**166**	**126**		**309.5**	
Percent	**39%** (104/264)	**39%** (160/407)	**51%** (166/328)	**40%** (126/318)		**48%** (309.5/650)	

Number of Pages in Ancillary Materials

Reference	K	Gr. 1	Gr. 2	Gr. 3	Gr. 4	Gr. 5	Total
Lesson Guide	419	421	456	450	944	961	**3651**
Appendices						56	**56**
Differentiation Handbook				155		155	**310**
Reference Manual	120			283		375	**778**
Assessment Handbook	109			305		314	**728**
Activity Cards	38						**38**
Math Masters	143		477	468	506		**1594**
Minute Math	258		168			254	**680**
Student Math Journal		209	310		343	431	**1293**
Student Reference Manual			165		345	442	**952**
Home Connection Handbook	22			102		110	**234**
Total	**1109**	**630**	**1576**	**1763**	**2138**	**3098**	**10,314**

- Appendices includes projects, grade-level goals, scope and sequence, glossary, and index.
- Math Masters includes teaching masters, game masters, project masters, and Home Links.

Comments

Kindergarten: Lots of counting songs (e.g., Ten Little Penguins, This Old Man, Five Little Monkeys) and books (e.g. *Anno's Counting Book*, *Five Little Monkeys Jumping on a Bed*, *The Doorbell Rang*), but no use of counting to work out the number facts.

Grades 1-3: Uses mental math (e.g., doubles plus or minus 1) and fact strategies to figure out the addition and subtraction facts.

Texas HSP Math

Harcourt School Publishers, 2009

Number of Instructional Pages

	K	Gr. 1	Gr. 2	Gr. 3	Gr. 4	Gr. 5	Total
Instruc. Pages	235	380	405	436	500	453	2409

Number of Instructional Pages on Whole Number Arithmetic

W. Numbers	K	Gr. 1	Gr. 2	Gr. 3	Gr. 4	Gr. 5	Total
No., Numeration	118	88	92	53	23	31	**405**
Addition Facts	19	71	21				**111**
Addition			45	23	16	10	**94**
Subt. Facts	19	73	15				**107**
Subtraction			45	21	16	10	**92**
Mult. Facts			10	55	17		**82**
Multiplication				17	55	21	**93**
Division Facts			10	53	19		**82**
Division					36	34	**70**
Total	**156**	**232**	**238**	**222**	**182**	**106**	**1,136**

- Number and numeration includes comparing, sorting, ordering, patterns, and place value.

Number of Instructional Pages on Fractions, Decimals, and Percent

Fractions	K	Gr. 1	Gr. 2	Gr. 3	Gr. 4	Gr. 5	Total
Concept	10	17	23				50
Equivalent				25	25	23	73
Addition						24	24
Subtraction						24	24
Multiplication							0
Division							0
Total	10	17	23	25	25	71	171

Decimals	K	Gr. 1	Gr. 2	Gr. 3	Gr. 4	Gr. 5	Total
Concept					23	14	37
Addition					10	10	20
Subtraction					10	10	20
Multiplication							0
Division							0
Total	0	0	0	0	43	34	77

Percent	K	Gr. 1	Gr. 2	Gr. 3	Gr. 4	Gr. 5	Total
Concept							0
Total	0	0	0	0	0	0	0

Number and Percent of Instructional Pages on Arithmetic

Topic	K	Gr. 1	Gr. 2	Gr. 3	Gr. 4	Gr. 5	Total
Whole Numbers	156	232	238	222	182	106	1136
Fractions	10	17	23	25	25	71	171
Decimals					43	34	77
Percent							0
Total	166	249	261	247	250	211	1,384
Percent	71% (166/235)	66% (249/380)	64% (261/405)	57% (247/436)	50% (250/500)	47% (211/453)	57% (1384/2409)

Number and Percent of Instructional Pages on Other than Arithmetic

Topic	K	Gr. 1	Gr. 2	Gr. 3	Gr. 4	Gr. 5	Total
Time	19	24	15	13	17		**88**
Money		15	34	15			**64**
Measurement	23	37	36	57	78	85	**316**
Geometry	10	23	34	51	61	51	**230**
Graphs	17	15	13	21	21	36	**123**
Data/Probability		17	12	17	21	28	**95**
Algebra				15	52	42	**109**
Total	**69**	**131**	**144**	**189**	**250**	**242**	**1,025**
Percent	**29%** (69/235)	**34%** (131/380)	**36%** (144/405)	**43%** (189/436)	**50%** (250/500)	**53%** (242/453)	**43%** (1025/2409)

Comments

Kindergarten: Count to 100 in chapter 8. Addition and subtraction facts withheld until the last two chapters.

Grade 1: Place value only to 99. Fractions limited to halves, thirds, and fourths. Patterns claimed to be algebra even though not generalized.

Grade 2: No multiplication and division until the *last* chapter. Patterns claimed to be algebra even though not generalized.

Grade 3: As much as 99 percent of the addition problems were "carry the 1" problems, as in solving 46 + 18 + 27 incorrectly in Chapter 4.

Grade 4: Probability chapter is about combinations. *Picture of the MOVE IT Math EquaBeam on p. 163!*

Texas Math
Houghton Mifflin, 2009

Number of Instructional Pages

	K	Gr. 1	Gr. 2	Gr. 3	Gr. 4	Gr. 5	Total
Instruc. Pages	200	395	424	457	468	404	2348

Number of Instructional Pages on Whole Number Arithmetic

W. Numbers	K	Gr. 1	Gr. 2	Gr. 3	Gr. 4	Gr. 5	Total
No., Numeration	90	108	92	51	51	34	426
Addition Facts	15	70	30			4	119
Addition			56	38	35	9	138
Subt. Facts	15	73	14			4	106
Subtraction			34	38	35	9	116
Mult. Facts			16	57	19	4	96
Multiplication			0	17	36	34	87
Division Facts			16	48		4	68
Division					38	17	55
Total	120	251	258	249	214	119	1,211

• Number and numeration includes comparing, sorting, ordering, patterns, and place value.

Number of Instructional Pages on Fractions

Fractions	K	Gr. 1	Gr. 2	Gr. 3	Gr. 4	Gr. 5	Total
Concept	5	14	36	17			72
Equivalent				19	17	36	72
Addition						10	10
Subtraction						10	10
Multiplication							0
Division							0
Total	5	14	36	36	17	56	164

Number of Instructional Pages on Decimals and Percent

Decimals	K	Gr. 1	Gr. 2	Gr. 3	Gr. 4	Gr. 5	Total
Concept					38	34	72
Addition					9	10	19
Subtraction					9	10	19
Multiplication							0
Division							0
Total	0	0	0	0	56	54	110

Percent	K	Gr. 1	Gr. 2	Gr. 3	Gr. 4	Gr. 5	Total
Concept							0
Total	0	0	0	0	0	0	0

Number and Percent of Instructional Pages on Arithmetic

Topic	K	Gr. 1	Gr. 2	Gr. 3	Gr. 4	Gr. 5	Total
W. Numbers	120	251	258	249	214	119	1211
Fractions	5	14	36	36	17	56	164
Decimals					56	54	110
Percent							0
Total	125	265	294	285	287	229	1,485
Percent	63% (125/200)	67% (265/395)	69% (294/424)	62% (285/457)	61% (287/468)	57% (229/404)	63% (1485/2348)

Number and Percent of Instructional Pages on Other than Arithmetic

Topic	K	Gr. 1	Gr. 2	Gr. 3	Gr. 4	Gr. 5	Total
Time	10	16	18	17	17	13	91
Money		16	16	15			47
Measurement	20	34	46	57	64	60	281
Geometry	35	30	16	49	70	19	219
Graphs	10	25	18	17	15	34	119
Graphing						21	21
Data/Probability		9	16	17	15	13	70
Algebra						15	15
Total	75	130	130	172	181	175	863
Percent	37% (75/200)	33% (130/395)	31% (130/424)	38% (172/457)	39% (181/468)	43% (175/404)	37% (863/2348)

Texas Mathematics

Macmillan McGraw - Hill, 2009

Number of Instructional Pages

	K	Gr. 1	Gr. 2	Gr. 3	Gr. 4	Gr. 5	Total
Instruc. Pages	260	391	390	528	526	580	2675

Number of Instructional Pages on Whole Number Arithmetic

W. Numbers	K	Gr. 1	Gr. 2	Gr. 3	Gr. 4	Gr. 5	Total
No., Numeration	132	87	58	79	72	20	448
Addition Facts	29	71	27				127
Addition			29	33	18	9	89
Subt. Facts	31	69	25				125
Subtraction			27	35	18	9	89
Mult. Facts			25	78	22		125
Multiplication				31	35	41	107
Division Facts				72	22		94
Division					43	37	80
Total	192	227	191	328	230	116	1,284

• Number and numeration includes comparing, sorting, ordering, patterns, and place value.

Number of Instructional Pages on Fractions

Fractions	K	Gr. 1	Gr. 2	Gr. 3	Gr. 4	Gr. 5	Total
Concept	11	23	25	39		78	176
Equivalent					35		35
Addition						20	20
Subtraction						20	20
Multiplication							0
Division							0
Total	11	23	25	39	35	118	251

Number of Instructional Pages on Decimals and Percent

Decimals	K	Gr. 1	Gr. 2	Gr. 3	Gr. 4	Gr. 5	Total
Concept					35	20	55
Addition					16	9	25
Subtraction					16	9	25
Multiplication							0
Division							0
Total	**0**	**0**	**0**	**0**	**67**	**38**	**105**

Percent	K	Gr. 1	Gr. 2	Gr. 3	Gr. 4	Gr. 5	Total
Concept							0
Total	**0**	**0**	**0**	**0**	**0**	**0**	**0**

Number and Percent of Instructional Pages on Arithmetic

Topic	K	Gr. 1	Gr. 2	Gr. 3	Gr. 4	Gr. 5	Total
Whole Nos.	192	227	191	328	230	116	1284
Fractions	11	23	25	39	35	118	251
Decimals					67	38	105
Percent							0
Total	**203**	**250**	**216**	**367**	**332**	**272**	**1,640**
Percent	**78%** (203/260)	**64%** (250/391)	**55%** (216/390)	**70%** (367/528)	**63%** (332/526)	**47%** (272/580)	**61%** (1640/2675)

Number and Percent of Instructional Pages on Other than Arithmetic

Topic	K	Gr. 1	Gr. 2	Gr. 3	Gr. 4	Gr. 5	Total
Time	25	25	33	19	10		112
Money		29	35				64
Measurement	21	35	50	65	78	113	362
Geometry	11	25	31	41	70	45	223
Graphs				9	9		18
Data/Probability		27	25	18	18	76	164
Statistics				9	9		18
Algebra						74	74
Total	**57**	**141**	**174**	**161**	**194**	**308**	**1,035**
Percent	**22%** (57/260)	**36%** (141/391)	**45%** (174/390)	**30%** (161/528)	**37%** (194/526)	**53%** (308/580)	**39%** (1035/2675)

Concepts Test

Name _____ Grade ____ Date _____

1. True or false: $5 = 2 + 3$

2. True or false: $7 = 7$

3. True or false: $4 + 6 = 8 + 2$

4. Add:

$$
\begin{array}{r}
2\ 4 \\
+\ 3\ 5 \\
\hline
\end{array}
\qquad
\begin{array}{r}
3\ 6 \\
+\ 4\ 9 \\
\hline
\end{array}
\qquad
\begin{array}{r}
1\ 2 \\
3\ 4 \\
+\ 2\ 1 \\
\hline
\end{array}
\qquad
\begin{array}{r}
4\ 6 \\
1\ 8 \\
+\ 2\ 7 \\
\hline
\end{array}
$$

5. Add:

    1 foot   8 inches
  + _____   7 inches

6. Subtract:

$$
\begin{array}{r}
8\ 5 \\
-\ 2\ 3 \\
\hline
\end{array}
\qquad
\begin{array}{r}
3\ 7\ 6 \\
-\ 1\ 4\ 5 \\
\hline
\end{array}
\qquad
\begin{array}{r}
8\ 6 \\
-\ 6\ 8 \\
\hline
\end{array}
\qquad
\begin{array}{r}
5\ 0\ 2\ 0 \\
-\ \ \ 4\ 6\ 3 \\
\hline
\end{array}
$$

Giving the MOVE IT Math Concepts Test

The MOVE IT Math concepts test determines children's perception of equality and their understanding of how to add and subtract in a base ten numeration system with place value. Allow about 10-15 minutes for its completion. Please refrain from saying anything about the items on the test until you have collected all of the tests. If giving the test to young children, you may read the first three items to them. To record the results, use the form provided. Tally the types of responses and then total the tally marks, as shown below for 15 true and 7 false responses to a true/false question.

True	False
///// ///// // /// 15	///// // 7

Items 1 through 3, Equals as Balanced

Amazingly (to adults), a large percentage of children do not know that the equal sign in arithmetic and algebra means "balanced" or "is the same as," not "get the answer," like on a calculator. The true/false questions for items 1 through 3 test for this.

Item 1: If a child thinks $5 = 2 + 3$ is false, they think it is written *backwards*. They think the answer is on the wrong side. That is, they think it isn't where it has been for just about *every* problem on *every* worksheet and textbook page they have worked. If asked to correct it, they will write $2 + 3 = 5$.

Item 2: If a child thinks $7 = 7$ is false, they think the equal sign requires a problem. If asked to correct it, they will write $7 + 0 = 7$ or $7 \times 1 = 7$. Children that answer this item false will be confused with equivalent fractions like $1/2 = 5/10$.

Item 3: If a child thinks $4 + 6 = 8 + 2$ is false, they think it is two problems or one big one written incorrectly. If asked to correct it, they will write $4 + 6 = 10$ and $8 + 2 = 10$ or $4 + 6 + 8 + 12 = 20$. Children that answer this item false are going to think that properties of numbers like $a + b = b + a$ are false, which will confuse them in algebra.

Item 4, Adding or Trading "Up" in Base Ten

All that matters here is the response to the fourth problem. A significant number of children who can do the first three problems "carry" the 1 or put the 1 up top for the fourth problem. This indicates a flawed understanding of place value in base ten numeration.

Item 5, Being a Trading Expert

This item is a check against how meaningfully children are working the fourth problem for item 4 in the event they get it correct. A surprising number of those who get it correct miss the next problem involving feet and inches. They add 7 and 8, get 15, and put the 5 in the answer and "carry" the 1. They do not realize that what is done with numbers is determined by the *context* in which they appear and that the "exchange rate" for this problem is 12, not 10.

Item 6, Subtracting or Trading "Down" in Base Ten

All that matters here are the responses to the last two problems. The typical (and common) mistake for the third one is to subtract the smaller number from the larger one and get 22. The fourth problem is simply difficult for many children because of the zeros in the minuend. Many leave it blank or subtract the smaller number from the larger one. Either indicates a flawed understanding of base ten place value numeration.

Concepts Test Results

School _____ Grade ____ Date _____

C = Correct, X = Incorrect, DT = Didn't try

	True	False
1. 5 = 2 + 3		
2. 7 = 7		
3. 4 + 6 = 8 + 2		

4.

24 + 35		36 + 49		12 34 + 21		46 18 + 27			
C	X	C	X	C	X	C	X		
59		85		67		91	82	Other	DT

5. 1 foot 8 inches
+ ____ 7 inches

C		X			
1 ft. 15 in. or 2 ft. 3 in.		2 ft. 5 in.	Other		Didn't Try

6.

85 − 23		376 − 145		86 − 68			5020 − 463		
C	X	C	X	C	X		C	X	
62		231		18	22	Other	4557	Other	DT

Addition Facts Algorithm

The Addition Facts Algorithm, attributed to L.B. Hutchings[67] (1976), is empowering. As its name suggests, it reduces addition to just the addition facts. Thus it eliminates the need to restrict the size of addition problems for young children. Moreover, it makes sense. It mirrors adding in base ten with arithmetic blocks or colored counters where ten the same must be traded for one of the next bigger thing.

Given the problem below, you may have been taught to try to simplify the columns by looking for number combinations that you knew or that made a ten, but if that did not help, you were probably taught to start at the top of each column, beginning with the ones column, and start adding: *"8 plus 6 is 14, plus 9 is 23, plus 5 is 28, plus 4 is 32, plus 7 is 39,"* as recorded below. The problem with adding as I just did is that it requires concentration and starting all over again if distracted while adding.

```
        3
        9    8
        8    6
        5    9
        2    5
        7    4
    +   6    7
        ─────────
             9
```

To introduce you to the Addition Facts Algorithm, I will add the numbers in the first column again but with this exception: When a sum is greater than or equal to ten, I will note the ten in the sum by putting a thumb or finger down and just add the *units* in the sum to the next number in the column. For example, for $8 + 6 = 14$, I will note the ten in 14 by putting a thumb or finger down and just add 4 to the 9 in the column.

So here goes with adding the numbers in the ones column again, but with the exception. $8 + 6 = 14$. I note the ten in 14 by putting my thumb down and add 4 and 9 (= 13). I note the ten in 13 by putting a finger down and add 3 and 5, which is 8, which I add to 4 (= 12). I note the 10 in 12 by putting down another finger and add 2 and 7 (= 9) and put the 9 in the ones column in the answer and a 3 at the top of the tens column, as shown, to account for the three tens I noted in adding the numbers in the ones column, the same as before. *Note that I only had to know or know how to work out the following addition facts: $8 + 6 = 14$, $4 + 9 = 13$, $3 + 5 = 8$, $8 + 4 = 12$, and $2 + 7 = 9$. In other words, I did not have to compute $14 + 9$, $23 + 5$, $28 + 4$, and $32 + 7$ while adding the column of numbers as before.*

---

[67] Hutchings, L. B. "Low-stress Algorithms: Measurement in School Mathematics," 1976 Yearbook, National Council of Teachers of Mathematics, 1976.

So here goes with adding the same numbers again, except with the Addition Facts Algorithm, as shown below. 8 + 6 = 14. I note the ten in 14 by "splitting" 14 about the 6 in the column and add 4 and 9. (Splitting 14 about the 6 in the column means putting the 1 in 14 to the left of the 6 and the 4 in 14 to the right of it.) Continuing, 4 + 9 = 13. I note the ten in 13 by splitting 13 about the 9 in the column and add 3 and 5. 3 + 5 = 8. To keep track of my place in the algorithm, I put the 8 to the right of the 5 in the column and add 8 and 4. 8 + 4 = 12. I note the ten in 12 by splitting 12 about the 4 in the column and add 2 and 7: 2 + 7 = 9. To finish with the column, I put the 9 in the ones place in the answer, count the number of tens that were noted, and put 3 at the top of the next column to the left.

```
   3                                3
   9   8                       1 9 2   8      3 + 9 = 12, split the 12
   8  1 6 4   8 + 6 = 14,      1 8 0  1 6 4   2 + 8 = 10, split the 10
              split the 14
   5  1 9 3   4 + 9 = 13,       5 5   1 9 3   0 + 5 = 5
              split the 13
   2   5 8    3 + 5 = 8         2 7    5 8    5 + 2 = 7
   7  1 4 2   8 + 4 = 12,      1 7 4  1 4 2   7 + 7 = 14, split the 14
              split the 12
 + 6   7 9    2 + 7 = 9      + 1 6 0   7 9    4 + 6 = 10, split the 10
        9                     4   0   9
```

Continuing with the second column, 3 + 9 = 12. I note the ten in 12 by splitting 12 about the 9 in the column and add 2 and 8. 2 + 8 = 10. I note the ten in 10 by splitting 10 about the 8 in the column and add 0 and 5. 0 + 5 = 5. To keep track of my place in the algorithm, I put the 5 to the right of the 5 in the column and add 5 and 2. 5 + 2 = 7. To keep track of my place in the algorithm, I put the 7 to the right of the 2 in the column and add 7 and 7: 7 + 7 = 14. I note the ten in 14 by splitting 14 about the 7 in the column and add 4 and 6: 4 + 6 = 10. I note the ten in 10 by splitting 10 about the 6 in the column. To finish with the column, I put the 0 in the tens place in the answer, count the number of tens that were noted, and put 4 in the hundreds place in the answer.

The Addition Facts Algorithm is like a magic wand.[68] It turns even horribly intimidating addition problems — *like adding ten 10-digit numbers to become certified in Monster Addition* — into manageable ones. Its only drawback is that it requires a lot of writing, thus more time, but the drawback is offset with the non-stressful nature of the algorithm. A user can take a break from adding a column of numbers anywhere in the process and pick up where they left off when they are ready to work on it some more. Also, the algorithm is diagnostic. If a mistake is made in adding a column of numbers, the addition fact that was worked incorrectly can be identified and corrected.

In MOVE IT Math, the Addition Facts Algorithm is not taught to elementary school children until they understand base ten numeration, which amounts to being able to add and subtract

---

[68] Dr. Hutchings, himself, taught me the Addition Facts Algorithm and its companion, the Multiplication Facts Algorithm. This occurred when he happened to be visiting Arizona State University when I was an Assistant Professor in the math department there.

in different bases. However, the algorithm is easy to teach by rote to older students. From middle school up, students can learn the algorithm, work the certification problem for Monster Addition, and receive their award certificates for their achievement in one class period.

A copy of the certification problem and the award certificate for Monster Addition is in the appendix. Why not certify yourself in Monster Addition? The feeling of accomplishment when your answer for the certification problem matches the one on the award certificate is worth experiencing, even for adults who are skilled in arithmetic.

Multiplication Facts Algorithm

With the Multiplication Facts Algorithm, one works from right to left as they multiply each digit in the multiplicand (top number) by each digit in the multiplier (bottom number), the same as with the standard algorithm for multiplication, called Tower Multiplication in MOVE IT Math because of the numbers that end up "towering" over the multiplicand, as shown below in the second working of each problem. The main difference between the two algorithms is how the Multiplication Facts Algorithm separates the addition from the multiplication.

To illustrate, to compute 5 x 47, 35 (from 5 x 7) is entered *diagonally* beneath the underline. The 3 is put a row down in the tens column, and the 5 is put in the row above it in the ones column, as shown. In contrast, in Tower Multiplication, the 3 is put above the 4 in the multiplicand, and the 5 is put in the ones column in the answer, as shown.

```
   47                  34                  157    12
 x  5    3          x 23    1           x 39    56
 ----              ----                ----
  0/5   47          9/2    34           95/    157
 2/3   x 5          0/    x 23          04/    x 39
 ----              6/    102           35/    1413
 235   235         0/     68           01/     471
                   ----               ----
                   782    782          6123   6123
```

Continuing, 20 (from 5 x 4) is entered diagonally beneath the underline. The 2 is put a row down in the hundreds column, and the 0 is put in the row above it in the tens column, as shown. To finish, <u>the columns so formed are added</u>, resulting in the answer 235. In contrast, in Tower Multiplication, the 20 is added to the 3 above the 4 in the multiplicand, and the sum (23) is put next to the 5 in the answer.

The underlined in the preceding paragraph indicates the connection between the Multiplication Facts Algorithm and the Addition Facts Algorithm. As you can see in the second and third examples, the Multiplication Facts Algorithm generates columns of numbers, and the Addition Facts Algorithm makes short work of adding them.

The Multiplication Facts Algorithm is as big a game changer as the Addition Facts Algorithm when it comes to computing with whole numbers and decimals. As revealed in Chapter 4, the standard multiplication algorithm is laden with cognitive distractions regardless of how well the multiplication facts are known. Every time the product of a digit in the multiplier and a digit in the multiplicand is greater than or equal to ten, the next product is followed by an addition problem.

In MOVE IT Math, three multiplication algorithms are taught: Lattice Multiplication (which you can view on YouTube), the Multiplication Facts Algorithm, and Tower Multiplication. Conventional wisdom warns that children will get them mixed up if taught all three, but I have not seen that. After children learn all three algorithms, they are told they can multiply with whichever one they choose. Unsurprisingly, they all choose Lattice Multiplication or the Multiplication Facts Algorithm, both of which are low-stress algorithms that separate the addition from the multiplication. Why, then, should we keep teaching Tower multiplication as if it were the only way to multiply?

Chunk It Division

Chunk It division is an extension of counting out the division facts by skip counting by the divisor to the dividend and keeping track of the count with tally marks, as demonstrated in Chapter 5. It is used with single-digit divisors and dividends of any size to answer the question *"How many divisors in the dividend?"* For example, to answer the question for 23 ÷ 9, skip count by 9 to 23 (9-18; 19, 20, 21, 22, 23) and note the number of 9s with tally marks (//) and the amount remaining with a dot for each count (.....), resulting in //....... Thus 23 ÷ 9 = 2, R 5.

In the first example below, the 0 above the 2 in the dividend is the answer to the question *"How many 9s in 2?"* Then, since the example is in base ten, the 2 in the dividend is traded for 20 of the next smaller thing, which are added to 3 of the same thing in the dividend, resulting in 23. To continue, skip count by 9 to 23, as done above and shown in the example, and put 2 above the 3 in the dividend to note the number of 9s in 23. Then, since the example is in base ten, the five dots are traded for 50 of the next smaller thing, which are added to 7 of the same thing in the dividend, resulting in 57. To finish, skip count by 9 to 57 (9-18-27-36-45-54; 55, 56, 57) and put 6 above the 7 in the dividend to note the number of 9s in 57 and write R 3 after the 6 to indicate the amount remaining.

Note: In using the algorithm, the long vertical lines between the digits of the dividend are drawn first to contain the work and keep it orderly.

Super Chunk It Division

Super Chunk It division is an algorithm for dividing whole numbers and decimals by multiple-digit divisors. Like Chunk It, it is based on skip counting to answer the question *"How many divisors in the dividend?* For example, in using it to solve 7623 ÷ 48, it answers the question *"How many 48s in 7,623"?*

Super Chunk It differs from Chunk It in that it requires making a skip counting table for the divisor, as shown below for the divisor 48. To make the table, 48 was added to itself eight times, but to get started, it is often enough to just add a divisor to itself five or six times. If it turns out that more times are needed, the table can be extended by adding the divisor to itself more times.

Making a skip counting table may seem like a time-consuming step, but not really. Making it amounts to constructing a handy list of answers to all of the multiplication problems that may arise in solving the problem with the standard division algorithm. Note, though, that regardless of how one divides, division with multiple-digit divisors is tedious, so once elementary school students have demonstrated proficiency with Chunk It and worked a few problems with Super Chunk It to see that it is just an extension of Chunk It, division with multiple-digit divisors should be relegated to calculators.

The two Chunk It algorithms were invented by Deanna Callahan, an elementary school teacher in New Braunfels, Texas. They are meaningful ways to divide for students who understand base ten numeration and can skip count. They demonstrate that division separates a quantity into equal amounts the size of the dividend.

MONSTER ADDITION PROBLEM

Name_____

Date_____

```
  9  9  9  3  6  1  7  3  2  1
  9  9  8  8  6  7  0  0  3  0
  9  9  7  5  9  8  4  2  1  3
  9  8  5  4  2  6  5  5  3  0
  9  5  9  2  7  4  3  1  1  0
  8  8  7  6  5  0  6  2  0  2
  8  6  8  7  6  5  2  0  4  1
  8  6  3  5  4  3  3  2  0  0
  7  9  6  9  0  2  0  4  0  0
+ 7  7  7  6  6  4  5  1  2  2
_____
```

MONSTER SUBTRACTION PROBLEM

NAME _____

DATE _____

```
  8 4 0 0 9 5 2 0 3 7
- 1 2 3 4 5 6 7 8 9 0
_____
```

MONSTER MULTIPLICATION PROBLEM

NAME _____

DATE _____

```
  2 5 8 6 7 1 0 3 9 4
x                 8 3
_____
```

MONSTER DIVISION PROBLEM

NAME _____

DATE _____

$$6\overline{)7\ 8\ 4\ 2\ 8\ 4\ 3\ 0\ 9\ 6}$$

MONSTER ADDITION AWARD

This certifies that

_____

can add any ten
10-digit numbers

_____
Witness/Date

Paul Shoecraft
President, Monster Math Club of America™

AWESOME
91,350,537,169
ADDER

MONSTER SUBTRACTION AWARD

This certifies that

_____

can subtract any two
10-digit numbers

_____
Witness/Date

Paul Shoecraft
President, Monster Math Club of America™

With Zeros!

SUBTRACTION
EXPERT
7,166,384,147

MONSTER MULTIPLICATION AWARD

This certifies that

_____

can multiply by any 1- or 2-digit number

_____
Witness/Date

Paul Shoecraft
President, Monster Math Club of America™

214,696,962,702
MULTIPLICATION ADEPT

MONSTER DIVISION AWARD

This certifies that

_____

can divide by any single-digit number

_____
Witness/Date

Paul Shoecraft
President, Monster Math Club of America™

1,307,140,516

Permission to duplicate.

316

Why, in Dividing Fractions, We Invert the Divisor and Multiply

When dividing fractions, we invert the divisor and multiply to divide by the number 1 instead of a fraction by utilizing the following special properties of the number 1:

- Any number times 1 is that number (N x 1 = 1 x N = N for all N).

- Any number divided by 1 is that number (N ÷ 1 = N/1 = N for all N).

- Any non-zero number divided by itself is 1 (N ÷ N = N/N =1 for all N ≠ 0).

Avoidance behavior drives the derivation of the "mysterious" rule for dividing fractions: *"Why divide by a fraction when dividing by 1 would be easier?"* For 2/3 ÷ 4/5, below, the solution amounts to pulling a rabbit out of a hat to turn the divisor 4/5 into the number 1 by utilizing the aforementioned special properties of the number 1.

$$\frac{2}{3} \div \frac{4}{5} = \frac{\frac{2}{3}}{\frac{4}{5}} = \frac{\frac{2}{3}}{\frac{4}{5}} \times 1 = \frac{\frac{2}{3}}{\frac{4}{5}} \times \frac{\frac{5}{4}}{\frac{5}{4}} =$$

$$\frac{\frac{2}{3} \times \frac{5}{4}}{\frac{4}{5} \times \frac{5}{4}} = \frac{\frac{2}{3} \times \frac{5}{4}}{\frac{20}{20}} = \frac{\frac{2}{3} \times \frac{5}{4}}{1} = \frac{2}{3} \times \frac{5}{4}$$

Then, by substituting the letter a for 2, b for 3 (b ≠ 0), c for 4, and d for 5 (d ≠ 0), the rule for dividing fractions is established for all fractions.

a/b ÷ c/d = a/b x d/c

Note: In dividing fractions, it matters which fraction is inverted. <u>The divisor must be inverted</u>, not the dividend. For example, 2/3 ÷ 1/3 = 2/3 x 3/1 = (2 x 3)/(3 x 1) = 6/3 = 2, *not 3/2 x 1/3 = (3 x 1)/(2 x 3) = 3/6 = 1/2,* which is the answer except upside down.

Student Survey

STOP

Do not turn this paper over until instructed to do so.

Please answer the questions below by filling in the blanks or circling the answers. Then, when told to begin, turn the paper over and, working from left to right and top to bottom, answer as many of the single-digit multiplication problems as you can in 60 seconds.

1. Where did you attend elementary school (city and state or country)?

   _____

2. Are you taking a math class now?   Yes   No   If no, when did you last take a math class?

   Last semester   About 1 2 3 4 5 year(s) ago   More than 5 years ago

3. Which word best completes this sentence for you? Math is _____.

   Easy   Fairly easy   Difficult   Very difficult

4. Which phrase best describes how much you like math?

   A lot   A little   Neutral   Not much   Not at all

5. How many years of math did you have in high school?   1   2   3   4

6. What was the name of your last high school math class (e.g., Algebra I, Honors, Geometry, etc.)? _____

7. What were your usual grades in high school math?

   As   As and Bs   Bs and Cs   Cs   Cs and Below

8. What is your current college level?

   1st Year   2nd Year   3rd Year   4th Year   Above 4th Year   N/A

9. What is your current/intended major? _____

10. How much math do you anticipate using in your future career?

    A lot   A little   None

Multiplication Facts

7 x 8	9 x 3	7 x 5	6 x 4	9 x 6	3 x 5	9 x 8	7 x 5	5 x 8	7 x 3
8 x 4	7 x 9	5 x 9	8 x 3	8 x 6	2 x 5	8 x 6	3 x 6	7 x 4	5 x 9
5 x 5	6 x 2	9 x 7	7 x 6	6 x 4	3 x 7	5 x 4	9 x 6	7 x 8	5 x 6
6 x 5	7 x 6	6 x 3	8 x 7	9 x 5	4 x 8	3 x 9	2 x 8	3 x 4	9 x 4
8 x 9	4 x 5	6 x 7	4 x 6	3 x 8	3 x 9	4 x 9	7 x 4	5 x 3	3 x 4
9 x 3	6 x 9	5 x 2	7 x 8	4 x 5	2 x 5	6 x 3	4 x 3	5 x 7	8 x 3
6 x 9	7 x 7	4 x 7	3 x 7	5 x 8	3 x 6	4 x 4	3 x 5	6 x 8	9 x 7
3 x 3	4 x 8	4 x 9	6 x 7	7 x 9	9 x 9	6 x 5	9 x 8	8 x 7	5 x 6
8 x 8	4 x 3	3 x 5	4 x 5	4 x 7	7 x 2	4 x 7	5 x 8	9 x 4	7 x 3
4 x 6	5 x 7	2 x 2	4 x 7	5 x 3	8 x 9	9 x 5	3 x 8	7 x 6	8 x 4
5 x 4	3 x 8	6 x 6	2 x 7	5 x 3	9 x 2	3 x 5	6 x 8	2 x 3	7 x 4
6 x 8	8 x 5	7 x 5	4 x 8	5 x 3	6 x 7	9 x 4	3 x 9	7 x 6	9 x 3

Kindergarten program in the *Mad Minute Primer* to Teach the Addition Facts, Subtraction Facts, Skip Counting by 2, 5, and 10, and the Multiplication and Division Facts for 2, 5, and 10.

Week	Day	Activity	Skill	Type	Page
1–7		**Teacher's choice of counting forward activities, songs, rhymes, games, and children's books**	**Count out the addition facts by counting all of the touchpoints on the numerals 1-9 or by counting on**	Instruction	**211-213, 223-224**
8	1	Figuring Addition Facts 1	Counting all to 10	Instruction	72
	2	Figuring Addition Facts 2	Counting all to 10	Instruction	73
	3	Jumbo Addition Facts 1	Counting all to 10	Practice	74
	4	Jumbo Addition Facts 2	Counting all to 10	Practice	75
9	1	Figuring Addition Facts 1	Counting all to 14	Instruction	76
	2	Figuring Addition Facts 2	Counting all to 14	Instruction	77
	3	Addition Facts Whales 1	Counting all to 14	Practice	78
	4	Addition Facts Whales 2	Counting all to 14	Practice	79
10	1	Figuring Addition Facts 1	Counting all to 18	Instruction	80
	2	Figuring Addition Facts 2	Counting all to 18	Instruction	81
	3	Addition Facts Ladybugs 1	Counting all to 18	Practice	82
	4	Addition Facts Ladybugs 2	Counting all to 18	Practice	83
11	1	Figuring Addition Facts 1	Counting on to 10 from 0–5	Instruction	84
	2	Figuring Addition Facts 2	Counting on to 10 from 0–5	Instruction	85
	3	Addition Facts Speckled Frogs 1	Counting on to 10	Practice	86
	4	Addition Facts Speckled Frogs 2	Counting on to 10	Practice	87
12	1	Figuring Addition Facts 1	Counting on to 14 from 0–5	Instruction	88
	2	Figuring Addition Facts 2	Counting on to 14 from 0–5	Instruction	89
	3	Figuring Addition Facts 3	Counting on to 14 from 6–9	Instruction	90
	4	Figuring Addition Facts 4	Counting on to 14 from 6–9	Instruction	91
13	1	Addition Facts Rams 1	Counting on to 14	Practice	92
	2	Addition Facts Rams 2	Counting on to 14	Practice	93
	3	Figuring Addition Facts 1	Counting on to 18 from 6-9	Instruction	94
	4	Figuring Addition Facts 2	Counting on to 18 from 6-9	Instruction	95
14	1	Addition Facts Dolphins 1	Counting on to 18	Practice	96
	2	Addition Facts Dolphins 2	Counting on to 18	Practice	97

Week	Day	Activity	Skill	Type	Page
15–18		**Teacher's choice of counting back activities, songs, rhymes, games, children's books**	**Counting out the subtraction facts by counting back from the minuend or counting on to the minuend**	Instruction	213-219, 225-226
19	1	Figuring Subtraction Facts 1	Counting back from 0–5	Instruction	124
	2	Figuring Subtraction Facts 2	Counting back from 0–5	Instruction	125
	3	Subtraction Facts Turtles 1	Counting back from 0–5	Practice	126
	4	Subtraction Facts Turtles 2	Counting back from 0–5	Practice	127
20	1	Figuring Subtraction Facts 1	Counting back from 6–9	Instruction	128
	2	Figuring Subtraction Facts 2	Counting back from 6–9	Instruction	129
	3	Subtraction Facts Pigs 1	Counting back from 6–9	Practice	130
	4	Subtraction Facts Pigs 2	Counting back from 6–9	Practice	131
21	1	Figuring Subtraction Facts 1	Counting back from 10–14	Instruction	132
	2	Figuring Subtraction Facts 2	Counting back from 10–14	Instruction	133
	3	Subtraction Facts Chameleons 1	Counting back from 10–14	Practice	134
	4	Subtraction Facts Chameleons 2	Counting back from 10–14	Practice	135
22	1	Figuring Subtraction Facts 1	Counting back from 15–18	Instruction	136
	2	Figuring Subtraction Facts 2	Counting back from 15–18	Instruction	137
	3	Subtraction Facts Moles 1	Counting back from 18	Practice	138
	4	Subtraction Facts Moles 2	Counting back from 18	Practice	139
23–24		**Teacher's choice of skip-counting activities, songs, rhymes, games, and children's books**	**Count out the multiplication and division facts for 2, 5, and 10 by skip counting by 2, 5, and 10**	Instruction	219-222, 227-229, 232, 237
25	1	Figuring Multiplication Facts	Counting by 2s	Instruction	152
	2	Figuring Multiplication Facts	Counting by 5s	Instruction	153
	3	Multiplying by 10	Counting by 10s	Instruction	154
	4	Multiplication Facts Bears 1	Counting by 2s, 5s, 10s	Practice	155
	5	Multiplication Facts Bears 2	Counting by 2s, 5s, 10s	Practice	156
26	1	Figuring Division Facts	Counting by 2s	Instruction	181
	2	Figuring Division Facts	Counting by 5s	Instruction	182
	3	Dividing by 10	Counting by 10s	Instruction	183
	4	Division Facts Goats 1	Counting by 2s, 5s, 10s	Practice	184
	5	Division Facts Goats 2	Counting by 2s, 5s, 10s	Practice	185

Toothpick Numerals (from the *Mad Minute Primer*)

Name

Make the shapes on the scrolls with toothpicks. Circle the numeral below that looks most like each shape.

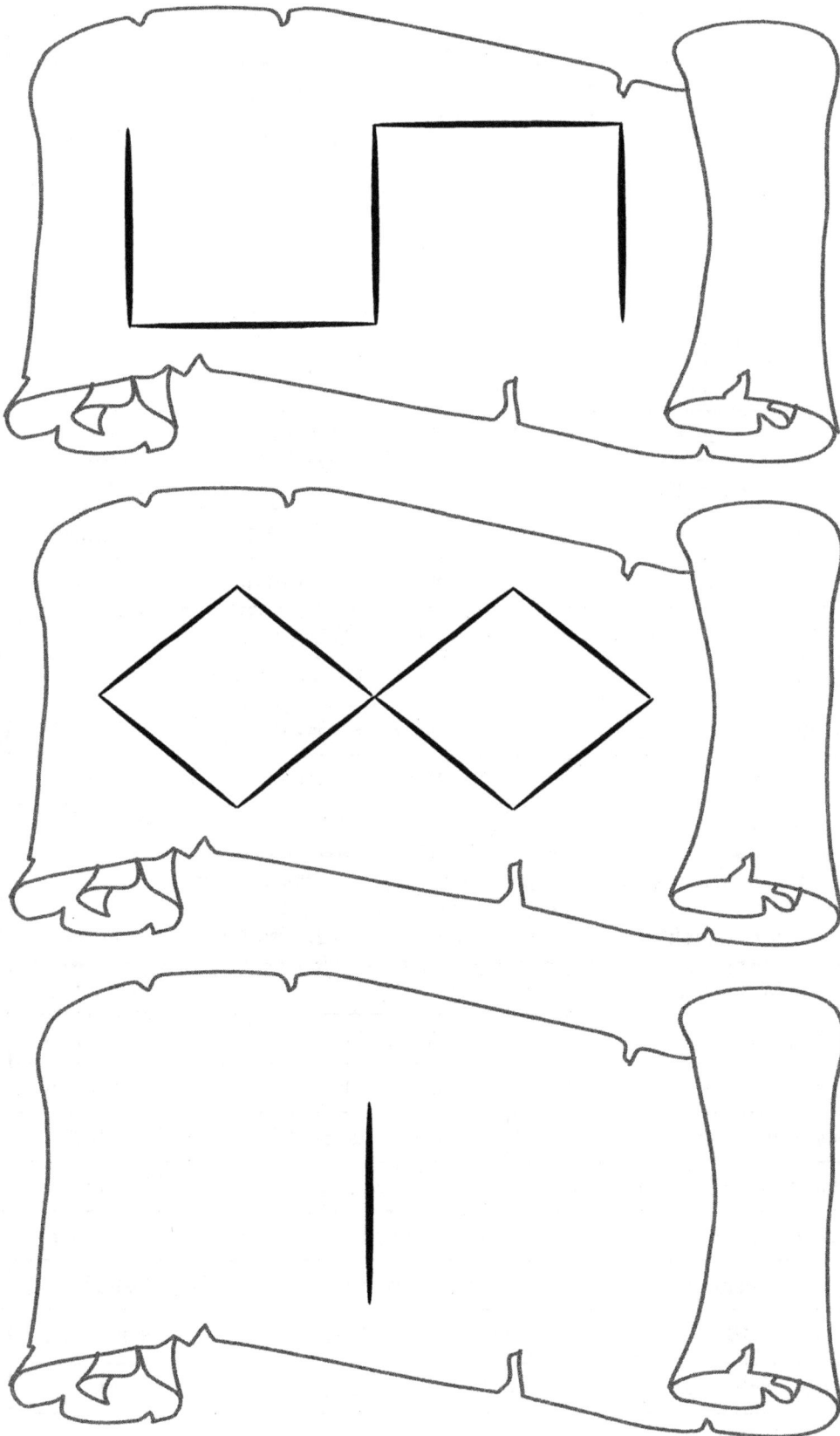

1 2 3 4 5 6 7 8 9

Toothpick Numerals (from the *Mad Minute Primer*)

Name _____

Make the shapes on the scrolls with toothpicks. Circle the numeral below that looks most like each shape.

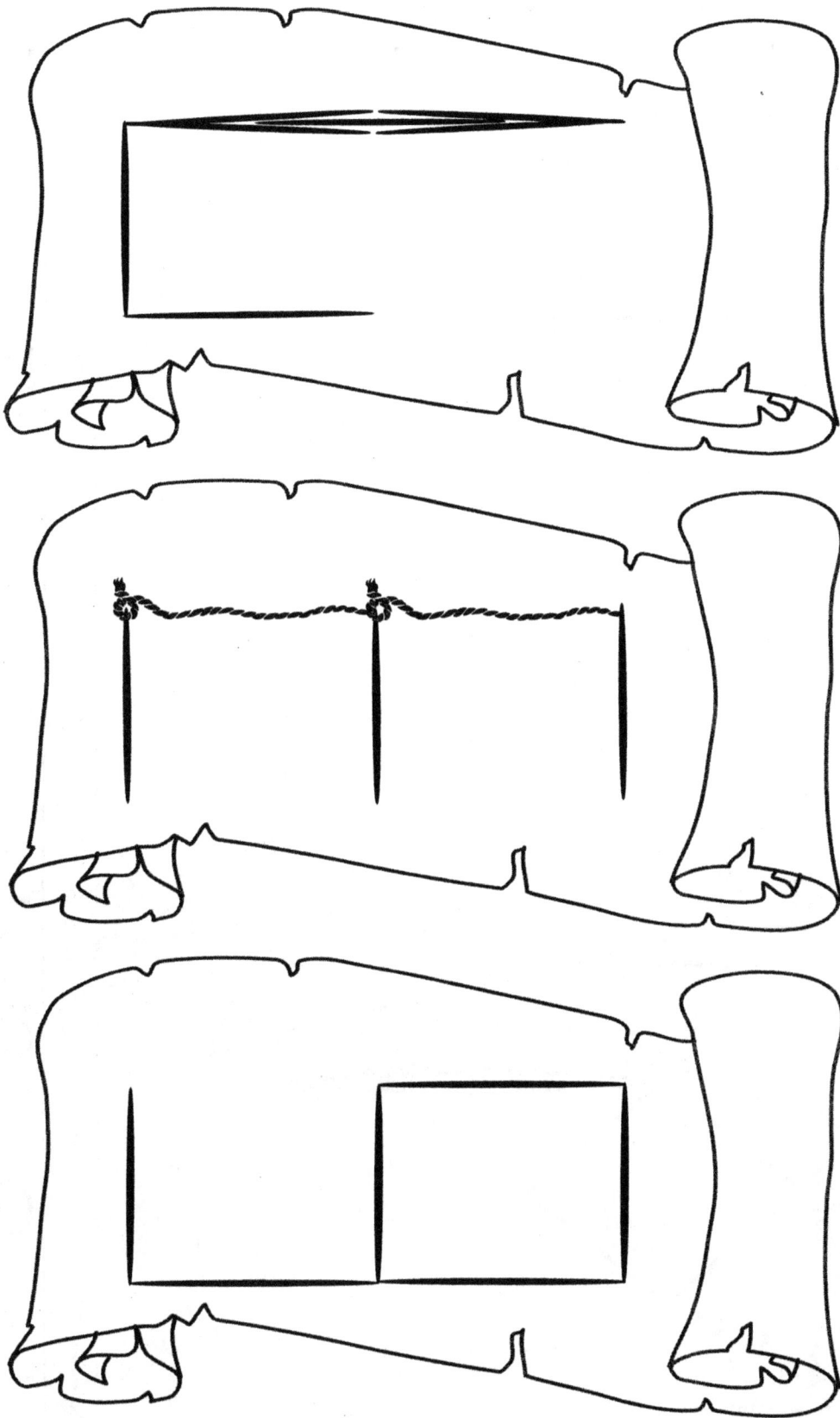

| 1 | 2 | 3 | 4 | 5 | 6 | 7 | 8 | 9 |

Toothpick Numerals (from the *Mad Minute Primer*)

Name _____

Make the shapes on the scrolls with toothpicks. Circle the numeral below that looks most like each shape.

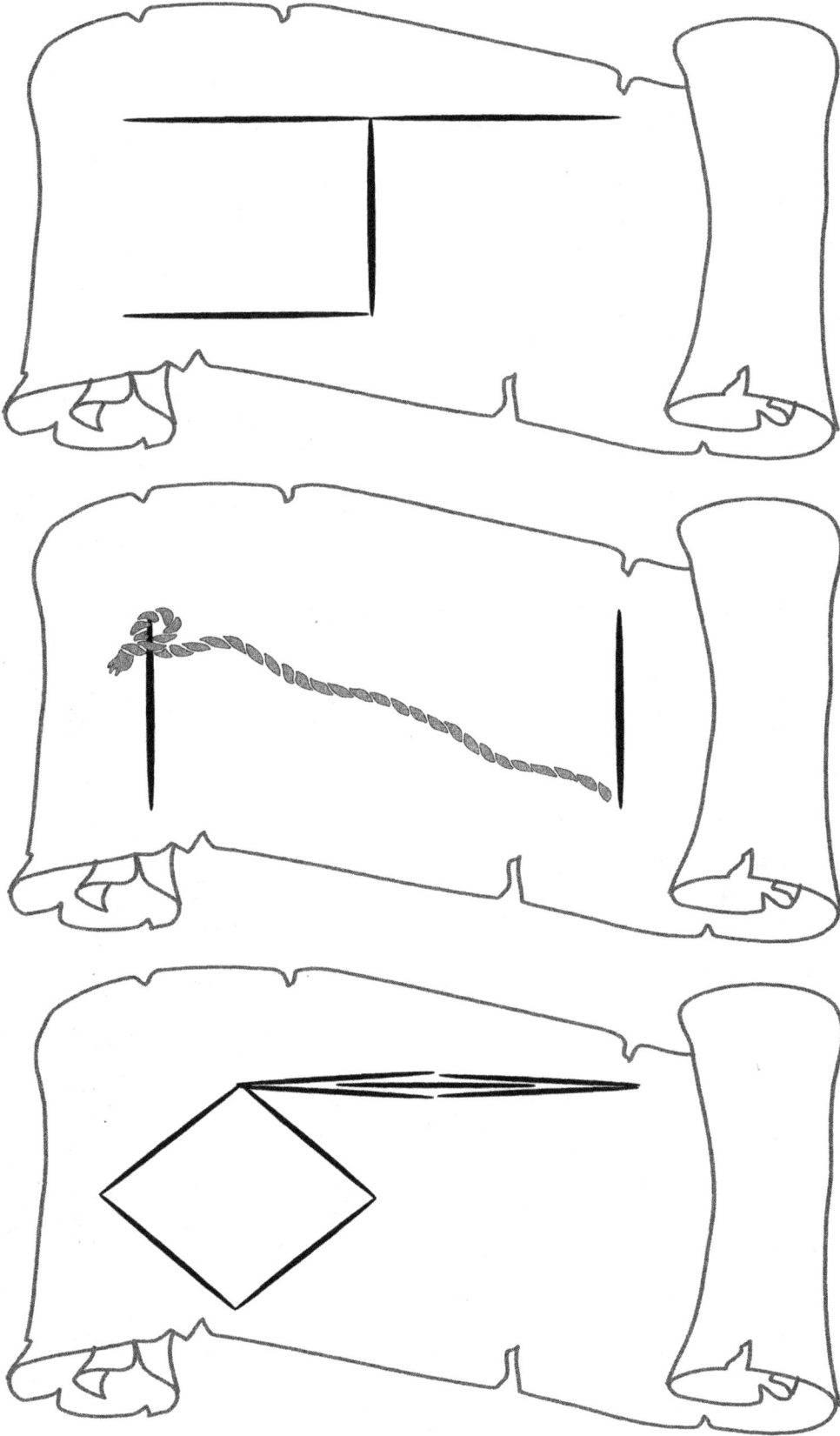

1 2 3 4 5 6 7 8 9

Adding and Subtracting in Base Two and Base Three with Arithmetic Blocks

Once children are adept at trading little blocks for big blocks and vice versa in Two Land (base two) and Three Land (base three), they are ready to add and subtract in those lands, called lands because they have laws: Two the same in Two Land and three the same in Three Land *must* be traded for the next bigger block to abide by the law.

For each of the activities that follow, children begin by loading a Fair Lands warehouse with the blocks shown in the warehouse. Then, if <u>buying</u> blocks, they load the truck with the blocks listed on the purchase order, after which they unload them in the warehouse and abide by the Law of the Land. The answer is the resultant number of blocks in the warehouse. If <u>selling</u> blocks, they load the truck with blocks <u>from the warehouse</u> to fulfill the purchase order. If they do not have enough blocks of a certain size to do that, they get more of them by trading the next bigger block for two of the next smaller block if in Two Land or three of the next smaller block if in Three Land. The answer is the number of blocks left in the warehouse.

Buying, no recording. Think blocks.

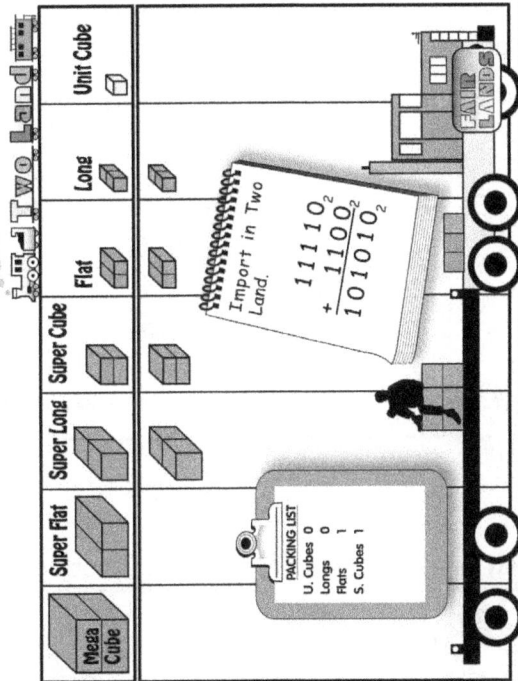

Import in Two Land.

$$11110_2$$
$$+ \; 1100_2$$

PACKING LIST
U. Cubes 0
Longs 0
Flats 1
S. Cubes 1

Selling, no recording. Think blocks.

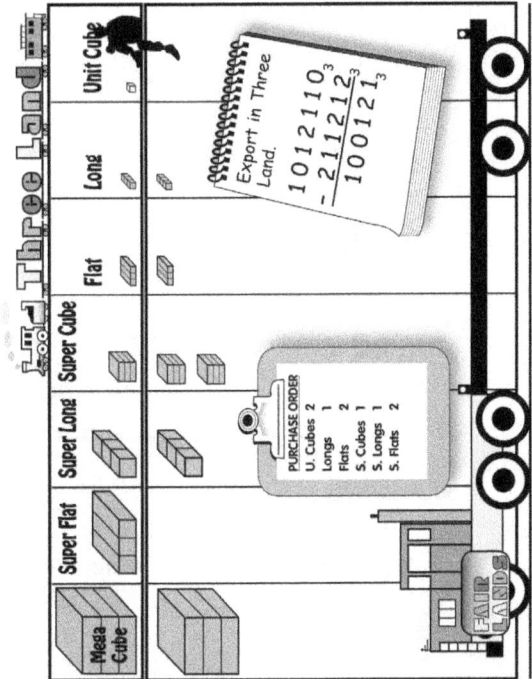

Export in Three Land.

$$1012110_3$$
$$- \; 211212_3$$
$$100121_3$$

PURCHASE ORDER
U. Cubes 2
Longs 1
Flats 2
S. Cubes 1
S. Longs 1
S. Flats 2

Buying, record the answer [1,001,001].

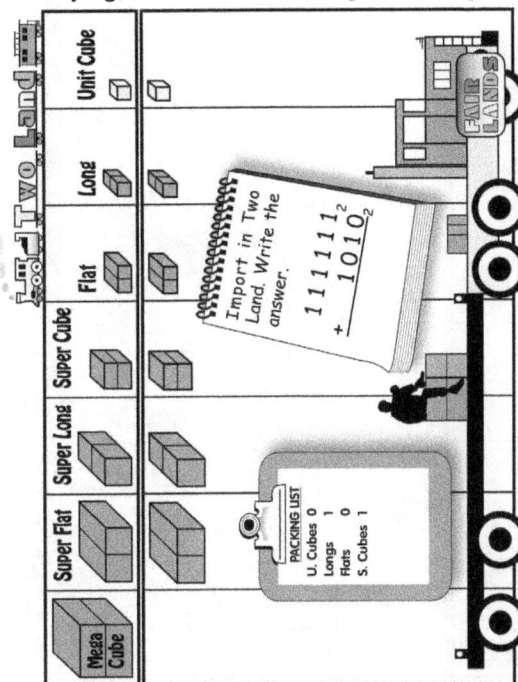

Import in Two Land. Write the answer.

$$111111_2$$
$$+ \; 1010_2$$

PACKING LIST
U. Cubes 0
Longs 1
Flats 0
S. Cubes 1

Selling, record the answer [221,221].

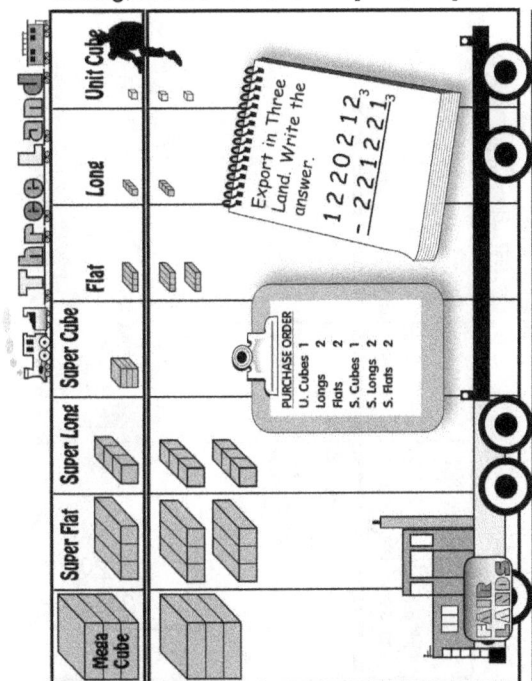

Export in Three Land. Write the answer.

$$1220212_3$$
$$- \; 212221_3$$

PURCHASE ORDER
U. Cubes 1
Longs 2
Flats 2
S. Cubes 2
S. Longs 2
S. Flats 2

Buying, recording column by column [1,002,000].

Three Land

Unit Cube	Long	Flat	Super Cube	Super Long	Super Flat	Mega Cube

Import in Three Land. Record column by column.
$$212120_3$$
$$+\ 12110_3$$

PACKING LIST
U. Cubes 0
Longs 1
Flats 1
S. Cubes 2
S. Longs 1

$$\begin{array}{r} 11111 \\ 212120_3 \\ +\ 12110_3 \\ \hline 1002000_3 \end{array}$$

Selling, recording column by column [10,110].

Two Land

Unit Cube	Long	Flat	Super Cube	Super Long	Super Flat	Mega Cube

Export in Two Land. Record column by column.
$$1001001_2$$
$$-\ 110011_2$$

PURCHASE ORDER
U. Cubes 1
Longs 1
Flats 0
S. Cubes 1
S. Longs 1
S. Flats 1

$$\begin{array}{r} 1 \qquad 1 \\ 0\ \cancel{2}\ 2\ 0\ \cancel{2}\ 2 \\ \cancel{1}\,0\,0\,\cancel{1}\,0\,0\,1_2 \\ -\ 110011_2 \\ \hline 10110_2 \end{array}$$

Buying, write and solve the addition problem.

Three Land

Unit Cube	Long	Flat	Super Cube	Super Long	Super Flat	Mega Cube

Write and solve the addition problem in Three Land.

PACKING LIST
U. Cubes 1
Longs 0
Flats 2
S. Cubes 2
S. Longs 1

$$\begin{array}{r} 200022 \\ +\ 12201 \\ \hline 220000 \end{array}$$

Selling, write and solve the subtraction problem.

Two Land

Unit Cube	Long	Flat	Super Cube	Super Long	Super Flat	Mega Cube

Write and solve the subtraction problem in Two Land.

PURCHASE ORDER
U. Cubes 1
Longs 0
Flats 1
S. Cubes 1
S. Longs 1

$$\begin{array}{r} 101010 \\ -\ 10101 \\ \hline 10101 \end{array}$$

Buying, no recording. Think blocks.

Ten Land

Thousand Cube	Hundred Flat	Ten Long	Unit Cube

Import in Ten Land.

$$\begin{array}{r} 1583 \\ +\ 786 \\ \hline 2369 \end{array}$$

PACKING LIST

U. Cubes 6

T. Longs 8

H. Flats 7

FAIR LANDS

Selling, record the answer [1,859].

Ten Land

Thousand Cube	Hundred Flat	Ten Long	Unit Cube

Export in Ten Land. Write the answer.

$$\begin{array}{r} 2708 \\ -\ 849 \\ \hline \end{array}$$

PURCHASE ORDER

U. Cubes 9

T. Longs 4

H. Flats 8

FAIR LANDS

Buying, record column by column [2,002].

Ten Land

Thousand Cube	Hundred Flat	Ten Long	Unit Cube

Import in Ten Land. Record column by column

$$\begin{array}{r} 1687 \\ +\ 315 \\ \hline \end{array}$$

PACKING LIST

U. Cubes 5

T. Longs 1

H. Flats 3

$$\begin{array}{r} \mathbf{1\ 1\ 1} \\ 1687 \\ +\ 315 \\ \hline 2002 \end{array}$$

FAIR LANDS

Selling, write and solve the subtraction problem.

Ten Land

Thousand Cube	Hundred Flat	Ten Long	Unit Cube

Write and solve the subtraction problem in Ten Land.

PURCHASE ORDER

U. Cubes 2

T. Longs 8

H. Flats 3

T. Cubes 1

$$\begin{array}{r} 1674 \\ -1382 \\ \hline 292 \end{array}$$

FAIR LANDS

327

2023 TIMSS Results for the U.S.

"Around the World in Math and Science: Scanning the Headlines on the Results of TIMSS 2023," *International Education News*, Thomas Hatch, December 11, 2024.

"Headlines touted gains in some countries like the United Arab Emirates, Turkey and Australia, but highlighted concerns about substantial declines in performance in countries like Israel and the US. <u>In the US, the drop in scores was particularly pronounced for the lowest performing students</u>, with one in five 8th graders [unable] to demonstrate even a basic level of proficiency. Adding to the concerns, OECD released the latest results of the 2023 administration of the Program for the International Assessment of Adult Competencies (PIAAC) which showed US adults are getting worse at reading and math as well."

"TIMSS Shows the Bottom Is Falling Out for US Test Scores," *American Enterprise Institute*, Nat Malkus, December 9, 2024.

"Since 2011, the score gap between the 75th and 25th percentiles in fourth-grade math grew by 35 points — over a year's worth of progress in learning for the average student. Similarly, the gap between the 90th and 10th percentiles grew by 58 points—about two years of learning — during that same period."

"U.S. Math Scores Drop on [TIMSS]," *Chalkbeat*, Erica Meltzer, December 4, 2024.

"<u>U.S. fourth graders saw their math scores drop steeply</u> between 2019 and 2023 on [TIMSS]. ... <u>Scores dropped even more steeply for American eighth graders</u>."

"U.S. Students Posted Dire Math Declines on [TIMSS]," *New York Times*, Dana Goldstein, December 4, 2024.

"On [TIMSS], American fourth and eighth graders posted results similar to scores from 1995 — <u>a sign of notable stagnation</u> — even as other countries saw improvements. ... The results found that since 2019, American fourth graders have declined 18 points in math, while eighth graders have declined 27 points."

"Math Scores [on TIMSS] Plummet, Progress 'Erased,' NCES Reports," *K-12 Dive*, Anna Merod, December 4, 2024.

"Average U.S. math scores for both 4th and 8th graders reverted to performance levels of 1995, the first year the TIMSS assessment was administered, meaning 'progress in prior years has been erased,'" said NCES Commissioner Peggy Carr." **[What progress?]**

"Insights into U.S. Students' Drop in Math & Science on International Test," *The 74*, Kevin Mahnken, December 4, 2024.

"In the United States, fourth and eighth graders performed much worse in math last year than students at the same age levels did in 2019; average scores in the subject fell to the level seen in 1995, the first time TIMSS was conducted."

2023 PIAAC Results for the U.S.

"Between 2017 and 2023, the overall average scores for U.S. adults <u>decreased</u> in numeracy." [69]

Average PIAAC Scores for U.S. Adults, Age 16-65, on Numeracy and Adaptive Problem Solving: 2012/2014, 2017, and 2023

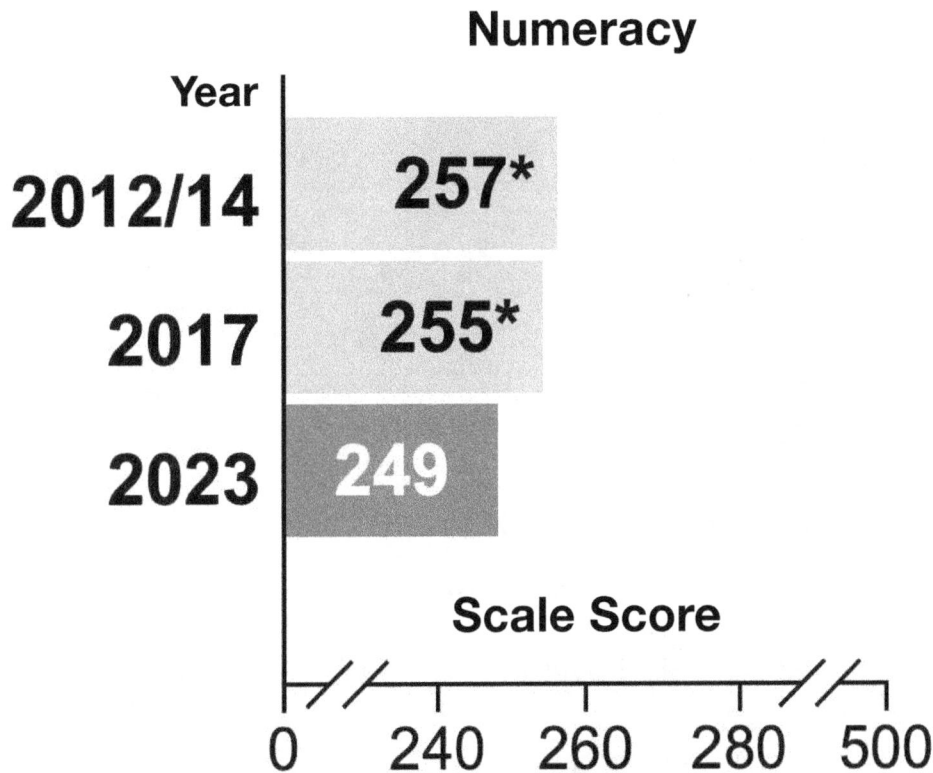

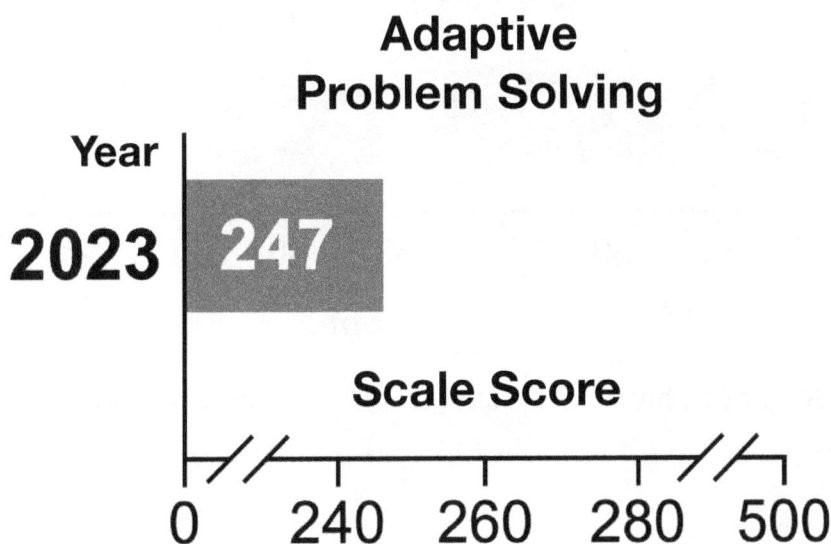

Numeracy

Year	Scale Score
2012/14	257*
2017	255*
2023	249

Scale Score

0   240   260   280   500

Adaptive Problem Solving

Year	Scale Score
2023	247

Scale Score

0   240   260   280   500

---

[69] "Highlights of the 2023 U.S. PIAAC Results," National Center for Education Statistics, U.S. Dept. of Education, December 2024.

"Between 2017 and 2023, the percent of adults performing at the lowest proficiency level (Level 1 or below) in numeracy <u>increased</u> from 29 to 34 percent. The percent of U.S. adults performing at the lowest level in adaptive problem solving in 2023 was 32 percent." [70]

Percent of U.S. Adults, Age 16-65, in PIAAC Proficiency Levels for Numeracy and Adaptive Problem Solving: 2012/2014, 2017, and 2023

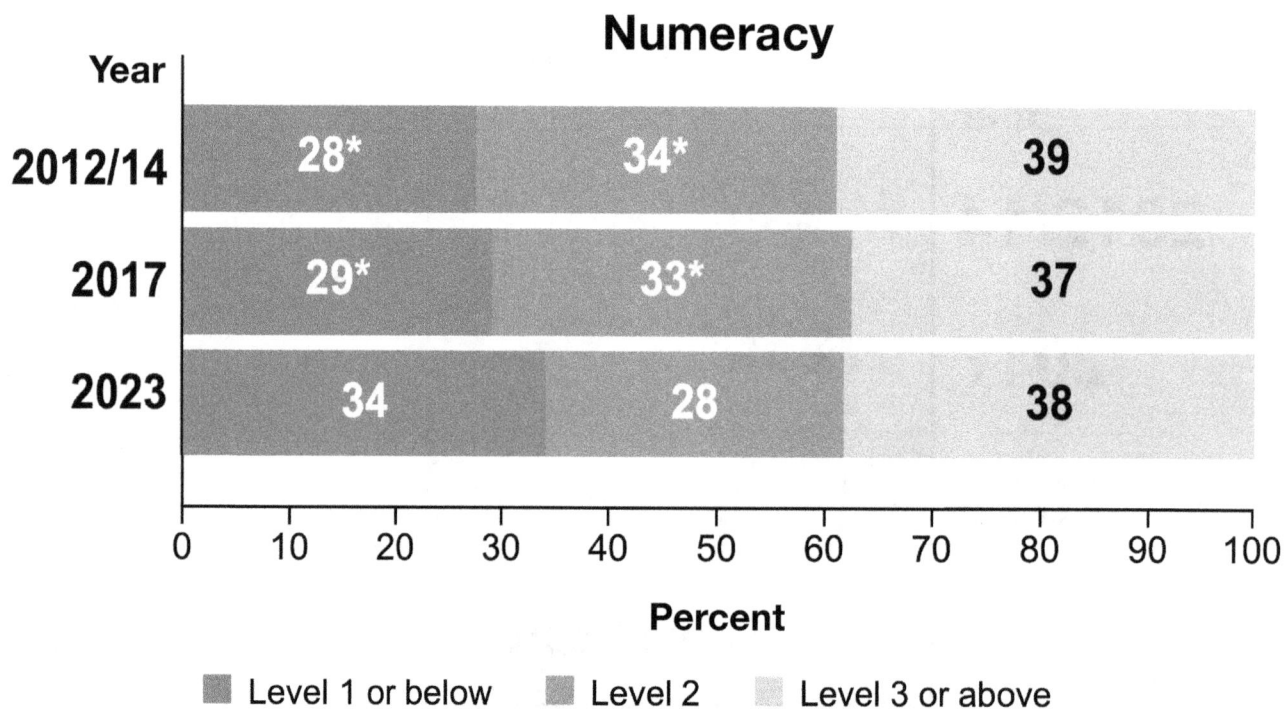

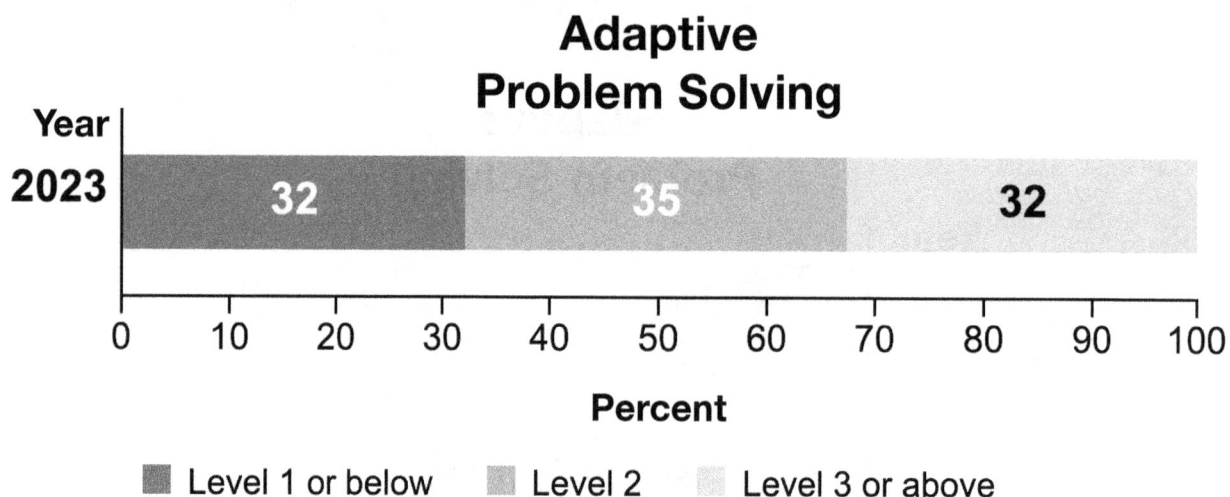

Numeracy

Year			
2012/14	28*	34*	39
2017	29*	33*	37
2023	34	28	38

Percent

■ Level 1 or below   ■ Level 2   ■ Level 3 or above

Adaptive Problem Solving

Year			
2023	32	35	32

Percent

■ Level 1 or below   ■ Level 2   ■ Level 3 or above

[70] "Highlights of the 2023 U.S. PIAAC Results," National Center for Education Statistics, U.S. Dept. of Education, December 2024.